Illegal Logging
in the Tropics:
Strategies
for Cutting Crime

Illegal Logging in the Tropics: Strategies for Cutting Crime has been co-published simultaneously as *Journal of Sustainable Forestry*, Volume 19, Numbers 1/2/3 2004.

Illegal Logging in the Tropics: Strategies for Cutting Crime

Ramsay M. Ravenel
Ilmi M. E. Granoff
Carrie A. Magee
Editors

Illegal Logging in the Tropics: Strategies for Cutting Crime has been co-published simultaneously as *Journal of Sustainable Forestry*, Volume 19, Numbers 1/2/3 2004.

CRC Press
Taylor & Francis Group
Boca Raton London New York

CRC Press is an imprint of the
Taylor & Francis Group, an **informa** business

Illegal Logging in the Tropics: Strategies for Cutting Crime

CONTENTS

Preface   xv
Francis E. Putz

Acknowledgments   xvii

INTRODUCTION

Introduction to Illegal Logging in the Tropics   1
Ramsay M. Ravenel
Ilmi M. E. Granoff

PROBLEM DESCRIPTION

Undercutting Sustainability: The Global Problem of Illegal
    Logging and Trade   7
    Wynet Smith

THEORETICAL APPROACHES TO UNDERSTANDING
FOREST GOVERNANCE

A System Dynamics Examination of the Willingness
    of Villagers to Engage in Illegal Logging   31
    Richard G. Dudley

Does Improved Governance Contribute to Sustainable Forest
    Management?   55
    Nalin Kishor
    Arati Belle

Illegal Logging and Local Democracy:
    Between Communitarianism and Legal Fetishism   81
    Antonio Azuela

You Say Illegal, I Say Legal: The Relationship Between 'Illegal'
Logging and Land Tenure, Poverty, and Forest Use Rights
in Vietnam   97
    Pamela McElwee

Can 'Legalization' of Illegal Forest Activities Reduce Illegal
Logging? Lessons from East Kalimantan   137
    Luca Tacconi
    Krystof Obidzinski
    Joyotee Smith
    Subarudi
    Iman Suramenggala

CASE STUDIES

Social and Environmental Costs of Illegal Logging in a Forest
Management Unit in Eastern Cameroon   153
    Philippe Auzel
    Fousseni Feteke
    Timothée Fomete
    Samuel Nguiffo
    Robinson Djeukam

Recent Trends in Illegal Logging and a Brief Discussion
of Their Causes: A Case Study from Gunung Palung
National Park, Indonesia   181
    Marc A. Hiller
    Benjamin C. Jarvis
    Hikma Lisa
    Laura J. Paulson
    Edward H. B. Pollard
    Scott A. Stanley

Community-Based Logging and *De Facto* Decentralization:
Illegal Logging in the Gunung Palung Area of West
Kalimantan, Indonesia   213
    Ramsay M. Ravenel

Combating Corruption and Illegal Logging in Bénin,
West Africa: Recommendations for Forest Sector Reform   239
    Ute Siebert
    Georg Elwert

INTERVENTIONS: THEORY AND PRACTICE

Illegal Actions and the Forest Sector: A Legal Perspective   263
    Kenneth L. Rosenbaum

The Role of Monitoring in Cutting Crime   293
    Wynet Smith

Illegal Logging and Deforestation in Andaman and Nicobar
    Islands, India: The Story of Little Andaman Island   319
    Pankaj Sekhsaria

Combating Forest Corruption: The Forest Integrity Network   337
    Aarti Gupta
    Ute Siebert

SYNTHESIS

Illegal Logging in the Tropics: A Synthesis of the Issues   351
    Ramsay M. Ravenel
    Ilmi M. E. Granoff

Index   373

ABOUT THE EDITORS

Ramsay M. Ravenel, AB, MF, MBA, holds Master of Forestry and Master of Business Administration degrees from Yale University and an AB in Environmental Science and Public Policy from Harvard College. He worked for the Gunung Palung Community-Based Forest Management Project in West Kalimantan, Indonesia in 1997-1998. His graduate work focused on forest management and financial engineering as tools for conservation. He currently works as an independent consultant in timberland investment and conservation finance.

Ilmi M. E. Granoff, BA, MESc, MA, holds masters degrees in Environmental Science and International Relations from Yale University and a BA in Biological Anthropology and Philosophy from Swarthmore College. His research interests include the relationship between governance-related development and environmental protection in the developing world, and conservation discourse as an instrument of political power. He has worked and studied throughout the tropics, and has a particular interest in Latin America. He is currently conducting research in conjunction with a Brazilian NGO on community forestry in an Amazonian extractive reserve.

Carrie A. Magee, BS, MF, holds a Master of Forestry degree from the Yale School of Forestry and Environmental Studies and a BS in Wildlife Ecology from the University of Wisconsin, Madison. She served as a Peace Corps Volunteer in Uganda, where she worked as a Forest Officer in the Kalinzu Forest Reserve area developing non-consumptive income generating activities in order to reduce illegal forest activities. She also conducted a study for the Kibale National Park on local wood use in forest patches outside the park. She is currently an Urban and Community Forester with the New Jersey Tree Foundation coordinating the Urban Airshed Reforestation Project in Camden City.

Preface

Although illegal logging has undoubtedly been going on since tree cutting was first outlawed by a local chieftain or clan leader, the practice has only very recently been scrutinized by scholars. Up until a very few years ago, when a poorly paid forest officer drove to his retirement party in his new Mercedes-Benz, people might have whispered disparaging remarks while others just sniggered, but seldom were graft, corruption, and outright theft discussed openly. When people in polite society did mention illegal logging, it was generally to condemn the tree fellers who, more often than not, were rural people being demonized for traditional forest practices that had been criminalized. In other social circles, the powerful beneficiaries of illegal logging were privately condemned but publicly accepted among governmental, business, and military leaders. Publication of this volume heralds increased awareness of the pervasiveness of illegal logging and the complexity of issues it entails. By comprehensively assessing this issue from a variety of disciplinary and philosophical perspectives, contributors to this volume suggest ways that this multidimensional problem might most reasonably be addressed.

Estimates vary, but it appears that about half of the trade in tropical timber involves illegal activities at one or more links along the market chain. Satellite imagery can help, but given the remote and difficult conditions under which most logging occurs in the tropics and the dangers of close association with people working outside of the law, it is not surprising that reliable estimates of the amount of illegal logging are hard to find. But whether the revenues involved amount to $1 billion or $10 billion per year, the problem is clearly huge, as are the windfall profits to some perpetrators of illegal forestry activities.

It becomes clear through reading this volume that illegal activities in the forestry sector need to be considered as expressions of elaborate and embedded social processes in which a diversity of actors operate under a variety of

[Haworth co-indexing entry note]: "Preface." Putz, Francis E. Co-published simultaneously in *Journal of Sustainable Forestry* (The Haworth Press, Inc.) Vol. 19, No. 1/2/3, 2004, pp. xxiii-xxiv; and: *Illegal Logging in the Tropics: Strategies for Cutting Crime* (ed: Ramsay M. Ravenel, Ilmi M. E. Granoff, and Carrie A. Magee) The Haworth Press, Inc., 2004, pp. xv-xvi. Single or multiple copies of this article are available for a fee from The Haworth Document Delivery Service [1-800-HAWORTH, 9:00 a.m. - 5:00 p.m. (EST). E-mail address: docdelivery@haworthpress.com].

constraints and take different risks for which they receive widely diverging rewards. Readers are also reminded that both forests and human futures suffer when forest laws are violated. Unfortunately, for many participants in illegal forest activities, the laws that are bent and broken are merely annoying impediments to be avoided with well-placed bribes. Others judge participation in what are classified as "illegal" activities as justifiable on the fundamental ethical ground that the government illegally expropriated the forests they are exploiting. In either case, after considering the diversity of perspectives presented in this volume, I will never again be content with simplistic and absolute notions of justice.

The focus of a number of chapters in this volume is on the mechanisms by which laws related to forests are violated. As a forest owner who was personally swindled by an unscrupulous forest operator, mostly due to my own ignorance mixed with a modicum of blind faith, I recognize the importance of understanding the details of the business of forestry. When the amount of business advice to which I had access is considered, my ignorance is particularly embarrassing. On the positive side of the ledger, the fates of my ventures into for-profit forestry may serve as an effective reminder that even where governance is modestly well developed, illegal activities are still likely where poorly informed forest owners deal with merchants who are more powerful, or at least more wily.

The contributors to this volume seem to agree that while social context should be considered, illegal logging generally undercuts efforts at sustainable forest management, denies benefits to rightful recipients, and otherwise undermines the rule of law. What is needed, they argue, are integrated solutions to this embedded social, environmental, economic and political problem. Hope derives from general societal condemnation of corruption and widespread support for transparency. It is less clear whether the cause will be served by harmonizing legislation with social norms, what roles international non-governmental organizations should play in forest monitoring and law enforcement, or how to best increase the costs of operating outside of the law. While corruption remains rife in tropical forestry and many government officials are reluctant or unable to enforce the laws of their lands, increased understanding of the patterns and processes of forest corruption should help to rein in illegal loggers for the benefits of tropical forests and the people they support.

Francis E. Putz
Department of Botany
University of Florida
Gainesville, FL 32611
E-mail: fep@botany.ufl.edu

Acknowledgments

This volume presents the papers delivered at the conference *Illegal Logging in Tropical Forests: Ecology, Economics and Politics of Resource Misuse*, held on March 29-30, 2002 at the Yale School of Forestry and Environmental Studies, New Haven, Connecticut. A few additional papers on the subject were also accepted for this publication after the conference. We wish to thank the presenters, moderators, participants, and volunteer organizers who helped make the conference a successful and engaging experience for all involved. We also thank the authors of these papers for their hard work and perseverance seeing this material through to publication. We also thank Paulo Adario, Anita Akella, and Wangari Maathai for their presentations at the conference. Wynet Smith was also kind enough to let us borrow her clever pun "cutting crime" for our title for this publication.

We would like to thank the many Yale faculty members who offered their support and participation in the conference: Dr. Lisa Curran, Dr. Graeme Berlyn, Dr. Mark Ashton, Dr. Chad Oliver, and Dr. Michael Dove. We also thank Dr. Lisa Curran, Dr. Benjamin Cashore, Dr. Andrew Shrank, and Professor Wangari Maathai for moderating discussions during the conference. Citlali Cortés-Montaño, Iona Hawken, Daniela Cusak, Abdalla Shah, and Barbara Bamberger, gave a great deal of their time and input as members of the Yale ISTF Conference Committee. We also thank the countless other student volunteers who helped us present the conference. Dr. Graeme Berlyn and Philip Marshall gave us invaluable encouragement and support during the editing process.

The conference and this publication of its proceedings would not have been possible without the financial support of a wide range of Yale University programs and offices. We thank Dean Gus Speth, Alan Brewster, Gordon Geballe, Charles Waskiewicz, Carmela Lubenow, and Kath Schoemaker of the School of Forestry and Environmental Studies; the Student Affairs Committee of the

[Haworth co-indexing entry note]: "Acknowledgments." Co-published simultaneously in *Journal of Sustainable Forestry* (The Haworth Press, Inc.) Vol. 19, No. 1/2/3, 2004, pp. xxv-xxvi; and: *Illegal Logging in the Tropics: Strategies for Cutting Crime* (ed: Ramsay M. Ravenel, Ilmi M. E. Granoff, and Carrie A. Magee) The Haworth Press, Inc., 2004, pp. xvii-xviii. Single or multiple copies of this article are available for a fee from The Haworth Document Delivery Service [1-800-HAWORTH, 9:00 a.m. - 5:00 p.m. (EST). E-mail address: docdelivery@haworthpress.com].

School of Forestry and Environmental Studies; Dr. Amity Doolittle of the Yale Tropical Resources Institute; Gary Dunning and Kate Kennedy of the Yale Global Institute of Sustainable Forestry; the Yale Center for International and Area Studies. We also thank Les Whitmore from our parent organization in Washington, DC, the International Society of Tropical Foresters for his introductory remarks at the conference.

INTRODUCTION

Introduction
to Illegal Logging in the Tropics

Ramsay M. Ravenel
Ilmi M. E. Granoff

SUMMARY. On March 29-30, 2002, the Yale Chapter of the International Society of Tropical Foresters convened social and natural scientists, resource managers, policy-makers, community leaders and other interested parties to share experiences, strategies, successes and failures in addressing illegal logging and corruption. The conference explored the framing of the illegal logging problem, the extent of the problem, its perceived causes, and potential solutions. Thirteen papers were presented at the conference entitled *Illegal Logging in the Tropics: The*

Ramsay M. Ravenel is an independent consultant in forestland investment (E-mail: rravenel@aya.yale.edu).

Illmi M. E. Granoff is a consultant in international conservation and development policy (E-mail: ilmi.granoff@aya.yale.edu).

Address correspondence to: Illmi M. E. Granoff at the above address.

[Haworth co-indexing entry note]: "Introduction to Illegal Logging in the Tropics." Ravenel, Ramsay M., and Ilmi M. E. Granoff. Co-published simultaneously in *Journal of Sustainable Forestry* (The Haworth Press, Inc.) Vol. 19, No. 1/2/3, 2004, pp. 1-6; and: *Illegal Logging in the Tropics: Strategies for Cutting Crime* (ed: Ramsay M. Ravenel, Ilmi M. E. Granoff, and Carrie A. Magee) The Haworth Press, Inc., 2004, pp. 1-6. Single or multiple copies of this article are available for a fee from The Haworth Document Delivery Service [1-800-HAWORTH, 9:00 a.m. - 5:00 p.m. (EST). E-mail address: docdelivery@haworthpress.com].

Ecology, Economics and Policy or Resource Misuse held at the Yale
School of Forestry and Environmental Studies on March 29-30, 2002.
This introduction presents a brief overview of the illegal logging issue
and describes the papers presented in this volume. *[Article copies avail-
able for a fee from The Haworth Document Delivery Service: 1-800-HAWORTH.
E-mail address: <docdelivery@haworthpress.com> Website: <http://www.
HaworthPress. com> © 2004 by The Haworth Press, Inc. All rights reserved.]*

KEYWORDS. Illegal logging, deforestation, tropical forest conservation, tropical forest management, forest policy

INTRODUCTION

On March 29-30, 2002, the Yale chapter of the International Society of
Tropical Foresters convened social and natural scientists, resource managers,
policy-makers, community leaders and other interested parties to share experiences, strategies, successes and failures in addressing illegal logging and corruption. This volume documents the presentations and discussions held at the
conference entitled *Illegal Logging in the Tropics: The Ecology, Economics
and Policy of Resource Misuse* at the Yale School of Forestry and Environmental Studies on March 29-30, 2002. The conference explored the framing of
the illegal logging problem, the extent of the problem, its perceived causes,
and potential solutions. This brief introduction presents an overview of the illegal logging issue and describes the papers presented in this volume.

ILLEGAL LOGGING IN TROPICAL FORESTS

Illegal logging has gained a great deal of attention in recent years. Local
and international conservation groups as well as governments and multilateral
institutions have called for action and initiated interventions (e.g., G8, 1998;
Callister, 1999; Global Witness, 1999; EIA/Telapak, 1999; Greenpeace, 2000;
UNFF, 2002; Global Forest Watch, 2003). Control over forest resources has
sparked social conflict for centuries. Illegal logging is best understood in this
context: as the latest manifestation of these conflicts. Today, however, they
occur in a highly interconnected social, political, and economic world.

Illegal logging involves a wide range of activities, beginning in the forest
and proceeding along the chain of transactions that delivers the forest products
to the end consumer. While at one level "illegal logging" boils down to the
acts of individual loggers violating national or sub-national laws in the forests,
the problem is much broader and more complex. These actions typically occur
in the context of a larger illegal logging trade system, and may be the conse-

quence of failed policies and laws at the national level. Adequate policy intervention thus entails considerations of both local and national drivers of the problem. The papers in this volume include analyses across scales, from Marc Hiller et al.'s examination of what motivates local actors to participate in illegal logging in Indonesia, to Wynet Smith's analysis of global trends. Perhaps most significantly, this publication discusses specific policy interventions aimed at curbing the problem, and all papers included are oriented toward identifying solutions to forest crime.

THE CONFERENCE AND PROCEEDINGS

Problem Description

This volume begins with an overview of available information by Wynet Smith in "Undercutting Sustainability: The Global Problem of Illegal Logging and Trade." Her literature review indicates that most countries with sizeable areas under forest cover experience at least some degree of illegal logging. Estimates include 80% for the Brazilian Amazon, over 90% in Cambodia, and over 70% in Indonesia. It is worth noting that although the geographic scope of the conference was limited to the tropics, the US and Canada also have documented problems with illegal logging. In the US, annual timber theft totals an estimated US$1 billion (valued at the mill) (Mendoza, 2003).

Theoretical Approaches

Having introduced the problem at the global level, this volume presents a variety of theoretical analyses of illegal logging. Although many forest policy problems occur at the national level, effective policy intervention requires a theoretical understanding of the drivers at the local scale. In "A System Dynamics Examination of the Willingness of Villagers to Engage in Illegal Logging," Richard Dudley provides an in-depth analysis of the incentive structures that bring local actors to commit forest crimes. Moving to the scale of national responses to the illegal logging problem, Nalin Kishor and Arati Belle analyze the degree to which illegal logging is reflective of larger governance failures in "Does Improved Governance Contribute to Sustainable Forest Management?"

Sometimes illegal loggers are not the source of the problem, and bad laws are what makes legitimate logging activities illegal. A nuanced policy approach should thus draw from a theoretical perspective that recognizes challenges of balancing between sometimes justifiable interests that drive illegal logging, and the legitimate exercise of state control over valuable resources. Antonio Azuela's "Illegal Logging and Local Democracy: Between Communitarianism and Legal Fetishism" explores the two sides of the "legality" question. In "You Say Illegal, I Say Legal," Pamela McElwee problematizes the issue of legality as perceived by local actors in Vietnam. McElwee demonstrates that while the state may view private logging in state-controlled forests as illegal, local actors may perceive their actions in quite different terms. Her analysis, like Azuela, helps address the utility of the legality discourse in terms of issues of effective forest management. Problematizing the issue of logging "illegality" begs the question of the effectiveness of legalization. Luca Tacconi et al. explore the implications of legalization on the ground in "Can 'Legalization' of Illegal Forest Activities Reduce Illegal Logging? Lessons from East Kalimantan."

Case Studies

After considering theoretical issues in illegal logging, this volume focuses on a number of case studies that describe how illegal logging transpires on the ground. Philippe Auzel et al. provide a detailed account of the nature and impacts of illegal logging in a specific site in their paper "Illegal Logging in a Forest Management Unit, Eastern Cameroon." Their research describes the means by which illegal logging occurs, its impacts on forests and on government revenues, along with the legal and political context in which it occurs.

In "Recent Trends in Illegal Logging and a Brief Discussion of Their Causes," Marc Hiller et al. investigate the proximate causes of illegal logging, including access to forests and equipment, and economic factors. Further, they describe the local ecological and socio-economic impacts of illegal logging in and around the Gunung Palung National Park, Indonesia. In "Community-Based Logging and De Facto Decentralization: Illegal Logging in the Gunung Palung Area of West Kalimantan, Indonesia," Ramsay Ravenel draws on experience in the same area and describes how community-based logging began to flourish in the mid-1990s as forest resources became exhausted and local networks of power emerged into the void left by a failing central government.

In "Combating Corruption and Illegal Logging in Bénin, West Africa: Recommendations for Forest Sector Reform," Ute Siebert and Georg Elwert analyze the various actors and forms of corruption in the forest sector in Bénin. They also provide a critical review of national forest policy and economy in the country and conclude that problems in the sector are not intractable. They report that a latent majority of people disapprove of forest corruption and suggest that they can be mobilized for sanctions.

Interventions: Theory and Practice

Having considered theoretical issues and case studies, this volume turns to the question of what can be done about illegal logging. Two papers suggest ways to address illegal logging at the national level. Kenneth Rosenbaum provides a systematic overview of the basis for effective forest crime law in "Illegal Actions and the Forest Sector: A Legal Perspective." In "The Role of Monitoring in Cutting Crime," Wynet Smith analyzes the critical issue of monitoring forest crime.

Additional insights can be garnered from the activities of non-governmental organizations involved in addressing forest crimes, and especially those that have already demonstrated their effectiveness. In the "Illegal Logging and Deforestation in Andaman and Nicobar Islands, India," Pankaj Sekhsaria provides a case study detailing the historical context of illegal logging and how action-oriented research by an Indian non-governmental organization Kalpavriksh halted all logging on the islands. The case study demonstrates that a combination of investigative reporting and judicial action in the Indian high courts proves a potent combination stemming illegal logging. In "The Forest Integrity Network," Aarti Gupta and Ute Siebert describe the efforts of Transparency International–a widely lauded international organization combating corruption–to address illegal logging at the international policy level. The paper details the institutional framework of this young "virtual" organization of experts cooperating to transform the forest sector.

Synthesis

This volume concludes with a synthesis of both the papers described above and the major issues discussed during the conference proceedings. In "Illegal Logging in the Tropics: A Synthesis of the Issues," Ramsay Ravenel and Ilmi Granoff describe how illegal logging "in the woods" is connected to a larger political economy of illegal logging. They review the various policy prescriptions proposed by other authors and emphasize that in most cases illegal logging is not so much a new problem as it is the latest manifestation of deeper historical problems. Some of the most promising strategies for addressing illegal logging involve demanding that retailers and financial institutions include legality and sustainability criteria to their due diligence procedures.

REFERENCES

Callister, D. J. 1999. Corrupt and Illegal Activities in the Forest Sector: Current Understandings and Implications for the World Bank Forest Policy. Draft for discussion. Forest Policy Implementation Review and Strategy Development: Analytical Studies. The World Bank Group, Washington, DC, USA.

EIA/Telapak Indonesia. 1999. The final cut: Illegal logging in Indonesia's orangutan parks. London: Environmental Investigation Agency and Telapak Indonesia.

G8. 1998. Action Programme on Forests. Released at the Foreign Ministers Meeting, London, England, May 9, 1998. Available at: *http://www.library.utoronto.ca/g7/foreign/forests.html*.

Global Forest Watch. 2003. Project description and resources. World Resources International, Washington, DC, USA. Available at: *http://forests.wri.org/project_text.cfm?ProjectID=58*.

Global Witness. 1999. *Made in Vietnam, Cut in Cambodia*. London: Global Witness.

Greenpeace. 2000. Spotlight on the illegal timber trade: Asia-Pacific. Amsterdam: Greenpeace.

Mendoza, M. 2003. Losing ground to timber thieves: Illegal logging chips away at forests, but one court puts foot down. *Associated Press*, May 26.

United Nations Forum on Forests. 2002. Report on the Second Session (22 June 2001 and 4 to 15 March 2002). United Nations Economic and Social Council. Official Records, 2002. Supplement No. 22. United Nations, New York. Available at: *http://www.un.org/esa/forests/documents-unff.html#2*.

PROBLEM DESCRIPTION

Undercutting Sustainability:
The Global Problem
of Illegal Logging and Trade

Wynet Smith

SUMMARY. Illegal logging and trade is a global problem that under-cuts attempts to sustainably manage forests. While data is not consistently or comprehensively available, existing studies highlight a large problem in both tropical and non-tropical timber producing countries.

Wynet Smith is now undertaking doctoral studies at the University of Cambridge. She can be contacted c/o Department of Geography, University of Cambridge, Downing Place, Cambridge, CB2 3EN, UK. The author conducted the initial research for this article while employed as a Senior Associate at the World Resources Institute (WRI).

Many thanks are due to reviewers of an earlier draft report: Janet Abramowitz, Arnoldo Contreras-Hermosilla, John Hudson, David Kaimowitz, Jan McAlpine, Kristof Obidzinski, Charles Palmer, John Spears and various WRI staff. A special thanks to Steve Johnston of ITTO for his contributions and support.

This paper is an expanded assessment of the global extent of illegal logging presented as a summary in a World Resources Institute report, Combating Illegal Logging: A Review of Initiatives and Monitoring Tools.

[Haworth co-indexing entry note]: "Undercutting Sustainability: The Global Problem of Illegal Logging and Trade." Smith, Wynet. Co-published simultaneously in *Journal of Sustainable Forestry* (The Haworth Press, Inc.) Vol. 19, No. 1/2/3, 2004, pp. 7-30; and: *Illegal Logging in the Tropics: Strategies for Cutting Crime* (ed: Ramsay M. Ravenel, Ilmi M. E. Granoff, and Carrie A. Magee) The Haworth Press, Inc., 2004, pp. 7-30. Single or multiple copies of this article are available for a fee from The Haworth Document Delivery Service [1-800-HAWORTH, 9:00 a.m. - 5:00 p.m. (EST). E-mail address: docdelivery@haworthpress.com].

This paper presents the results of a literature review on the extent of the problem. Results indicate that most nations with significant portions of forest cover suffer some level of illegal logging. Estimates range from 80 percent for the Brazilian Amazon, to over 90 percent for Cambodia and over 70 percent for Indonesia. Further studies at both national and local levels will help to identify the complex set of factors that contribute to the problem and appropriate solutions. *[Article copies available for a fee from The Haworth Document Delivery Service: 1-800-HAWORTH. E-mail address: <docdelivery@haworthpress.com> Website: <http://www.HaworthPress. com> © 2004 by The Haworth Press, Inc. All rights reserved.]*

KEYWORDS. Illegal logging, sustainable forest management, forest industry

INTRODUCTION

Illegal logging and associated trade is being addressed at international fora such as the United Nations Forum on Forests and action plans of the G8 as well as through bilateral agreements between producing and consuming countries. An historic meeting on the topic of illegal logging was held in Bali, Indonesia in September 2001 and resulted in a declaration that was endorsed by all attending governments. Similar ministerial meetings are planned for Africa in early 2003 and for Latin America. A number of private sector companies and associations have made commitments or developed codes of conduct to improve their buying practices in the case of retailers or to improve their lending practices, in the case of financial institutions.

This flurry of activity is underway as illegal logging is now considered to be a major global problem (G8, 1998). But just how big is this problem? While there is no certainty as to its extent, some estimates do exist and cases of illegal logging activities are well documented in many countries. For example, the World Bank estimates that $5 billion in revenues is lost globally each year by governments through failure to collect forest taxes and that the market value of losses from illegal logging is over $10 billion per year (World Bank, 2001). One overview indicates that studies exist for over 70 tropical and non-tropical wood producing countries (Toyne, O'Brien, and Nelson, 2002).

This paper presents the results of a review of the literature on the problem in key forest producing countries, as well as some other more minor timber producing countries. The objective is to provide an overview of the pervasiveness of the problem of illegal logging and its wide reaching impacts. This background research is important because the extent and scope of the problem can guide the choice of effective solutions both at international and national levels.

It is important to note up front that the topic of illegal logging is incredibly complex. What is considered illegal is usually restricted to what is prohibited under national (or sub-national) legislation. Additionally, activities can span the entire commercial timber production chain, from the actual felling of trees without authorization or without meeting required operating standards, to the transport, processing and export of products. Fraudulent reporting practices often designed to reduce forestry charges and taxes, are also considered illegal. Non-timber harvesting forest crimes can include illegal forest clearing for agriculture, wildlife poaching, and fuelwood gathering.

METHODS

Research was undertaken to examine the extent of the problem and existing documentation. Information was collected over many months through literature reviews, web searches, and references passed on by other researchers, government officials and nongovernmental organization (NGO) staff. The priority was to look for data and information on major forest countries with significant industrial wood production. Forest cover, percent of forest area, and deforestation data were collected. The 2000 FAO (Food and Agricultural Organization) FRA (Forest Resource Assessment) statistics are used for all official forest cover, percent of land area and deforestation rates. Roundwood and industrial roundwood production and export statistics were collected from FAO's on-line database, FAOSTAT.

There are known limitations and problems with these various data. For example, FAO FRA data are not necessarily accurate and may over or under state amount of forest cover in given jurisdictions (Matthews, 2001). Official production numbers for roundwood (fuelwood and industrial roundwood) and industrial roundwood (all officially traded commodities) do not capture most illegally cut and processed timber and so inherently underestimate total production for countries with significant illegal logging problems. The use of a consistent data set, however, provides a common reference point. ITTO (International Tropical Timber Organization) production and trade statistics were also examined. All FAO and ITTO data quoted are year 2000. Other statistics are also included were readily available.

Key information sources on illegal logging include reports by NGOs, newspaper articles, journal articles, government reports, and forestry sector reports by multinational development banks and bilateral agencies. Overview documents of the problem were also key sources of background information (Brack, Gray, and Hayman, 2002; Callister, 1999; Contreras-Hermosilla, 2001). Collected reports and articles on illegal logging were reviewed for data on amounts of illegal logging, value of losses and impacts on revenues. All economic values have been converted to US$. The findings are summarized here.

RESULTS

Estimates of illegally harvested and traded timber are available for many key tropical timber-producing countries and range from 80% for Brazil to 50% for Cameroon and 33% for the Philippines. Illegal logging, however, is not just a problem in tropical countries. Canada, Estonia, Georgia and Russia all have documented cases of illegality and non-compliance in the forestry sector. Some countries, such as Indonesia and Cambodia, have received a tremendous amount of coverage. Others, such as the Democratic Republic of Congo and Peru, are the subjects of few reports.

This section summarizes results of the literature review by four major regions: Asia-Pacific, Africa, the Americas, and Europe-Russia. Tables provide key forest statistics and indicate the main type of illegal logging and/or existing estimates, as well as the types of data sources used.

Asia-Pacific

The Asia-Pacific region has been the largest source of tropical timber for many years. Deforestation rates have been high in many countries and forest degradation is also a problem (Dudley, Jeanrenaud, and Sullivan, 1995). It is also the most widely studied region in terms of illegal logging. Reports on the region and on specific countries date back to the early 1990s (Callister, 1992).

The region suffers from large-scale illegal logging and trade. An estimate from the early 1990s was that hundreds of thousands of acres were illegally logged annually in Asia-Pacific (Callister, 1992). Timber flows across borders, is often cut without authorization, transported in defiance of log export bans, and processed without licenses (Environmental Investigation Agency and Telepak, 2001b). The military in many countries is an integral part of these dealings, including Burma (Brunner, Talbott, and Elkin, 1998), Cambodia (Global Witness, 2001a; Le Billon, 2000) and Indonesia (Barber and Talbott, 2003).

The analysis for Asia-Pacific includes the eight countries in Asia with more than 10 million hectares of forest, as well as some smaller key timber producers (Vietnam, Cambodia and the Philippines) with smaller amounts of forest cover. Papua New Guinea (PNG) and the Solomon Islands are included as significant Pacific Island forest countries (Table 1). A number of countries, including Cambodia and Indonesia, have received a large share of attention, largely due to the scale of illegal logging. One early study estimated that in the decade preceding 1992, billions of dollars in foreign exchange, uncollected taxes, and foregone revenues were lost in the Asia-Pacific region (Callister, 1992).

China has the most forest cover in the region and is the largest producer of industrial roundwood. It has problems with illegal logging and instituted a log ban in 1998 to try to stop related environmental effects, such as flooding and

TABLE 1. Summary of known information on illegal logging and trade in the Asia-Pacific region.

Country	Forest Cover 2000		Deforestation Rate/year 1990-2000 (percentage)	Industrial Roundwood Production 2000 (000 m³)	Industrial Roundwood Export 2000 (000 m³)	Data sources on illegal logging
	Area (000 ha)	Percentage of land area (%)				
China	163,480	17.5	1.2	96,421	781.0	Articles
Indonesia	104,986	58	−1.2	31,357	1,503.7	Reports Articles
India	64,113	21.6	0.1	1,575	3.1	Articles
Myanmar	34,419	52.3	−1.4	3,574	949.0	Reports
Japan	24,081	64	n.s	17,987	3.8	Articles
Malaysia	19,290	58.7	−1.2	24,468	6,845.0	Articles
Thailand	14,800	28.9	0.7	2,894	0.2	Articles
Laos People's Democratic Republic	12,610	54.4	−0.4	866	204.8	Report
Vietnam	9,819	30.2	0.5	4,560	35.1	Report Articles
Cambodia	9,340	53	−0.6	809	0.1	Reports Articles
The Philippines	5,789	19.4	1.4	3,079	0.0	Articles Reports
Papua New Guinea (PNG)	30,601	67.6	−0.4	3,064	1,901.5	Articles Report
The Solomon Islands	2,536	88.8	−0.2	734	424.0	Mentioned in reports

soil erosion (Murray, 2001). China sources its wood from a number of countries where illegal logging activities are well documented (Toyne, O'Brien, and Nelson, 2002). Studies already indicate that their need for wood is fueling illegal logging in surrounding countries (Pearce, 2001), such as Russia (Morozov, 2000; Newell and Lebedev, 2000), Myanmar and Indonesia (Toyne, O'Brien, and Nelson, 2002).

Indonesia, with official forest cover of over 104 million hectares, has suffered high rates of deforestation (FWI/GFW, 2002). Indonesia ranks second after China in terms of total amount of official industrial roundwood production and has been one of the major tropical timber producers for the past 30 years. It also still ranks first in Asia in terms of export values and in total amounts exported of industrial roundwood. The country suffers from high levels of illegal logging. Research by various organizations, from NGOs (see, for example, Environmental Investigation Agency and Telepak, 1999; 2001a), research organizations such as CIFOR (Center for International Forestry Research), and government departments, all point to major levels of illegal forestry activities (see, for example, Barr, 2000; Brown, 1999; Obidzinski and Palmer, 2002). The existing studies complement each other and indicate that levels of illegal logging are at least 50% (Scotland, 2000), but more likely 75% (Scotland, Fraser, and Jewell, 1999) or greater (Palmer, 2001).

International demand for Indonesia's wood is considered a significant factor in these high levels of illegal activities (Environmental Investigation Agency and Telepak, 2001b). Estimates of economic loss include unpaid taxes and royalties of $600 million per year for the whole country (Baird, 2001). Other research, however, estimates unpaid taxes of $300 million in 2001 for one particular type of logging concession in just three districts of East Kalimantan (Obidzinski and Palmer, 2002). Indonesia's responses to growing awareness of the problem include institution of a log export ban and hosting the Bali Ministerial Forest Law Enforcement conference in September 2001. They have also signed an agreement with the United Kingdom that commits both countries to work together to address the problem of illegal logging.

India ranks as the country with the third largest amount of forest cover and yet is only ninth in terms of industrial roundwood production. Much of the wood is consumed domestically and India is also a major tropical timber importer (ITTO, 2001). Available information on illegal logging in India highlights the degree to which illegal logging in India is a village level problem, rather than an issue of industrial level activity. India also buys wood from many tropical timber producers with known illegal logging problems, including Indonesia and Myanmar. Unfortunately, India does not provide import statistics to ITTO (International Tropical Timber Organization), which makes it difficult to estimate how much illegal wood may be flowing to India (Smith, 2003).

In last 30 years, analysis shows that Myanmar–the seventh largest industrial roundwood producer–has undergone a substantial increase in unsustainable logging and that most of it is illegal (Brunner, Talbott, and Elkin, 1998). Myanmar's log exports to its two other major trading partners (Thailand and China) have consistently been reported as lower than import reports from these countries (Johnson, 2002). In 1995, Myanmar reported a volume of exports of logs that was 276,000 cubic meters less than that reported by importing countries. By 1994, Thailand's reported log imports from Myanmar were four times higher than Myanmar's declared exports (80% illegal), although this was only twice as much in 1995 (50% illegal) (Brunner, Talbott, and Elkin, 1998). Myanmar declared no log exports to China in 1995 but China declared imports of 500,000 cubic meters. Unfortunately, a lack of data does not allow for an assessment of trade with India, Myanmar's main reported log customer (Johnson, 2002). Other reports, however, show the types of activities that are taking place. For example, rebel groups and Thai loggers established at least 200 illegal sawmills around 1990 to process logs taken from outside concession areas (Callister, 1992).

Japan is the fifth ranked country in terms of forest cover. It has recently suffered some high profile forestry corruption cases (Anonymous, 2002d, 2002e, 2002f, 2002h, 2002j). As noted below in the Discussion, Japan purchases wood products from many of the other Asian countries and Russia, many of which have high levels of illegal logging.

Malaysia is a significant wood product producer and is the region's number one forest product exporter, both in terms of amount and value. In the early 1990s, estimates indicated 33% of all timber logged was illegally harvested (Callister, 1992). During the 1996-1999 period, an average of 93.5 cases per year of illegal logging were reported, of which twenty cases were considered to be very serious (Mohd and Yaman, 2000). Recent trade data analysis shows a widening gap in the import and export reports between Malaysia and China, its new largest customer. These discrepancies require further assessment (Johnson, 2002). Malaysia's impact on surrounding countries such as Indonesia is great. Although they did not attend the Bali Ministerial meeting in September 2001, they have recently bowed to international pressure and announced they will cease importing Indonesian logs (Anonymous, 2002c; Harrabin, 2002).

Newspaper reports indicate a significant occurrence of illegal logging and trade activities in Thailand (e.g., see Kasem, 1998; Lebel, 1998). Thailand's need for wood, due to the depletion of its own forests, is impacting surrounding countries, including Laos and Cambodia (see below). Thailand's log ban is not considered to have been an effective deterrent to illegal logging and deforestation (Macan-Markar, 2001).

In the Peoples Republic of Laos, the volume of illegal logging has been estimated to be at least one-sixth of the legal harvest (TFP, 2000). A comparison of export and import statistics between Laos and Thailand revealed twice as much reported as entering Thailand as leaving Laos (Thongleua and Castren, 1999 as cited by TFP, 2000). In another example of irregular reporting, Laos had an annual export volume of 200,000 cubic meters from Saravan province but submitted a proposal to Vietnam that stated harvests were 1.5 million cubic meters (Greenpeace, 2001b). Other documented irregularities include downgrading of logs. One report states that the royalties paid to the government were less than 50% of what was due in every year but one between 1995 and 1998 (TFP, 2000). A lack of transparency in the forest management process makes any on-going assessment of illegal activities difficult.

Vietnam has lost about one-third of its forest cover between 1985 and 2000, partly as a result of illegal logging (Linh, 2000). Illegal logging is also blamed for flooding in areas of the country (Anonymous, 1999). Vietnamese papers report that some 55,000 cases of illegal logging were uncovered during 2001, leading to the imposition of more than 7,600 fines (Anonymous, 2002i). Vietnam also imports frequently from surrounding countries and a significant proportion of trade is considered illegal (Global Witness, 1999a). Government attempts to crackdown on the internal illegal logging problem are resulting in attacks on forest rangers (Associated Press, 2002). For example, between 1996 and 2000, a total of 12 forest rangers were killed and 490 injured in attacks by loggers (Linh, 2000).

Cambodia lost 2.6 million hectares or 14.4% of its forest cover in 30 years. Unchecked logging–much of it illegal–has played a significant part in this deforestation (Global Witness, 2000). Cambodia's processing capacity was 2.5 million cubic meters in 1997 while the sustainable yield of the forests was probably only 500,000 cubic meters (Global Witness, 1999b). The World Bank reported that 4 million cubic meters of timber were harvested in 1997 (as cited by Global Witness, 2000). This level of harvest exceeds the sustainable yield of Cambodia's forests, which is in the order of 0.5 to 1.5 million cubic meters according to government sources (Ouen, Sokhun, and Savet, 2001; Savet, 2000).

Government losses include $77 million in 1997 from failure to capture revenues (Magrath and Grandalski, 2000). From 1995 to 1997, forest exploitation formally contributed an average of $15 million per year to Cambodian government revenue whereas "informal" payments were estimated at $200 million in 1997 alone (Global Witness, 1999b). The estimated 3 to 4 million cubic meters of illegally logged timber, as noted above, equaled a revenue loss to the government of $60 million (World Bank as cited by Global Witness, 1999b). In 1999, the government cracked down and there was a dramatic decrease in illegal logging. In 2000/2001, Global Witness contends that there has been an increase again in incidences of illegal logging (Global Witness, 2001a). In January 2002, the government again imposed a temporary ban on logging and in July 2002, a major company had its concessions cancelled (Carmichael, 2002).

In the Philippines, illegal logging has been blamed for wiping out the forests (Bengwayan, 1999). There is no authoritative source on the measurement of the magnitude of illegal logging, but it is publicly acknowledged as having been significant, especially during the 1960s and 1970s. There is a large gap (2.3 million cubic meters) between the legal supply of industrial round wood (2.7 million cubic meters) and the demand (5 million cubic meters). This deficit–about 45 percent of domestic requirements–comes from illegal sources (Acosta et al., 2000). Illegal logging and trade was so widespread that estimates are that the Philippines lost $800 million annually in the early 1990s (Dauvergne, 1997). In summer of 2002, the Philippine Department of Natural Resources stepped up enforcement action on illegal logging and confiscated hundreds of thousands of dollars worth of timber (for example, see DENR, 2002a, 2002b, 2002c; DENR, 2002d).

In the Pacific, two important forest producing countries are PNG and the Solomon Islands. In the Solomon Islands, there have been reports of transfer pricing (Sizer and Plouvier, 2000). In PNG in the 1990s, there was an in-depth inquiry to investigate illegal activities in the forestry sector. After two years of study, the Commission produced a report that documented pervasive forest crime in the sector, from illegal permits granted to companies, collusion of regional and national level politicians with foreign companies, to massive tax evasion and transfer pricing. These fraudulent activities were estimated to have led to losses in national income equivalent to the annual aid the country received (Glastra, 1999). There are also anomalies in trade data. Official PNG log export data for 1996 shows exports of 1,390,000 cubic meters at an average price of $140 per cubic meter. Japan's records show imports for the same period to be 1,852,000 cubic meters at an average price of $175 per cubic meter. This amounts to a discrepancy of $130 million in 1998 (Greenpeace, 2001b).

Africa

Many African countries have suffered extensive deforestation and forest degradation in recent years (FAO, 2001). International investments and transnational logging companies active in Africa increasingly are Asian, competing with (and in some cases replacing) the European companies whom long dominated the logging sector. A number of countries have known problems with illegal logging and trade, some more egregious then others. Systemic corruption exists in many of these countries. In Africa, 17 countries have more than 10 million hectares of forest cover (see Table 2). Sixteen countries had industrial roundwood production of over 1 million cubic meters in 2000 (FAO, 2002). Results are shown for most major forest regions that have significant industrial roundwood production and reported incidents of illegal logging. A few other countries, such as Tanzania and Benin, where most wood is consumed domestically, are also included. This section first presents results for the Congo Basin countries, followed by results for West African and then East African countries.

The Democratic Republic of Congo (DRC) production numbers are far below those of other central African countries (ITTO, 2001). A United Nations Expert panel on the Illegal Exploitation of Natural Resources and Other Forms of Wealth in the DRC found that timber extraction and its export are rife with unlawfulness and illegality (UN, 2000). Activities include logging without authorization and in violation of legislation, and export without appropriate certificates. Wood flows from the DRC through Uganda and other East African

TABLE 2. Summary of known information on illegal logging and trade in Africa.

Country	Forest Cover 2000		Deforestation Rate/year 1990-2000 (percentage)	Industrial Roundwood Production 2000 (000 m^3)	Industrial Roundwood Export 2000 (000 m^3)	Data sources on illegal logging
	Area (000 ha)	Percentage of land area (%)				
Democratic Republic of Congo	135,200	60	−0.4	3,727	18.6	Report
Tanzania	38,800	43.9	−0.2	2,314	12.1	Articles
Zambia	31,246	42	−2.4	834	0.8	
Mozambique	30,601	39		1,319	74.0	
Cameroon	23,858	51.3	−0.9	2,960	575.0	Reports Articles
CAR	22,907	36.8	−0.1	1,011	250.0	Reports Articles
Congo, Republic of	22,060	64.6	−0.1	646	757.0	Reports
Gabon	21,800	84.7	n.s	2,584	2,584.0	Reports Articles
Zimbabwe	19,040	49.2	−1.5	1,136	0.0	
Kenya	17,000	30	−0.5	1,977	0.0	Articles
Nigeria	13,500	15	−2.6	9,418	7.3	Report summaries
Ghana	6,340	27.8	−1.7	1,087		Reports Articles
Liberia	3,480	31.3	−2.0	337	637.0	Reports
Benin	2,650	24	−2.3	332	55.0	Report

Results shown for all countries with over 10 million hectares of forest cover and over 300,000 m^3 industrial roundwood production, as well as for a few additional countries of interest.

states (UN, 2000). One company alone has been exporting approximately 48,000 cubic meters of timber per year since 1998. This is a significant number, given official export numbers of 50,000 cubic meters for 1999 and 2000 and 46,000 cubic meters for 1998.

Studies for Cameroon indicate that at least 50% of the concessions/licenses are illegal (Global Forest Watch Cameroon, 2000). The country is one of the most important wood producers and exporters in Africa, and ranks forth in ITTO producers after Malaysia, Gabon and PNG (ITTO, 2001). A study undertaken in 2000 indicates that up to 90% of logging licenses are operating illegally (Greenpeace, 2000). Illegal activities include logging outside of concessions, logging above limit, logging inside of inactive concessions, logging inside allocated concessions but outside annual harvest areas, and logging inside of protected areas. Analysis indicated that over half of operating licenses in 1997-98 failed to comply with current regulations and that less than a third of concessions allocated that year fully complied with guidelines set out in the new forestry legislation (Global Forest Watch Cameroon, 2000). Internal Ministry of Environment and Forests (MINEF) reports from a series of field inspections conducted in December 1999 indicate major illegalities in almost all concessions visited (Greenpeace, 2000). Undeclared timber volume was close to one-third of production in the Eastern Province in 1992-93 (Glastra, 1999). Cameroon has appointed Global Witness as an independent monitor to help audit compliance with forestry legislation (Global Witness, 2001b). In 2002, the government signed an agreement with Global Forest Watch to undertake satellite monitoring of concessions (WRI, 2002).

There are also problems in Gabon, another important African wood producer. The country is the largest exporter of logs in Africa (ITTO, 2001). There are widespread problems with the implementation of forestry laws and regulations. As of early 2000, only 5 of 200 logging companies had initiated work on a management plan (Global Forest Watch Gabon, 2000). Analysis of ITTO trade data shows imbalances between Gabon's reported log exports and China's reported log imports for both 1998 and 2000 (27% difference in 1998 and 26% difference in 2000 between exports and imports). Gabon's log export figures for 1998 and 2000 are significantly lower than China's import figures for the same years (Johnson, 2002).

There are reports of illegal logging for both the Republic of Congo and the Central African Republic. Reports indicate there may be as many as 400 illegally granted concessions in the Republic of Congo (Vidal, 2002a). Illegally logged timber is believed to flow to many European countries, including the United Kingdom (Vidal, 2002b).

Forestry is an important part of Ghana's economy; it has traditionally been the third-largest export commodity (Glastra, 1999). At least 90% of its forests have been logged since 1940s (Sayer, Harcourt, and Collins, 1992). Ghana is also known to have had a major illegal logging problem. The scale of illegal exports became clear to the Ghanian authorities in 1987 when several shiploads of illegally harvested wood were halted and fraud involving export documents was revealed (Glastra, 1999). In 1994, the Ghanian Forestry Department estimated about 34% of logs were illegally harvested (Glastra, 1999). Corruption is also a problem, with district forestry officials, concessionaire, communities and law enforcement agencies involved. During the worst period, total losses from illegal logging were approximately $28.97 million, equivalent to about 2% of the GDP and foreign exchange losses were estimated at $600,000 each month. The government instituted reforms that appear to be having an effect (Glastra, 1999).

Other countries, such as Liberia, are identified as countries with severe governance issues, as well as resource management problems. Basic forest management standards have not been adopted in law. There are also major concerns that revenues from timber harvesting help fund regional conflict (Global Witness, 2001c). Logging is not illegal in the sense that it contravenes national laws. Logging appears, however, to be unsustainable and to be funding regional conflict

The few remaining closed forests in Nigeria suffer from illegal exploitation and fraudulent practices have been well documented (Glastra, 1999). Little information exists for Benin. The existing research indicates that there is widespread illegal logging. Most of Benin's annual demand for wood of 7,779,000 cubic meters is being met by the exploitation of natural forests. Approximately 90% of timber is felled illegally with the consent or encouragement of state forest officers (Siebert, 2001). In one area, 29,500 cubic meters of wood have been illegally logged between 1996 and 1999, with a considerable increase between 1998 (6,707 cubic meters) and 1999 (approx. 11,000 cubic meters). Siebert (2001) also estimates that 80 to 90% of timber "imports" from the neighboring countries are, in fact, re-imports of timber illegally felled in Benin.

In Tanzania, an estimated 500,000 hectares of forests disappear each year due to various causes (Glastra, 1999). The few remaining closed forests suffer from illegal exploitation and fraudulent practices abound (Odhiambo, 1999). Government officials were implicated in the illegal timber trade in Tanzania (World Rainforest Movement, 1999). In Kenya, logging inside Mount Elgon Forest Reserve has continued despite a presidential ban since 1996 (Glastra, 1999).

The Americas

North America is a significant timber producer. South America has begun to be a more important tropical timber producer internationally. Companies from Asia and North America are logging in many countries there. Illegal logging problems occur in many countries across the Americas. The problem appears to be especially significant in some Central and South American countries, although Canada and the United States also have problems with timber theft and other illegal activities. This section summarizes results for South America and North/Central America (see Table 3).

Brazil is the most significant country in South America both in terms of forest cover and timber production. The timber industry has grown extensively in the last 20 years. Illegal logging numbers in the Brazilian Amazon range from 80 to 95%. IBAMA, the Brazilian environmental agency, released figures that indicate nearly 30 million cubic meters of wood was harvested in 2000 even

TABLE 3. Summary of known information on illegal logging and trade in the Americas.

Country	Forest Cover 2000		Deforestation Rate/year 1990-2000 (%)	Industrial Roundwood Production 2000 (000 m^3)	Industrial Roundwood Export 2000 (000 m^3)	Data sources on illegal logging
	Area (000 ha)	Percentage of land area (%)				
Brazil	543,905	64.3	−0.4	102,994	751.8	NGO and Government reports
Peru	65,200	50.9	−0.4	927		Report, articles
Bolivia	53,608	48.9	−0.3	468	3.2	Reports, articles
Colombia	49,601	47.8	−0.4	3,783		
Venezuela	49,506	56.1	−0.4	1,549		
Argentina	34,648	12.7	−0.8	6,652		
Paraguay	23,372	58.8	−0.5	4,044		NGO research
Guyana	16,879	78.5	−0.3	308	54.0	Noted in reports
Chile	15,536	20.7	−0.1	24,437	681.0	NGO research
Suriname	14,113	90.5	n.s	184	10.0	
Ecuador	10,557	38.1	−1.2	5,719		
Central and North America						
Canada						Reports, articles
United States						Articles
Mexico	55,205	28.9	−1.1	8,105	9.6	Articles
Honduras	5,383	48.1	−1.0	759	40.0	Research
Nicaragua	3,278	27.0	−2.3	228		Research
Panama	2,876	38.6	−1.6	77		
Guatemala	2,850	26.3	−1.7	466.6		
Costa Rica	1,968	38.5	−0.8	1,687	0.2	Report
Belize	1,348	59.1	−3.0	61.6		

Results only shown for countries with over 10 million hectares of forest cover for South America and 1 million ha for Central America.

though only 4.5 million cubic meters of wood were authorized (Greenpeace, 1999). This implies that 85% of all logging in 2000 was illegal (Greenpeace, 2001a). In 1997, the Secretaria de Assuntos Estrategicos (SAE) estimated that 80% of all wood logged in the Amazon was extracted illegally (Greenpeace, 2001a). Government audits of concession management plans revealed that at least 70% of them did not comply with regulations and that in some municipalities, 95% of the concessions were illegal (Glastra, 1999). Recent investigative work by Greenpeace Brazil (see Smith, this volume) has helped to create awareness and provide political support to IBAMA. In late 2001, IBAMA undertook raids and seized timber worth millions of dollars. The government of Brazil passed a decree suspending all logging and trade in mahogany. Many importing countries have promised to not import illegally harvested wood (Anonymous, 2002b), though shipments were still imported into the United States in 2002 (Anonymous, 2002a; Heilprin, 2002).

Reports of illegal logging in Peru exist but there is no current estimate for total amounts or percents of wood harvested. Trade data for Peru indicates there may be a problem as import and export numbers vary considerably. In 2000, the Peruvian government announced unauthorized logging by an American company. The estimated value of mahogany being harvested in this area was $37 million to $40 million (Associated Press, 2000). In July 2002, clashes took place between indigenous peoples and loggers, who were allegedly illegally invading indigenous reserves (Powers, 2002; Webber, 2002).

Bolivia had a major illegal logging problem in the 1990s. A new law was passed in 1996 in an effort to reduce extensive illegal acts and corruption in the management of the forestry sector (Contreras-Hermosilla and Vargas Rios, 2002). The new law included various "textbook" provisions to improve prevention, detection, and control of such acts. A recent assessment by government, however, shows that while illegal practices have been reduced substantially, at least 80% and perhaps as much as 90% of all forest clearing is still illegal (Contreras-Hermosilla, 2002).

A one million hectare illegal colonization scheme, involving top officials in Paraguay, was reported in the late 1990s. An illegal trafficking problem on the border with Brazil exists and illegal trade in wildlife and wildlife products is high (Glastra, 1999).

In Chile, data from NGO reports indicates that there are problems with illegal logging of Alerce (*Fitzroya cupressoides*), a species declared a national monument in the country. Any logging of this species is illegal (Lara, Lobos, and Thiers, 1999).

Minimal reports are also available for Guyana and Suriname. For example, an Asian company was allegedly logging without approval where the value of illegally logged trees was estimated at $6.75 million (Forest Peoples Programme, 1998). In 1997, reports of large-scale illegal logging in the interior began to circulate (Sizer and Plouvier, 2000).

Although relatively small both in terms of overall size and amounts of forest cover, reports of illegal logging exist for many Central America countries. Illegal logging has accounted for at least 26 to 58% of the total market of timber in Costa Rica. Estimates derived from a survey indicate that illegally cut timber accounts somewhere between 28 and 41% of all timber cut. Only 4% of the illegal wood comes from areas under management; most come from pasturelands (Arce et al., 2000). An illegal logging study is underway in Honduras and Nicaragua. Current estimates indicate that 80% of logging is illegal in Honduras (ODI, 2002).

There are also many reports for North American countries. In Mexico, community and indigenous forests have been subject to illegal logging. Community members have protested against these illegal activities (Dillon, 1999) and there are on-going conflicts over forest resources (Klooster, 2000). Canada and the United States both have documented problems with illegal activities in the forestry sector.

There is no overall estimate of illegal logging or corruption in the forest sector in Canada. A number of reports exist, however, that indicate at least some degree of timber theft, irregular-scaling and reporting practices by forestry companies, as well as problems with non-compliance with existing regulations in terms of both logging and processing standards (e.g. see SLDF, 2001; SLDF and Wildlands League, 2001). In 1998, The Royal Canadian Mounted Police estimated that at least US$130 million was lost annually through timber theft in British Columbia alone (Weatherbe, 1998).

Europe-Russia

During the early 1990s, there was not much documented literature on illegal logging for Europe and Russia. This has changed in the late 1990s and early 2000s. NGOs have published a number of studies in recent years. Many World Bank documents address the issue as well. In general, a number of major problems in the region exist. These include substantial quantities of illegal Russian wood flowing both east and west, and illegal wood flowing out of many former eastern bloc countries (Table 4).

Russia contains the most forests in the region. It is also a significant timber producer. A crumbling state infrastructure and rampant corruption, however, have put the forests at risk. Estimates of illegal logging in Russia include an overall figure of 20%–5 million cubic meters–for the country as a whole (Morozov, 2000; Newell and Lebedev, 2000). WWF Russia estimates illegal logging exceeds 30% (Kotlobay and Ptichnikov, 2001). It is thought that Siberia and the Far East suffer levels around 50%. For example, Greenpeace estimates that 50% of logging in the Far East is considered illegal (Morozov, 2000). In certain forest districts, 80% of timber in storage is illegal (Kotlobay and Ptichnikov, 2002). WWF Russia estimates the Russian governments lose over $1 billion annually in uncollected taxes, fees and other revenues (Kotlobay and Ptichnikov, 2001).

After independence in 1991, lands in Estonia were transferred back to the private domain. Widespread rumors of illegal harvesting of wood exist. Landowners, who are often infrequent visitors to their land, have sometimes discovered that the trees have been cut down without their permission (pers.

TABLE 4. Summary of known information on illegal logging and trade in Europe and Russia.

Country	Forest Cover 2000		Deforestation Rate/year 1990-2000 (percentage)	Industrial Roundwood Production 2000 (000 m³)	Industrial Roundwood Export 2000 (000 m³)	Data sources on illegal logging
	Area (000 ha)	Percentage of land area (%)				
Russia	851,392	50.4	n.s.	478,699	30,835.0	Articles
Romania	6,448	28.0	0.2	10,116	531.0	Report
Georgia	2,988	43.7	n.s.		56.0	Reports, articles
Estonia	2,060	48.7	0.6	7,270		Report, articles
Albania	991	36.2	−0.8	119	0.4	Report

Georgia is included in FAOSTAT in Asia grouping. It is included here as part of Europe.

comm., 2000). NGOs in Estonia are now undertaking their own study on the topic. To date, their results indicate that of all the timber felled 5% involves outright forest theft, 20% involves violation of felling regulations, 20% do not have or fabricate documentation, 15% use off-shore firms or secret identities, and around 50% do not pay employer's taxes and income tax (Ahas, 2001).

Concern about illegal logging in Georgia is high and there are many allegations about the problem (Chelidze, 1999). Government statistics vary widely and range up to 2.5 million cubic meters per year but trade data indicates a significant problem (Siry, 2000). In May 1998, Parliament adopted a resolution forbidding commercial cutting and logging in entire territory of Georgia until the new Forestry Code is adopted. The resolution is in effect but is not enforced. Corruption appears to be widespread through government and various sectors and the Government has set up an anti-corruption bureau to help address the issue (Ugulava, 2001).

In Albania, data from the Illegal Logging Monitoring System indicates that the volume of illegal logging in 1999 amounted to 72,606 m³ of logs. This figure is 7.3 times less than the volume of logs illegally removed in 1997. Despite claims that illegal logging is subsiding, concerns are still high (ACER, 2001).

DISCUSSION

A Global Trade Problem

Recent studies indicate illegal logging accounts for a significant portion of tropical forest products sold in international markets worldwide, including up to one-third of timber-based products imported by G8 and China (Toyne, O'Brien, and Nelson, 2002). FOE-UK estimates that over 62% of tropical forest products entering the United Kingdom may be illegal (Matthew, 2001a, 2001b). FERN (2001) estimates additionally that 27% of logs and 18% of sawnwood entering Germany and 25% of logs and 33% of sawnwood entering Spain are illegal. Illegal imports into France likely include 10% of logs and 48% of sawnwood while 50% of imports into the Netherlands is likely illegal (FERN, 2001).

Illegal logging also accounts for an important component of trade in forest products from temperate and boreal forests. At least 20 percent of wood from the Russian Far East is thought to be illegally cut and traded with China (Morozov, 2000; Newell and Lebedev, 2000). China's log ban is a factor in increased illegal logging in many countries, including Russia and Myanmar (Pearce, 2001). Analysis for the Republic of Georgia indicates that a major portion of the forest products trade to Turkey may be illegal (Siry, 2000).

Some countries play significant roles in terms of their purchasing power (for further details see Toyne, O'Brien, and Nelson, 2002). China's increasing demands for timber and wood products is impacting many countries, including Russia and Myanmar. Japan's demand for forest products also influences trade and harvesting activities in many of the same countries (Dauvergne, 1997). European countries purchase significant amounts of wood from Asia and Africa (FERN, 2001; Matthew, 2001b). The United States imports large quantities of timber from Brazil, including the now banned mahogany (Berman, 2002; Heilprin, 2002; Jordan, 2001)

Good Numbers Are Hard to Find

There are a number of limitations, however, with current estimates (Smith, 2002). The data available for each country and region come from a number of sources, and are often of variable age and quality. Figures are collected using a variety of methods, which makes comparing data between and even within countries difficult. For many countries, data does not exist or is not accessible to the public.

Reasons for the lack of comprehensive comparable data include the inherent difficulty of trying to document illegitimate activities. As noted earlier, these activities often take place in remote regions, in situations that are not transparent, and often with the involvement of key officials. The range of illegal activities that can take place is great, making detection and monitoring difficult and placing a heavy burden on law enforcement agencies, which often lack the necessary capacity for monitoring and assessment. Even where officials and civil servants observe problems, the political will to report real conditions is rare due to the potential risks (career and personal safety) involved.

The absence of data for a particular country does not necessarily mean that there are no illegal or corrupt activities in its forest sector. A broad lack of transparency or freedom of inquiry could prevent citizens or outsiders from undertaking research and publicly identifying problems. Alternatively, no one may have yet paid particular attention to that country (Callister, 1999). More attention has been paid, for example, to tropical than temperate forests in the past by the media and NGOs (FAO, 2001). Additionally, the results reported here are generally restricted to English language reports, articles and books and may not have unearthed all relevant source materials.

While the current figures are problematic and can only be considered general assessments and often 'best guesses,' they are the best available and do provide sufficient evidence that action must be taken. The scale and consequences of the problem are tremendous. Current estimates underscore how important it is for governments, the private sector and civil society to address the issue and try to solve (or substantially reduce) the problem of illegal logging. Further monitoring and study are needed, however, to address the problem of illegal logging (see Smith this volume).

CONCLUSION

An analysis of existing literature shows a major global problem. Illegal logging activities–of various types–happen in many forest-rich and timber producing countries. Levels in some countries–including Brazil, Cambodia, and Indonesia–are almost as great as legal production. Illegal logging activities are contributing to deforestation and forest degradation, as well as having significant economic and social impacts. These high levels of illegal logging undercut attempts to improve forestry and achieve sustainable forest management.

Fortunately, signs of hope and political will do exist. There is movement at both the international and national levels. As noted, the G8 developed an action plan on forests that included recommendations related to illegal logging and trade. The Ministerial meeting in Bali, Indonesia in September 2001, resulted in all countries in attendance endorsing a Ministerial declaration to address the problem. After years of stalemate at the International Tropical Timber Council, decisions were passed in November 2001 that called for action.

There are also signs of progress at national levels. For example, there have been significant decisions made in Cameroon. The government appointed Global Witness as the independent monitor in June 2001 and the government signed an agreement with Global Forest Watch in June 2002 to have it undertake remote sensing monitoring of logging activities. In Cambodia, the government, at the request of donors, set up the Forest Crimes Monitoring Project in 1999. Although opinion is mixed on its impact to date, the Cambodian government instituted a temporary ban on logging in January 2002 and cancelled one company's concessions in June 2002 (Carmichael, 2002). In Indonesia, the government has begun to take steps to address the problem. Besides hosting the Ministerial meeting, the government has instituted a log ban and signed a MOU with the United Kingdom in April 2002 (Anonymous, 2002g).

These initial steps must be followed by concrete action plans that will help address the problem of illegal logging and associated trade. Besides far-reaching policy and legislative reform, monitoring is a key element of any framework to reduce illegal logging (see Smith this volume). More substantive research on the problem in specific countries is needed as the range of activities and underlying causes and contributing factors are so broad.

REFERENCES

ACER. 2001. *Illegal Logging Independent Study.* Tirana, Albania: Albanian Center for Economic Research.

Acosta, R. T., E. S. Guiang, L. A. Paat, and W. S. Pollisco. 2000. "The Control of Illegal Logging: The Philippine Experience." *World Bank/WWF workshop, Controlling Illegal Logging, Jakarta, Indonesia, 2000.*

Agence France Presse. 2002. "Vietnam PM Orders Crackdown on Violence by Illegal Loggers." Forest Conservation Portal. January 19. Available online at *http://forests. org/articles/reader.asp?linkid=6865.*

Ahas, R. 2001. "Illegal Logging in Estonia." *Taiga News*: 11.

Anonymous. 1999. "Logging Blamed in Vietnam's Second Wave of Floods." Environment News Service. December 9. Available online at *http://ens-news.com/ens/ dec1999/1999-12-09-01.asp.*

Anonymous. 2002a. "Brazil Seizes Record Haul of Illegal Mahogany." *Reuters News Service*, June 26.

Anonymous. 2002b. "Europe Rejects Brazilian Mahogany Imports." Environment News Service. March 29. Available online at *http://ens-news.com/ens/mar2002/ 2002-03-29-01.asp.*

Anonymous. 2002c. "Malaysia Bans Indonesian Log Imports." BBC News. June 25. Available online at http://www.news.bbc.co.uk/1/hi/world/asia-pacific/2065596.stm.

Anonymous. 2002d. "Prosecutors Indict Suzuki, Aide on Charges of Accepting Bribes." *Japan Times*, July 11.

Anonymous. 2002e. "Prosecutors Nab Suzuki: Lawmaker Arrested After Lower House OK." *Jakarta Post*, June 20.

Anonymous. 2002f. "Prosecutors Seek Warrant for Suzuki." *Japan Times*, June 18.

Anonymous. 2002g. "RI, UK to Curb Illegal Logging." *Jakarta Post*, April 25.

Anonymous. 2002h. "Suzuki Served Fresh Arrest Warrant Over Bribery." *Japan Times*, August 2.

Anonymous. 2002i. "Yamarin Escaped Illegal Logging Penalties: Link Suspected Between Hokkaido Firm and Alleged Bribes to Suzuki." *Japan Times*, June 21.

Arce, J. J. C., M. C. Calvo, R. V. Soto, C. M. Rodriguez, and M. G. Flores. 2000. *La Tala Ilegal en Costa Rica: un analisis para la discusion. Executive summary available in English.* San Jose: CATIE.

Associated Press. 2000. "Peru Seeks to Stem Amazon Logging." Forest Conservation Portal. July 7. Available online at *http://forests.org/recent/2000/perseeks.htm*.

Associated Press. 2002. "Illegal Loggers Attack Police, Forest Rangers in Northern Vietnam." Environment News Network. November 8. Available online at *http:// enn. com/news/wire-stories/2002/11/11082002/ap_48915.asp*.

Baird, M. 2001. "Forest Crime as a Constraint on Development." *Presentation at the Forest Law Enforcement and Governance Conference, Bali, Indonesia, 2001.*

Barber, C. and K. Talbott. 2003. The Chainsaw and the Gun: The Role of the Military in Deforesting Indonesia. *Journal of Sustainable Forestry* 16: 137-166.

Barr, C. 2000. *Profits on Paper: The Political-Economy of Fiber, Finance and Debt in Indonesia's Pulp and Paper Industries.* Bogor, Indonesia: Center for International Forestry Research.

Bengwayan, M. 1999. "Illegal Logging Wipes Out Philippines Forests." Environment News Service. October 11. Available online at *http://ens-news.com/ens/oct1999/ 1999-10-11-01.asp*.

Berman, D. 2002. "U.S. Continues to Hold Brazilian Imports; Ibama Forest Chief Sacked," Greenwire.

Brack, D., K. Gray, and G. Hayman. 2002. *Controlling the International Trade in Illegally Logged Timber and Wood Products.* London: Royal Institute for International Affairs.

Brown, D. W. 1999. *Addicted to Rent: Corporate and Spatial Distribution of Forest Resources in Indonesia; Implications for Forest Sustainability and Government Policy.* Jakarta: Indonesia-UK Tropical Forest Management Programme.

Brunner, J., K. Talbott, and C. Elkin. 1998. *Logging Burma's Frontier Forests: Resources and the Regime.* Washington, DC: World Resources Institute.

Callister, D. 1992. *Illegal Tropical Timber Trade: Asia-Pacific.* Cambridge: TRAFFIC International.

Callister, D. 1999. *Corrupt and Illegal Activities in the Forestry Sector: Current Understandings and Implications for World Bank Forest Policy.* World Bank.

Carmichael, R. 2002. "GAT Axed for Illegal Logging." *Phnom Penh Post*, June 21.

Chelidze, N. 1999. "Illegal Logging Damages Caucasus Mountain Forests." Environment News Sevice. May 7. Available online at *http://forests.org/archive/europe/ illlogge.htm*.

Contreras-Hermosilla, A. 2001. *Law Compliance in the Forestry Sector: An Overview.* Washington DC: World Bank.

Contreras-Hermosilla, A. 2002. *Policy and Legal Options to Improve Law Compliance in the Forest Sector.* Rome: FAO.

Contreras-Hermosilla, A. and M. T. Vargas Rios. 2002. *Social, Environmental and Economic Dimensions of Forest Policy Reforms in Bolivia.* Washington, DC: Forest Trends.

Cotton, C. and T. Romine. 1999. *Facing Destruction: A Greenpeace Briefing on the Timber Industry in the Brazilian Amazon.* Amdsterdam: Greenpeace.

Dauvergne, P. 1997. *Shadows in the Forest: Japan and the Politics of Timber in Southeast Asia.* Cambridge and London: MIT Press.

DENR. 2002a. "4 Pump Boats Caught with P3.3M-Worth of Hot Lumber as DENR Region 5 Official Bares Harassment." Department of Environment and Natural Resources, the Republic of the Philippines. August 14. Available online at *http://www1.denr.gov.ph/article/articleview/370/1/138/.*

DENR. 2002b. "Alvarez Orders Crackdown on Illegal Logging in Quezon." Department of Environment and Natural Resources, the Philippines. June 27. Available online at *http://www1.denr.gov.ph/article/articleview/241/1/106/.*

DENR. 2002c. "DENR Nabs P3.5M Hot Logs in Quezon." Department of Environment and Natural Resources, the Republic of the Philippines. August 23. Available online at *http://www1.denr.gov.ph/article/articleview/383/1/138/.*

DENR. 2002d. "DENR Nabs P7-M 'Hot' logs in Agusan Del Norte as Anti-Illegal Logging Operations Goes Hi-Tech." Department of Environment and Natural Resources, the Philippines. June 19. Available online at *http://www1.denr.gov.ph/article/articleview/236/1/106/.*

Dillon, S. 1999. "Mexicans Protest Illegal Logging." *San Jose Mercury News*, May 3.

Dudley, N., J.-P. Jeanrenaud, and F. Sullivan. 1995. *Bad Harvest? The Timber Trade and the Degradation of the World's Forests*. London: Earthscan and WWF.

Environmental Investigation Agency and Telepak. 1999. *The Final Cut: Illegal Logging in Indonesia's Orangutan Parks*. Washington, DC: Environmental Investigation Agency.

Environmental Investigation Agency and Telepak. 2001a. *Illegal Logging in Tanjung Puting National Park: An Update on the Final Cut Report*. London: Environmental investigation Agency.

Environmental Investigation Agency and Telepak. 2001b. *Timber Trafficking: Illegal Logging in Indonesia, South East Asia and International Consumption of Illegally Sourced Timber*. London: Environmental Investigation Agency.

FAO. 2001. *State of the World's Forests 2001*. Rome: Food and Agricultural Organization.

FAO. 2002. "FAOSTAT." Food and Agricultural Organization. Available online at *http://www.fao.org.*

FERN. 2001. *Special Report: EU Illegal Timber Imports*. Moreton-in-Marsh, Gloucestershire: Forests and the European Union Resource Network.

Forest Peoples Programme. 1998. "Illegal Logging Discovered in Guyana." April 20, 1998. Available online at *http://forests.org/recent/1998/illogdi.htm.*

FWI/GFW. 2002. *The State of the Forest: Indonesia*. Bogor and Washington, DC: Forest Watch Indonesia and World Resources Institute.

G8. 1998. *Action Programme on Forests*. G8.

Glastra, R. Editor. 1999. *Cut and Run: Illegal Logging and Timber Trade in the Tropics*. Ottawa: International Development Research Centre.

Global Forest Watch Cameroon. 2000. *An Overview of Logging in Cameroon*. Washington, DC: World Resources Institute.

Global Forest Watch Gabon. 2000. *A First Look at Logging in Gabon*. Washington, DC: World Resources Institute.

Global Witness. 1999a. *Made in Vietnam, Cut in Cambodia*. London: Global Witness.

Global Witness. 1999b. *The Untouchables: Forest Crimes and the Concessionaires-Can Cambodia Afford to Keep Them?* London: Global Witness.

Global Witness. 2000. *Chainsaws Speak Louder Than Words*. London: Global Witness.

Global Witness. 2001a. *The Credibilty Gap-And the Need to Bridge It: Increasing the Pace of Forestry Reform.* London: Global Witness.

Global Witness. 2001b. "Global Witness Takes Up Its Position as the Independent Observor of the Forest Sector in Cameroon." Global Witness. July 13. Available online at *http://www.globalwitness.org.*

Global Witness. 2001c. *Taylor-Made: The Pivotal Role of Liberia's Forests and Flag of Convenience in Regional Conflict.* London: Global Witness.

Greenpeace. 2000a. *Spotlight on the Illegal Timber Trade: Asia-Pacific.* Amsterdam: Greenpeace.

Greenpeace. 2000b. *Spotlight on the Illegal Timber Trade: Cameroon.* Amsterdam: Greenpeace.

Greenpeace. 2001. *The Santarem Five and Illegal Logging-A Case Study.* Amsterdam: Greenpeace.

Harrabin, R. 2002. "Malaysia Targets Illegal Timber." BBC. June 11. Available online at *http://www.news.bbc.co.uk/1/hi/world/asia-pacific/2039125.stm.*

Heilprin, J. 2002. "US Details Brazil Mahogany Shipments." *Associated Press,* May 3.

ITTO. 2001. *Annual Review and Assessment of the World Timber Situation 2000.* Yokohama: International Tropical Timber Organization.

Johnson, S. 2002. Documenting the Undocumented. *Tropical Forest Update* 12: 6-9.

Jordan, M. 2001. "Brazilian Mahogany: Too Much in Demand." *The Wall Street Journal,* November 14.

Kasem, S. 1998. "Over 5,000 Logs Found in Tak Seized: Evidence Shows They Came from Salween." *Bangkok Post,* September 16.

Klooster, D. 2000. "Community Forestry and Tree Theft in Mexico: Resistance or Complicity in Conservation?," in *Forests: Nature, People, Power.* Edited by M. Doornbos, A. Saith, and B. White, pp. 275-298. Oxford: Blackwell.

Kotlobay, A. and A. Ptichnikov. 2001. *Illegal Harvesting and Problems of Control in Russia's Forest Sector: Summary.* Moscow: WWF.

Kotlobay, A. and A. Ptichnikov. 2002. *Illegal Logging in the Southern Part of the Russian Far East: Problem Analysis and Proposed Solutions.* Moscow: WWF.

Le Billon, P. 2000. The political ecology of transition in Cambodia 1989-1999: War, Peace and Forest Exploitation. *Development and Change* 31: 785-805.

Lebel, L. 1998. "Illegal logging in Thailand: Salween Logging Scandal." *Bangkok Post,* July 29.

Linh, T. T. T. 2000. "Vietnam's Besieged Forestry Workers Prepare to Fight Back." Trin Tuc Thanh Hoa (Thanh Hoa News). September 20. Available online at *http://www.geocities.com/tranthithuylinh/10-2000/TTTL-vanderung.html.*

Macan-Markar, M. 2001. "Logging Bans in Asia Prove Ineffective." Malaysiakini. November 18. Available online at *http://www.earthisland.org/borneo/news/articles/011112article.html.*

Magrath, W. and R. Grandalski. 2000. "Forest Law Enforcement–Policies, Strategies, and Technologies." *World Bank/WWF Workshop on Controlling Illegal Logging, Jakarta, Indonesia, 2000.*

Matthew, E. 2001a. *European League Table of Imports of Illegal Tropical Timber.* London: FOE-UK.

Matthew, E. 2001b. *Import of Tropical Timber into the UK.* London: FOE-UK.

Matthews, E. 2001. *Understanding the FRA 2000.* Wshington, DC: World Resources Institute.

Mohd, R. and A. R. Yaman. 2000. "A Country Report on Forest Law Enforcement in Peninsular Mayalsia." *World Bank/WWF Workshop on Illegal Logging, Jakarta, Indonesia, 2000.*

Morozov, A. 2000. *Survey of Illegal Forest Felling Activities in Russia: Forms and Methods of Cuttings.* Moscow: Greenpeace Russia.

Murray, G. 2001. "China Combats Illegal Logging." *Kyodo News,* March 10.

Neiro, E., H. Verscheure, and C. Revenga. 2002. *Chile's Frontier Forests: Conserving a Global Treasure.* Santiago, Chile and Washington, DC: World Resources Institute, Comite Nacional Pro Defensa de la Fauna y Flora, and University Austral of Chile.

Newell, J. and A. Lebedev. 2000. *Plundering Russia's Far Eastern Taiga: Illegal Logging Corruption and Trade.* Vladisvostok: Bureau for Regional Oriental Campaigns and PERC.

Obidzinski, K. and C. Palmer. 2002. *"How Much Do You Wanna Buy?"–A Methodology for Estimating the Level of Illegal Logging in East Kalimantan.* Bogor, Indonesia: CIFOR.

Odhiambo, N. 1999. "Illegal Logging Rips up Tanzanian Forests." Environment News Service. August 26. Available online at *http://ens-news.com/ens/aug1999/1999-08-26-04.asp.*

ODI. 2002. "Illegal Logging in Honduras." *Promoting Transparency in the Forest Sector: Best Practices for Detecting Illegal and Destructive Commercial Logging, Washington, DC, 2002.*

Ouen, M., S. T. Sokhun, and E. Savet. 2001. "Cambodia Paper on Forest and Wildlife Law Enforcement Experience." *Forest Law Enforcement and Governance Conference, Bali, Indonesia, 2001.*

Palmer, C. E. 2001. *The Extent and Causes of Illegal Logging: An Analysis of a Major Cause of Tropical Deforestation in Indonesia.* London: University College London.

Pearce, F. 2001. "Logging Ban Backfires." *New Scientist Magazine,* February 28.

Powers, M. 2002. "Illegal Loggers Invade Primordial Peruvian Natives." Environment News Service. August 9. Available online at *http://www.ens-news.com/ens/aug2002/2002-08-09-01.asp.*

Savet, E. 2000. "Country Paper on Forest Law Enforcement in Cambodia." *World Bank/WWF Workshop, Controlling Illegal Logging, Jakarta, Indonesia, 2000.*

Sayer, J., C. Harcourt, and N. M. Collins. Editors. 1992. *The Conservation Atlas of Tropical Forests: Africa.* London: Macmillan.

Scotland, N. 2000. "Indonesian Country Paper on Illegal Logging-DRAFT," pp. 45. Jakarta.

Scotland, N., A. Fraser, and N. Jewell. 1999. *Roundwood Supply and Demand in the Forest Sector in Indonesia, Draft Report.* Jakarta: Indonesia-UK Tropical Forest Management Programme. PFM/EC/99/08.

Siebert, U. 2001. "Benin's Forest Corruption." *Forest Integrity Network, Washington, DC, 2001.*

Siry, J. 2000. "Annex A: Georgia Timber Exports Assessment Based on Turkey Import Data," in *Total Economic Valuation of Georgian Forests: Under the Current Re-*

source Management Regime. Edited by T. Arin and J. Siry. Washington, DC: World Bank.

Sizer, N. and D. Plouvier. 2000. *Increased Investment and Trade by Transnational Logging Companies in Africa, the Caribbean and the Pacific: Implications for the Sustainable Management and Conservation of Tropical Forests.* Brussels: World Wild Fund for Nature, WRI, and European Commission.

SLDF. 2001. *Stumpage Sellout: How Forest Company Abuses of the Stumpage System Is Costing B.C. Taxpayers Millions.* Vancouver: Sierra Legal Defense Fund.

SLDF and Wildlands League. 2001. *Improving Practices, Reducing Harm.* Toronto: Sierra Legal Defense Fund and the Wildlands League.

TFP. 2000. *Aspects of Forestry Management in the LAO PDR.* Amsterdam: Tropical Forest Programme.

Toyne, P., C. O'Brien, and R. Nelson. 2002. *The Timber Footprint of the G8 and China: Making the Case for Green Procurement by Government.* London: WWF.

Ugulava, V. 2001. "Presentation on Corruption in Georgia." *Forest Integrity Network, Washington, DC, 2001.*

UN. 2000. *Report of the Panel of Experts Appointed Pursuant to UN Security Council Resolution 1306(2000) Paragraph 19 in Relation to Sierra Leone.* New York: United Nations.

Vidal, J. 2002a. "Big Five Seek to Save Congo Forests." *The Guardian,* August 27.

Vidal, J. 2002b. "UK Plays Key Role in Illegal Logging: Multimillion Trade Violates Environment and Defrauds West African Countries." *The Guardian,* April 19.

Weatherbe, S. 1998. "Canada Log Thefts Tally Seen C$500 Million Yearly." *Reuters,* June 3.

Webber, J. 2002. "'Naked' Natives Block Illegal Loggers in Peru." *Reuters,* August 3.

World Bank. 2001. *A Revised Forest Strategy for the World Bank Group.* Washington DC: The World Bank Group.

World Rainforest Movement. 1999. "Tanzania: Where Illegal Logging Is Almost Legal." World Rainforest Movement. 27 Sept./Oct. Available online at *http://www.wrm.org.uy.*

WRI. 2002. "Cameroon Ink Pact to Monitor Forests, Curb Illegal Logging." World Resources Institute. June 6, 1992. Available online at *http://www.wri.org* and *http://www.globalforestwatch.org.*

THEORETICAL APPROACHES TO UNDERSTANDING FOREST GOVERNANCE

A System Dynamics Examination of the Willingness of Villagers to Engage in Illegal Logging

Richard G. Dudley

SUMMARY. Much of the work of illegal logging in Indonesia is carried out by villagers. Several factors determine villagers' willingness to

Richard G. Dudley is a fisheries and natural resources consultant. During 2002-2003 he was a visiting fellow at Cornell University (E-mail: drrdudley@compuserve.com).

Much of the work reported here was carried out while he was working as a consultant with the Center for International Forestry Research in Bogor, Indonesia. Among the many people who provided advice and discussion during this work were: Joyotee Smith, Krystof Obidzinski, Carol Colfer, Patrice Levang, Herry Purnomo, Doddy Sukadri, Subarudi, Agus Purnomo, Agus Setyarso, Graham Applegate, and Eva Wolenberg.

Modeling used Vensim software available from Ventana Systems, Inc., 60 Jacob Gates Road, Harvard, MA 01451 USA (E-mail: vensim@vensim.com) (Web site: http://www.vensim.com).

[Haworth co-indexing entry note]: "A System Dynamics Examination of the Willingness of Villagers to Engage in Illegal Logging." Dudley, Richard G. Co-published simultaneously in *Journal of Sustainable Forestry* (The Haworth Press, Inc.) Vol. 19, No. 1/2/3, 2004, pp. 31-53; and: *Illegal Logging in the Tropics: Strategies for Cutting Crime* (ed: Ramsay M. Ravenel, Ilmi M. E. Granoff, and Carrie A. Magee) The Haworth Press, Inc., 2004, pp. 31-53. Single or multiple copies of this article are available for a fee from The Haworth Document Delivery Service [1-800-HAWORTH, 9:00 a.m. - 5:00 p.m. (EST). E-mail address: docdelivery@haworthpress.com].

participate in such activities. Chief among these are: (1) the need for income, (2) the fact that other villagers (and non-villagers) are already illegally logging, and (3) the realization of loss of community control over traditional forest areas. These factors form the basis of feedback loops, which trap villagers in illegal logging systems, which will likely result in the disappearance of a major source of livelihood. Ideas for system dynamics model structure were obtained from field reports and interviews with stakeholders. These ideas were examined using causal loop diagrams to represent different views of illegal logging. One village level view was formulated as a quantified system dynamics model using Vensim software. The model allows examination of scenarios, which might alter system behavior. The model is a tool for understanding consequences of various proposed strategies to control illegal logging. These strategies include enforcement of laws, strengthening of community rights, the prevention of outside labor in local forests, and the provision of alternate sources of income. This is part of a larger effort to describe and analyze illegal logging using system dynamics modeling. *[Article copies available for a fee from The Haworth Document Delivery Service: 1-800-HAWORTH. E-mail address: <docdelivery@haworthpress.com> Website: <http://www.HaworthPress.com> © 2004 by The Haworth Press, Inc. All rights reserved.]*

KEYWORDS. System dynamics modeling, illegal logging, Indonesia, villagers

INTRODUCTION

The decline of Indonesian forests is well documented. In 1997 and 1998, during the first phases of Indonesia's economic crisis, between 3 and 50% of Indonesian timber harvest was unaccounted for in official statistics (Palmer, 2001; Scotland et al., 2000). Illegal logging was thought to account for a large portion of this shortfall. Prior to 1997 Indonesian forests were disappearing at the rate of almost 1.6 million ha per year, equivalent to an annual decline of 1.5% (World Bank, 2001). Annual rate of decline was more rapid within Sumatra (2.8%) and Kalimantan (1.9%). Several reports indicate significant increases in illegal logging since 1998 (e.g., McCarthy, 2000; Obidzinski and Suramenggala, 2000; Casson, 2000), so the rate of forest loss has presumably increased considerably.

The nature of illegal logging activity has rapidly changed. Prior to 1998, much of the "illegality" was confined to large scale timber operations owned by a well-connected business and political elite. To a significant extent these operations were technically legal because laws and regulations establishing them were created and used by the same elites. Some small-scale illegal logging took place as local people tried to gain access to traditional lands that were within the extensive forest concessions granted to powerful business and government interests. These attempts were suppressed by military and police who had ties to concession holders.

After the fall of President Soeharto in 1998, the situation changed dramatically. Large scale concession holders, timber tycoons, and their political backers no longer had the political power to control what happened in the provinces. The effect of this failure of central government was magnified by an official, and inevitable, move toward decentralization during 2000 and 2001, which was backed by foreign donors.

A two-step change occurred. First, from 1998 to 2000, there was an explosion of illegal logging at the local level whereby entrepreneurs came in to cut trees illegally, often making deals directly with villagers and village heads. Second, by early 2000 newly empowered local governments started to assert their new authority. They created laws to permit local logging concessions. These new laws allowed local officials and entrepreneurs to create corrupt business arrangements more easily. Many local communities saw this as a challenge and an opportunity. They claimed blocks of land as traditional forest, requesting approval from local government, which then allowed them to also make deals directly with entrepreneurs. Using this approach, traditional lands can even be claimed inside established protected areas and forest concessions. Sustained yield forest management has little to do with this type of quasi-legal over-exploitation. In any case, modern forestry expertise is largely non-existent at the local level (see Obidzinski et al., 2001; Casson and Obidzinski, 2001).

When we consider the involvement of villagers in illegal logging we need to consider the above context. Villagers live in a world of uncertain laws, with the knowledge that existing laws have always been manipulated by powerful individuals for their own ends. Laws have little meaning if they are widely ignored and un-enforced. Nevertheless, the emergence of local, as opposed to centralized, political power makes the use of local and traditional laws a seemingly attractive option for natural resource management. To use such an approach effectively, a better understanding of the involvement of local people in illegal logging is needed.

The purpose of the research reported here is to develop a conceptual model of village level aspects of illegal logging that explains basic causal relationships leading to villagers' willingness to engage in illegal and/or destructive activities that appear to be against their own long term interests. A system dynamics model of this sort can be considered a theory about how a system works, and why it produces particular results. It can then be used to gain insights into the workings of the actual system. The theory can also be compared to reality and modified as better information is obtained.

This paper represents part of a larger effort to develop system dynamics models of various aspects of illegal logging (e.g., Dudley, 2002).

METHODS

Data and information used as the basis for model building were obtained from field reports (e.g., Casson, 2000; McCarthy, 2000; Obidzinski and Suramenggala, 2000; Obidzinski et al., 2001; and Wadley, 2001) and interviews with various stakeholders. Several qualitative conceptual models, from different stakeholder perspectives, were earlier formulated as causal loop diagrams (see Dudley 2001 for details). Subsequently, some of these conceptual models were used as the basis for quantified system dynamics models. The model presented here is based on one of those conceptual models (Figure 1) that deals with one aspect of the local area view of illegal logging. Other local area aspects include the views of entrepreneurs, and the relationship between entrepreneurs and local officials. The conceptual model was used as the basis for construction of a system dynamics model. See Sterman (2000) for a discussion of the system dynamics modeling approach.

FIGURE 1. Conceptual framework for modeling was based on this diagram from Dudley (2001) which describes factors affecting villagers' perception of illegal logging.

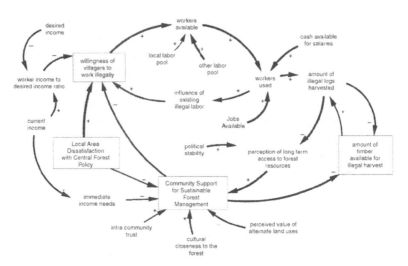

Figures presented herein are of two types: causal loop diagrams and parts of the actual model. The loop diagrams (Figures 1 and 7) show only major causal relationships and do not specifically show flows. The other figures illustrate parts of the actual model. The following conventions are used in labeling model components: stocks (sometimes called state variables or levels) are capitalized and are enclosed in boxes, flows are shown as hollow arrows with a valve and are labeled in lowercase. Auxiliary variables are also in lowercase. Constants are shown in all uppercase. Arrows are illustrated with a plus (+) or a minus (−) to indicate the general trend of the relationship between the two connected variables. A plus indicates a change in the same direction. That is, if X increases then Y also increases if this relationship is taken by itself. A minus indicates a change in the opposite direction. The actual relationship is described by the model equations. In general thicker arrows were used to illustrate more important relationships. When used in the text, names of model components are italicized.

MODEL STRUCTURE

The model describes a theoretical group of small communities with a total labor pool of 1,000 villagers available for logging work. The communities have 5,000 ha of well-forested traditional land holding 200 m^3 of merchantable timber per ha. Normal harvests from this forest start at about 5,000 m^3 per year (i.e., 1.0 m^3 ha^{-1}), which is calculated as a fixed fraction of available timber. The model also assumes a baseline illegal harvest of 1,400 m^3 per year for a total annual harvest of 6,400 m^3 (1.2 m^3 ha^{-1}), a reasonable sustainable harvest from these forests (see, for example, Bruenig 1996, p. 173). This creates an income of $143 per village forest laborer per year ($125 from normal logging and $1 from illegal logging). It is assumed that, initially, one percent of villagers engage in illegal logging. Villagers have other income sources so that the total annual income for members of the above labor pool averages $1,800.

The model consists of 5 interacting feedback loops that describe the following relationships:

1. An increase in need for money increases villagers' willingness to work illegally.
2. As more villagers work illegally they influence others to work illegally.
3. As the amount of forest still intact decreases, community support for sustainable management gradually declines. This increases willingness of villagers to work illegally.
4. As community support for sustainable management declines the amount of forest made available to entrepreneurs for exploitation increases, raising entrepreneurs' desire and ability to provide illegal salaries.
5. As the forest eventually disappears the funds from entrepreneurs decline, and jobs from logging also disappear.

Loop 1: Need for Income Forces Villagers to Work Illegally

According to the model, if villagers have adequate income then they have no need to work illegally. If their income level drops for any reason their willingness to participate in illegal logging will rise, other things being equal, and this *willingness based on income need* can be viewed as a function of the ratio of income to desired income, called the *desired income ratio*. The *desired income ratio* could also change if *desired income* changes. This might happen, for example, if villagers became aware of new desirable consumer goods, or if school fees were increased.

If the income needs of villagers increase, for example, the *current willingness of villagers to work illegally* will gradually increase as well. If this happens then some villagers will start earning illegal income. This, in turn, will raise the income of the villagers as a whole. It is assumed that the money is shared with other villagers so the average income of villagers seeking work is raised even when only some participate in illegal activities. For example, food may be purchased from neighbors. Eventually, the *desired income ratio* is raised enough so that willingness does not rise any further unless disturbed by some other factor. This is a negative feedback, or stabilizing, loop (Figure 2).

Loop 2: Illegal Workers Create More Illegal Workers

As illegal workers (villagers or outsiders) become more common they have a significant influence on others to participate in illegal logging. This creates a positive feedback that can spiral out of control in the absence of other controlling factors. If the *current willingness of villagers to work illegally* increases, then the number of *local villagers available for work* increases. If illegal work is available and enforcement of laws is weak then the *number of illegal workers actually used* for labor will increase, and will include both villagers and outside workers if they are available. It is likely that in any village there are some *people normally working illegally* and other villagers are accustomed to this. However, at some point the *number of illegal workers actually used* rises above the normal number. As this *illegal worker ratio* increases it causes an increase in the willingness of villagers to work illegally as indicated in Figure 3. This is a positive, or reinforcing, feedback loop, which will lead to all villagers participating in illegal logging if no other factors influenced the outcome.

Loop 3: Disappearing Forest Decreases Community Support for Good Forest Management

As forests disappear in relation to what villagers see as *normal forest cover*, there will be a weakening of the community's *strength of perception of long*

FIGURE 2. One factor affecting the willingness of villagers to work illegally is the need for income. This is a negative feedback, or stabilizing, loop, forming part of the model. As income from illegal logging rises the willingness stabilizes. While enforcement can limit illegal workers, it will also prevent the rise of income levels so the willingness to work illegally will remain high. For clarity some model components have been omitted in this view.

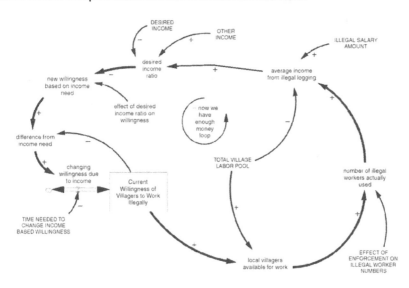

term access to resources. As a community's sense of control over these resources dwindles, *community support for sustainable forest management* also decreases. If *community support for sustainable forest management* is strong then community norms, local customs and rules tend to discourage villagers from working on illegal timber operations. If *community support for sustainable forest management* weakens, then, other things being equal, the *current willingness of villagers to work illegally* will increase, causing a further increase in illegal logging and a decrease in forest cover. If no other factors come into play, this positive feedback loop will spiral out of control as forest cover disappears (Figure 4).

Loop 4: Decreasing Community Support Makes More Forest Available for Illegal Operations

Decreasing *community support for sustainable forest management* causes an overall weakening of traditional community control over its lands. If that happens, community leaders may become more willing to make illegal or cor-

FIGURE 3. A positive feedback loop which illustrates the effect that existing illegal workers have on the willingness of villagers to work illegally. As villagers realize that others are carrying out illegal logging, they gradually lose remaining inhibitions to do so. Over time illegal logging becomes the normal thing to do if other forces do not act on the system. For clarity some model components have been omitted in this view.

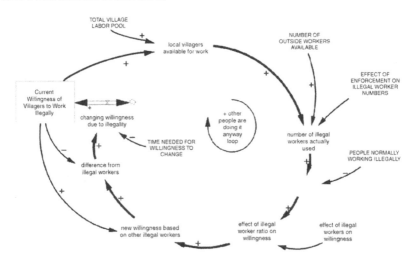

rupt arrangements with outsiders, or outsiders may become more willing to ignore community rules and regulations that are no longer considered important by community members. As a consequence, the amount of community lands available for illegal or inappropriate exploitation will increase or decrease depending on the direction of change in *community support for sustainable forest management.* If more community lands are believed to be open for exploitation, and the risks are considered acceptable by entrepreneurs, then they will provide operating capital for equipment, salaries and other needs. Workers will be hired and more illegal logging will occur, further degrading *community support for sustainable forest management* (Figure 5).

The *fraction of forest that can be logged illegally* also depends on the *strength of community rights* under the legal system (Figure 5). In some cases, community rights have been eroded as new laws give rights to other government authorities (e.g., see McCarthy, 2000). As presented here, *strength of community rights* is a constant (which can be manipulated by the model user), but other model formulations are possible. For example, we may wish to assume that as *community support for sustainable forest management* decreases, the legal support for such control would also change after some delay. That is, there is a distinction between the actual legal backing for community rights

FIGURE 4. The effect of changes in *forest cover* on *community support for sustainable forest management* and, therefore, also on the willingness of villagers to work illegally. If forests disappear a community's perception of its access to resources weakens, weakening community support for long term management approaches. This, in turn, tends to weaken or remove community sanctions or restrictions on villagers working illegally. For clarity some model components have been omitted in this view.

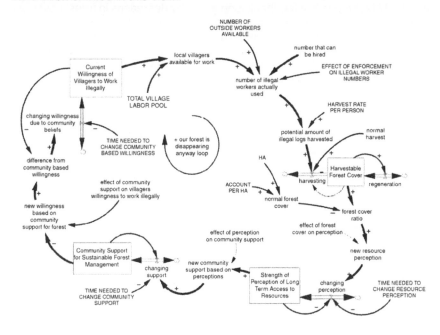

and the desire to apply those rights to manage and control logging activities on community lands.

Loop 5: Disappearing Forests Cause Disappearing Jobs

Harvest from the forest provides jobs. If forest cover becomes degraded not only will sustained yield forest productivity decline, but the temporary benefits of over-harvest will also disappear. While significant *harvestable forest cover* remains there is a large *potential illegal harvest* based on the *fraction of forest that can be logged illegally*. Eventually, *harvesting* becomes self limiting: as over-harvest occurs potential illegal harvest will decline if no other factors interfere. Unfortunately, this self limitation does not stop over-harvest. It merely decreases the rate at which timber is removed. Forest cover will gradually approach zero as will forest related jobs (Figure 5, upper loop). It is impor-

FIGURE 5. Two additional feedback loops make up the remainder of the model. The lower loop (which includes most of the upper loop) illustrates how weakening *community support for sustainable forest management* can lead to increasing illegal log harvest. As community support drops, the community is more likely to allow lands to be accessed by questionable operators. Operators see the weakening community control in terms of larger *potential illegal log harvest*, and thus are willing to risk larger amounts of money in the necessary investments. This is a positive feedback loop whereby decreases in forest cover lead to lower community support for good forest management leading to more illegal operations leading to more illegal harvest. Note that *community support for sustainable forest management* also affects the *willingness of villagers to work illegally* (see Figure 4). The upper loop is a negative feedback loop reflecting the idea that loss of *forest cover* will eventually limit the *harvesting* of timber.

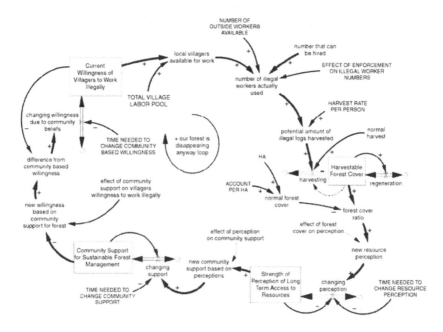

tant to remember that illegal harvest is not necessarily over-harvest, and legal harvest is not necessarily sustainable.

One minor difficulty arises in the selection of a simple approach for depicting the forest that is being harvested. It is not my intention here to provide a detailed forest vegetation model, but the approach used should reflect the ability of the forest to regenerate (grow and reproduce), and *regeneration* should be somewhat higher at intermediate stand densities. I have chosen a biomass approach that disregards stems per ha, size of the trees, and species composition

(Figure 6). This is not to say these components lack importance, but that such a level of detail is not necessary to describe the illegal logging dynamics discussed here.

The Full Model–Additional Comments

The full model incorporates all of the above feedback loops, which are interlinked as illustrated, in simplified form, in Figure 7. A complete model diagram and equations are presented at the end of the document.

MODEL OUTCOMES

Presented here are some model outcomes that might be expected if an increase in illegal logging initially started in response to significant drops in income levels of rural communities. A drop of 50% in income over a two year period (1997-1998) was used as a triggering mechanism. To accomplish this,

FIGURE 6. Harvestable forest is described by a simple biomass model. Net gain (*regeneration*) of harvestable forest biomass is a fraction of existing biomass. When forest cover is very low that fraction is highest (equal to *max regen rate*). As harvestable forest cover increases, stock ratio increases, and because of the shape of the regeneration function (*rgn function*) the *effect of stock on regeneration* lowers *regeneration*. *Regeneration* reaches is zero if *harvestable forest cover* equals or exceeds *max per ha*.

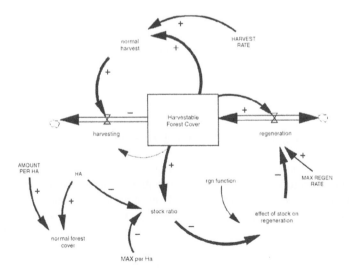

FIGURE 7. The full model is shown here as a causal loop diagram. All five major feedback loops are represented in general form. A detailed model diagram, and model equations, are presented at the end of the paper.

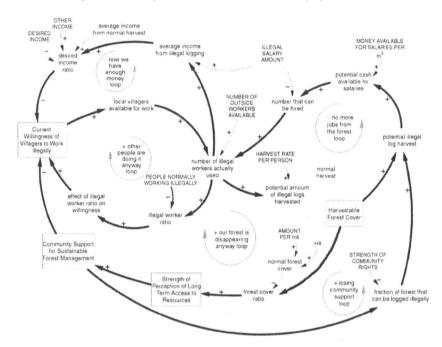

other income was decreased with a ramp function at a rate of $450 per year over the two years. The model was started in approximate equilibrium with the sum of income sources equaling the annual *desired income* of $1,800. This represents the effects of the Asian monetary crisis of those years. Primary factors examined for their effect on illegal logging were *number of outside workers available*, *strength of community rights*, and *effect of enforcement on illegal worker numbers*.

General Pattern of Willingness to Work Illegally

A typical response of villagers' willingness to engage in illegal logging is illustrated in Figure 8. Here we see three phases of willingness while forest still exists. First, willingness rises rapidly as incomes drop and people become willing to work illegally. During this period, as more people work illegally the willingness of others to work illegally also increases. At the same time *community support for sustainable forest management* remains strong and, as income needs are partially satisfied, is able to limit a further rise of willingness.

FIGURE 8. This figure represents the general pattern, over time, of villagers' willingness to engage in illegal logging. The triggering even during 1997-98 is a 50% drop in income. There is a rapid rise in willingness in response to income need, but this need is addressed somewhat as income rises. During this same period the existence of illegal workers stimulates more illegal workers. After this first growth in willingness ends in late 1999 it remains relatively stable because of residual community desire for long term management. By the end of this period forests are largely gone and remaining community control collapses. By the end of this second period incomes have jumped again and willingness starts to stabilize at a higher level. Near the end of this period income from the forest is depleted and willingness due to income need rises rapidly but no additional income is forthcoming. Eventually willingness drops as illegal logging disappears due to the disappearance of the forest.

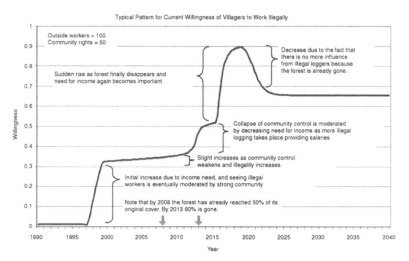

After this period of rapid increase, willingness increases more slowly. During this period, the effects of *community support for sustainable forest management* almost balance the effects of income need and other workers' illegality. Overall, willingness rises only slightly. Even so, the effect on the forest is extreme because the willingness to work illegally remains between 0.3 and 0.4; over 30% of villagers in the labor pool are willing to work illegally. These villagers, plus outside workers, are hired by entrepreneurs. By 2008 (in the scenario shown) 50% of the forest is gone, and by 2013 it has been reduced to 0% of its original amount.

Toward the end of the previous phase the strength of *community support for sustainable forest management* collapses as communities finally realize that the forest is disappearing anyway, regardless of their efforts. This causes a third phase where willingness jumps again. This jump is reinforced by the effect of illegal workers as more illegal workers take to the field. However, this period of increase is short-lived because it also increases income flowing into the village. Increased income limits further increases in willingness as income levels approach the desired income level causing willingness to remain at just above 0.5. At the end of this phase the forest is essentially gone, but further changes in willingness occur.

Following the disappearance of the forest income levels drop precipitously causing a large jump in willingness to work illegally. However, there is no forest left to cut. Willingness jumps to over 0.9 but then drops as illegal loggers, and the idea of illegal logging, disappear along with other forest related jobs (Figure 8 and Figure 9).

FIGURE 9. This figure details the changes in willingness illustrated in the previous figure. Here the left-hand axis gives the change in willingness. Three sources of changes are shown. When the sum of these three is zero willingness (right-hand axis) will not change. When the sum is positive willingness will rise and when negative willingness will decline. We can think of the dotted or dashed lines above zero as pulling willingness up and those that are negative as pulling willingness down. *Changing willingness due to community beliefs* refers to changes caused by strength of community support for sustainable forest management. *Changing willingness due to illegality* refers to the effect of other illegal workers, and *changing willingness due to income* refers to the effect of income need.

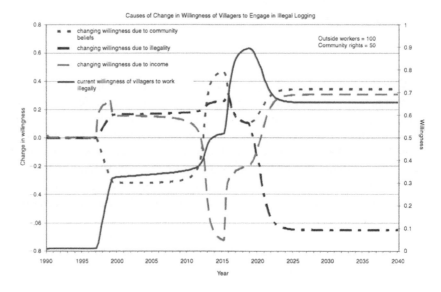

Causes of Change in Willingness of Villagers to Engage in Illegal Logging

The Effect of Outside Workers

Not all illegal logging is done by villagers themselves. Some is carried out by migrants who move to forested areas specifically to find such jobs. As the number of outsiders increases, the forest disappears faster (Figure 10). Outsiders take illegal jobs that locals don't want, but their presence also influences locals to start working illegally (Figure 11). An increase in *the number of outside workers available* causes an increase in *harvesting* because there are more outside workers and also because more villagers are influenced to work illegally.

In addition, outside workers take salaries that would ultimately go to locals. Assuming that little money paid to outsiders is spent locally, the overall amount of money reaching villagers is lower if more outside workers are present. However, shortly prior to the collapse of illegal logging, villagers actually bring slightly higher amounts of money into the village when more outsiders are present. This is because more of the available village labor force is stimulated to work illegally under those conditions creating a peak in illegal incomes that precedes the collapse (Figure 12).

Effect of the Strength of Community Rights

Strong community rights can limit the amount of forest that illegal entrepreneurs can log. If entrepreneurs expect to have difficulty making illegal ar-

FIGURE 10. Influence of outside workers on forest disappearance. Outside workers work illegally and also influence villagers to work illegally. In addition, outsiders absorb some of the cash that would otherwise have gone to villagers.

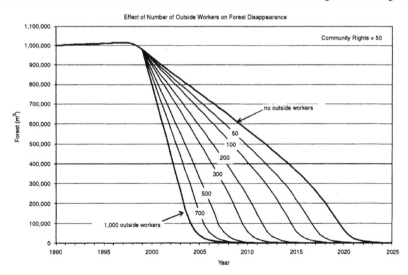

FIGURE 11. The effect of outside illegal workers on villagers' willingness to engage in illegal activities. Outside workers stimulate more villagers to participate in illegal logging earlier than they would in the absence of outsiders.

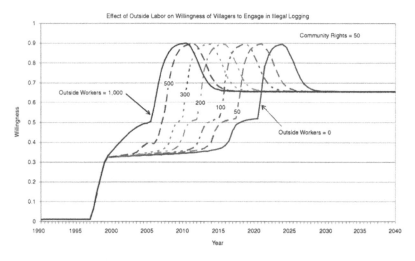

FIGURE 12. Average salary flowing to the village labor pool under situations with different numbers of outside workers. Ultimately, the more outside workers there are the less money goes to village workers. (The area under each curve can be considered the average amount earned per labor pool member.)

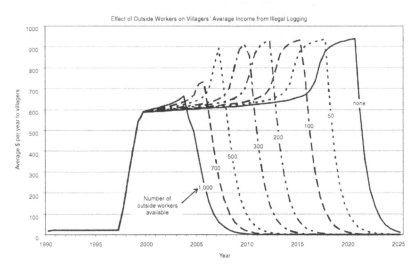

rangements to get logs then they will be less willing to invest in illegal operations. When community rights are strong we would expect to see less illegal logging, assuming that communities want to maintain a long term management approach.

Two problems tend to counteract the role that community rights can play in preventing illegal logging. Firstly, unless rights are very strong there is always some fraction of the forest that can be logged illegally. As that forest is logged community resolve will weaken allowing a larger fraction to be logged. Secondly, logging is to some extent limited by worker availability. If labor is limited then illegal entrepreneurs will be satisfied to use what labor is available on that portion of the forest not currently under community protection. In cases where labor is limited, the effects of moderate levels of community rights may not be noticeable.

Because there is a fairly large amount of forest available to be cut, community rights do not have a significant impact unless the strength of community control is high. For example, if *strength of community rights* is only 50% then entrepreneurs can work on harvesting the remaining 50% until community resolve weakens (Figure 13).

FIGURE 13. The strength of community rights can have a strong effect on forest integrity, but this effect is dependent on the proportion of outside workers available and has little effect if rights are not reasonably strong. Community rights which protect only a small portion of forest are irrelevant from the entrepreneurs' point of view since they can merely harvest other areas of the forest until community resolve weakens.

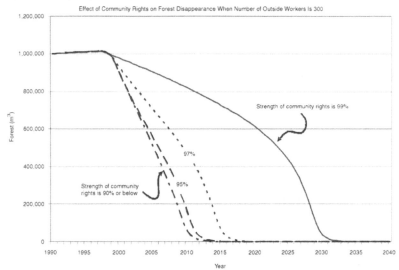

If community rights are weak but the number of workers is limited, then the number of workers available will limit entrepreneurs' operations. Consequently, if *number of outside workers available* is high then *strength of community rights* plays a bigger role (compare Figure 13 and Figure 14).

Enforcement

If enforcement is feasible it can have a major effect on forest protection. This is especially true if *strength of community rights* is weak and the *number of outside workers available* is high. That is, under conditions where illegal logging is most likely to occur enforcement is most likely to have a positive effect. Figure 15 illustrates this situation with 1,000 outside workers and 50% community control. At low levels of enforcement the forest disappears quickly, but at higher levels the disappearance is delayed considerably. Preventing villagers from working illegally prevents their income needs from being satisfied, which raises their willingness to work illegally. On the other hand, by maintaining forest cover enforcement tends to reinforce community support for long-term management. (Factors limiting the effectiveness of enforcement present an excellent subject for another model.)

Under the conditions of the model, enforcement did not significantly lower the effect that other illegal workers in the forest had on willingness. This is be-

FIGURE 14. If the number of outside workers is high then the effect of community rights will be more important because labor to harvest available forest is less limiting.

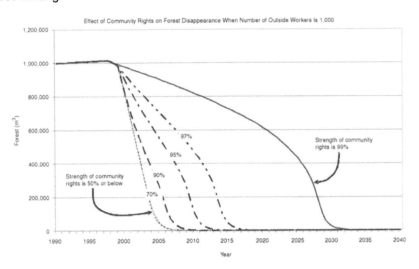

FIGURE 15. Enforcement has a major effect on protecting forest, especially if community rights are weak and the number of outside workers is high. However, with strong enforcement, villagers' willingness to engage in illegal activities due to income need remains high. Thus, if enforcement is suddenly weakened (as illustrated here in year 2010) then illegal activities will rapidly increase.

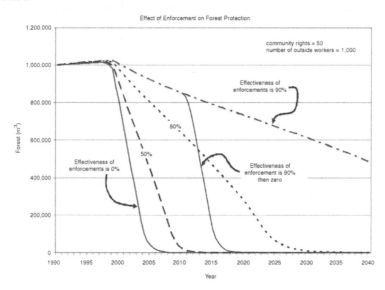

cause for most model conditions used (especially with 1,000 outsiders), regardless of enforcement effectiveness, there were always more illegal workers than the number of *people normally working illegally.* This situation tends to raise willingness. Even in a case with no outsiders, unless enforcement is very strong, enough illegal workers will be working (initially due to income need) that other workers will be influenced to want to join the ranks of illegal workers. This is, of course, dependent on how easily villagers are so influenced. Note that willingness to work illegally, by itself, does not necessarily mean the villagers work illegally. Both availability of salary money and enforcement affect whether a willing villager will actually work illegally.

DISCUSSION

The Model and the Real World? What Is Missing?

In considering the usefulness of the modeling process we need to consider if the model is good enough for its intended purpose. That is, can this model help to explain basic causal relationships leading to villagers' willingness to engage in illegal logging? The model represents one attempt to describe and investigate these causal relationships. It simultaneously considers five feedback relationships connecting villagers' willingness to work illegally, their need for income, the availability of forest to support that income, the communities' role in good forest management, and illegal entrepreneurs' role in hiring villagers.

Within limitations the model does help to elucidate these relationships. We see more clearly that although income levels are raised by illegal logging, other factors also serve to stimulate additional willingness to work illegally even when income levels are already reasonably high. As forests disappear, willingness increases even further because community support for long-term management disappears as well. In examining the model, we are stimulated to consider other elements not included in the model's current structure. Some aspects of illegal logging at the village level are not specifically addressed by the model. Would addition of such factors improve the model, or would they merely make the model more confusing and harder to understand?

We may wonder, for example, if the amount that villagers consider as *desired income* might also rise as incomes from illegal logging rise. If this were the case, then as income expectations rose, it would be harder for *desired income* to be matched by actual income, and illegal logging levels would grow more rapidly. Also, the model includes no additional sources of income that might emerge as forests disappear. One example might be labor income from work on mono-crop plantations–typically oil palm in Indonesia. Alternate income sources would tend to lower the need for illegal employment, especially if these alternate sources grew as forests declined.

In the model, communities' views of long term forest management remain strong until the forest is obviously disappearing. In the real world, community views may be influenced, for example, by possibilities for other uses of community land such as plantation development. Most probably, as the value of alternate land uses rises, the communities' desire for long term forest management would weaken more rapidly.

In the model, forest disappearance does not generate more concern for forest protection, and this seems to agree with reality, at least in practical terms. That is, there is not a resurgence of meaningful community desire to protect forest resources as they disappear. We might use the model to help us consider possible changes to this portion of the real world system. As forest cover disappears, is there a feedback mechanism whereby the desire to protect and rehabilitate forest can be strengthened? Are communities willing to merely accept conversion of forest to non-forest, or can new mechanisms stimulate the desire for long-term forest management? Can such mechanisms be self-reinforcing?

Except for competition for timber, the model does not examine specific relationships between legal and illegal logging. Where laws and regulations limit access to forest, increases in illegal logging from those same forests will remove future timber harvest originally destined for the legal operators. The model reflects this case. If, however, illegal logging is carried out in protected areas, for example, legal and illegal operations would not be in competition for timber. The model also does not reflect the situation whereby cheaper illegal timber and timber products provide competition for legally produced items. This is a subject for a related modeling effort.

Once illegal logging starts, its development in a given area seems to accelerate. This acceleration may be partly related to competition among illegal entrepreneurs and their corrupt colleagues in local government. In such an environment, those who can tend to scramble for resources creating a gold rush mentality. The urgency created by such activity would create additional pressures on communities, and community leaders may be bribed or tricked into illegal agreements. This would clearly hasten the spread of illegal logging. The model does not address such competition. Entrepreneurs in the model are motivated only by money available in standing timber, and there is no direct influence of entrepreneurial desires on weakening community resolve.

Although the model examines the role of outside workers, it is assumed that outsiders and villagers share work in proportion to their numbers in the labor pool. It is also assumed that money paid to villagers remains among villagers, and that paid to outsiders leaves the area. What proportion of outsiders become de facto villagers? Some of their money may stay within the communities and provide income for others. In the model the number of outsiders is determined externally. It is possible to allow the number to be determined by demand for labor. In this case labor would be less limiting and illegal logging would proceed more quickly.

The preceding paragraphs present a number of areas where the model may be inadequate or unfinished. However, the correction of these weaknesses could make the model less understandable and less able to provide insights. The purpose of the model is to provide a framework for thinking about illegal logging at the local level. It is not intended that this model will provide detailed management strategies. However, it can be used as a first step to examine larger, as well as more detailed, issues. Such examination can expose those secondary issues that might provide avenues for modification, not only to the model itself, but more importantly to the real illegal logging system that is our ultimate target.

MODEL 1. Willingness of Villagers to Engage in Illegal Logging

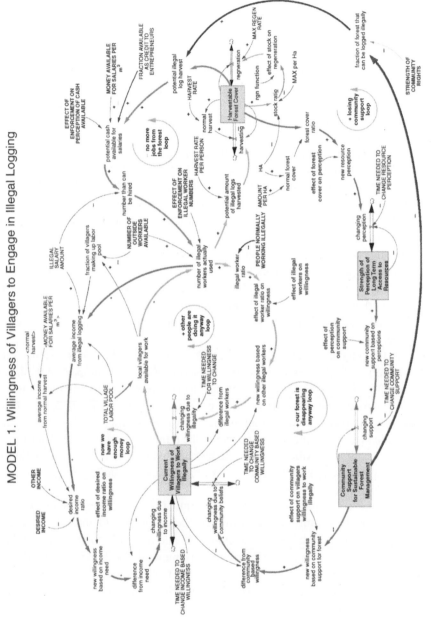

REFERENCES

Bruenig, E. F. 1996. Conservation and management of tropical rainforests: An integrated approach to sustainability. CAB International. Oxon. 339p.

Casson, A. 2000. Illegal tropical timber trade in Central Kalimantan, Indonesia. Draft report prepared for the Center for International Forestry Research, Bogor, Indonesia.

Casson, A. and K. Obidzinski. 2001. From new order to regional autonomy: Shifting dynamics of 'illegal logging' in Kalimantan, Indonesia. Resource Management in Asia-Pacific Conference on Resource Tenure, Forest Management and Conflict Resolution: Perspectives form Borneo and New Guinea, Australian National University, Canberra.

Dudley, R. G. 2001. Dynamics of illegal logging in Indonesia. In: pp. 35-32. Colfer, C. J. P., and I. A. P. Resosudarmo (eds.). Which Way Forward? Forests, People and Policymaking in Indonesia. Resources for the Future. Washington, DC.

Dudley, R. G. 2002. An application of system dynamics modeling to the question of a log export ban for Indonesia with comments on illegal logging. Paper presented at the 20th International Conference of the System Dynamics Society, July 2-August 1, 2002, Palermo, Italy.

McCarthy, J. F. 2000. "Wild logging": the rise and fall of logging networks and biodiversity conservation projects on Sumatra's rainforest frontier. Center for International Forestry Research, Occasional Paper No. 31. Bogor, Indonesia.

Obidzinski, K. and I. Suramenggala. 2000. Illegal logging in Indonesia–A contextual approach to the problem. Draft paper prepared for the Center for International Forestry Research, Bogor, Indonesia.

Obidzinski, K., I. Suramenggala and P. Levang. 2001. L'exploition forestière illégale en Indonésie: Un inquiétant processus de légalization (Illegal logging in Indonesia: A disturbing legalization trend). *Bois et Forets des Tropique* 270(4):5-97.

Palmer, C. E. 2001. The extent and causes of illegal logging: An analysis of a major cause of tropical deforestation in Indonesia. CSERGE Working Paper. Centre for Social and Economic Research on the Global Environment, University College, London.

Scotland, N. (with J. Smith, H. Lisa, M. Hiller, B. Jarvis, C. Kaiser, M. Leighton, L. Paulson, E. Pollard, D. Ratnasari, R. Ravenel, S. Stanley, Erwidodo, D. Currey, A. Setyarso). 2000. Indonesia country paper on illegal logging. Report prepared for the World Bank-WWF Workshop on Control of Illegal Logging in East Asia. Jakarta, 2 August 2000. Edited by William Finlayson and Neil Scotland.

Sterman, J. D. 2000. Business dynamics: Systems thinking and modeling for a complex world. Irwin/McGraw-Hill, Boston. 92p.

Wadley, R. L. 2001. Histories of natural resource use and control in West Kalimantan, Indonesia: Danau Sentarum National Park and its vicinity (100-2000). A Report for the CIFOR Project "Local People, Devolution, and Adaptive Co-Management." Center for International Forestry Research, Bogor, Indonesia.

World Bank. 2001. Indonesia: Environment and natural resource management in a time of transition. 129p.

Does Improved Governance Contribute to Sustainable Forest Management?

Nalin Kishor
Arati Belle

SUMMARY. This paper explores the hypothesis that improving governance is beneficial in reducing deforestation. The hypothesis is tested by including six objectively constructed measures of governance as explanatory variables in an econometric model of the causes of deforestation. Analysis of cross-section data for 90 countries shows an indirect but strong impact of governance on deforestation, working through per capita income. However, the evidence to support a *direct* beneficial impact of improved governance on deforestation is quite weak. The paper argues that if the main objective is to reduce deforestation–especially in the short run–then undertaking reforms directly related to the forest sector, such as in the areas of forest policy, scientific forest management, and forest law enforcement and compliance are likely to be the most effective both in terms of cost and outcomes. However, improving overall governance should not be ignored as it is likely to yield long term bene

---

Nalin Kishor and Arati Belle are both affiliated with The World Bank.

The views expressed in this paper are strictly the authors' and do not necessarily reflect those of the World Bank or any of its affiliated agencies. Helpful comments from the participants at the Yale conference and an anonymous referee are acknowledged.

Address correspondence to: Nalin Kishor, The World Bank, 1818 H Street, Room# J4-005MC5-771, Washington, DC 20433 (E-mail: nkishor@worldbank.org).

[Haworth co-indexing entry note]: "Does Improved Governance Contribute to Sustainable Forest Management?" Kishor, Nalin, and Arati Belle. Co-published simultaneously in *Journal of Sustainable Forestry* (The Haworth Press, Inc.) Vol. 19, No. 1/2/3, 2004, pp. 55-79; and: *Illegal Logging in the Tropics: Strategies for Cutting Crime* (ed: Ramsay M. Ravenel, Ilmi M. E. Granoff, and Carrie A. Magee) The Haworth Press, Inc., 2004, pp. 55-79. Single or multiple copies of this article are available for a fee from The Haworth Document Delivery Service [1-800-HAWORTH. 9:00 a.m. - 5:00 p.m. (EST). E-mail address: docdelivery@ haworthpress.com].

fits including enhanced effectiveness of reforms within the forestry sector. *[Article copies available for a fee from The Haworth Document Delivery Service: 1-800-HAWORTH. E-mail address: <docdelivery@haworthpress.com> Website: <http://www.HaworthPress.com> © 2004 by The Haworth Press, Inc. All rights reserved.]*

KEYWORDS. Deforestation, illegal logging, forest policy, scientific forest management, tenure, governance, corruption, rule of law

THE PROBLEM:
THE NATURE AND IMPACT OF FOREST CRIMES

The problem of illegal logging and corruption and other forest crimes, such as wildlife poaching, trade in endangered species, arson and theft, is global and pervasive. Examples of illegal practices in the forestry sector include: unlawful occupation of forest land by rural families or corporations, international trade in protected species, logging outside concession boundaries, logging in protected areas, undergrading and misclassifying species, timber smuggling, transfer pricing in timber trade, timber processing without a license, etc. (Contreras, 2002). Furthermore, illegal acts and forest crimes of various kinds are common everywhere, in developing as well as developed nations, and in all major forest types–tropical, temperate and boreal (Callister, 1999, Contreras, 2002).

Recent estimates indicate that as much as 15% of global timber trade involves illegalities and corrupt practices. A case study of timber trade shows that illegally harvested ramin can be purchased in Indonesia for about US$20/ m^3 and (after being routed through Malaysia and Singapore) is sold to high-end users in the USA for US$1,000/m^3 (EIA, 2001). Illegal logging in public lands worldwide is estimated to cause losses in assets and revenue in excess of US$10 billion annually (Baird, 2001).

Some examples of forest crimes at the country level include (Contreras, 2002, Glastra, 1999):

- Canada, where violations were detected in 55% of areas designated for protection,
- Brazil, where a presidential commission concluded that fully 71% of the management plans in concessions did not comply with the law,
- Russia, where 20% of timber logged is in violation of the law,
- PNG, where $20 million a year were being lost to illegal practices,
- Cambodia, where only 10% of logging was legal (estimate for 1997),
- Cameroon, where one-third of the timber cut was undeclared (estimate for 1992-93).

Forest crimes pose a threat to the sustainable management of forest resources everywhere. They also lead to a leakage of resources (tax revenues in particular) that legitimately belong in the government treasury. A low-end estimate of the royalties, reforestation fund and export taxes payments that are not being paid to the Government of Indonesia on stolen timber amount to US$ 600 million per annum. This amount is more than twice what the government spent on subsidized food programs for the poor in 2001 (Baird, 2001). In addition, illegal logging and other forest crimes put at risk the livelihoods of the poor and directly threaten ecosystems and biodiversity in protected areas and parks across the world (Contreras, 2002, Thomas et al., 2000). In overall terms, poor governance and weak law and order is likely to contribute to accelerated deforestation and forest degradation.

Efforts by international development agencies at assisting client countries to control illegal logging and improve governance in the sector have had limited success in slowing down deforestation. The constraints to effective action stem from several factors: a weak and resource constrained forest department; vested interest groups, particularly commercial ones; weak or nonexistent voice/participation of critical stakeholders; poor rule of law and ineffective judicial and police systems; paucity of useful information, and the lack of high-level political commitment, regional cooperation and collaboration among the producing and consuming nations.

Effectiveness of measures within the forestry sector is likely to be compromised if the quality of overall governance in the economy is poor, particularly if the forestry sector is large and has strong linkages with the rest of the economy. However, regardless of whether the forestry sector is large or small, one would expect that the probability of success of reforms aimed at improving governance in the sector is highest if they are accompanied by economy-wide reforms aimed at improving overall governance. Otherwise, following Gresham's Law–which states that if counterfeit money is not controlled, it will supplant legal tender and lead to monetary anarchy–bad practices will likely drive away the good policies (Johnston and Doig, 1999, World Bank, 1997).

The basic aim of this paper is to quantitatively explore some of these ideas by analyzing cross-section data on deforestation for 90 countries. The paper will estimate an econometric model of the causes of deforestation, but will depart from traditional models in an important way, by testing six different measures of governance as explanatory variables. Here governance is "unbundled" into six main measurable components: Rule of law, Control of corruption, Government effectiveness, Lack of regulatory burden, Voice and accountability, and Political stability and lack of violence. Governance measures are defined in detail in section III of the paper.

There are several advantages to this approach. First, the model will enable a valuable insight into the relative contribution of "traditional" variables (such as per capita incomes, population density, and roads) vis-à-vis governance and corruption, to deforestation. This will be useful in prioritizing policy reform options. Second, the model will indicate which among the six governance measures have the most impact on deforestation. In turn this can identify areas of governance most requiring improvement and attention. Finally (apart from the information revealed from the empirical regularities), the statistical approach will help identify the "outliers." Analysis of the outliers will yield valuable insights in both types of countries–those characterized by–poor quality of governance and low levels of deforestation and those characterized by high quality of governance with high deforestation.

THE FRAMEWORK AND THE "DRAMATIS PERSONAE"

Forests provide multiple benefits to a multitude of users. If not carefully managed, this multiplicity of users can create situations of conflict leading to resource misuse. Thus, Sustainable Forest Management (SFM) (defined from the broad perspective of preserving all ecosystem services for the present and all future generations) requires attention to a range of issues (legislation, property rights capacity to implement, etc., for example) and the need to involve a number of actors (government, local stakeholders, NGOs, private sector, etc.). This has two important implications for the design of strategic approaches. First, the overall state of governance is one of several critical sets of factors determining outcomes in the sector (Kaimowitz, 2001, Kaimowitz and Angelsen, 1999). In other words, while improvement in governance is a necessary condition for SFM it cannot be sufficient by itself. But, by the same token, governance issues cannot be ignored otherwise the success of other reform efforts in forestry will likely be short-lived. And second, several stakeholders will need to be involved in any realistic strategy aimed at improving governance. In other words, any effective strategy to promote lasting SFM will likely require a multi-pronged approach (Kishor, 2002).

Figure 1 illustrates conceptually the 4 major sets of factors–forest policy, scientific forest management, overall governance, and specific forest law enforcement–which collectively influence SFM.

A comprehensive strategy to promote SFM will therefore require action on 4 fronts:

   i. Actions to promote sustainable forest management via forest policy reforms;
  ii. Actions to encourage adoption of scientific forest management techniques;
 iii. Actions to promote a better overall quality of governance in the economy; and,

iv. Actions to improve law enforcement and promote specific anticorruption efforts in the forestry sector.

Key actions in the area of *forest policy* include: Clarification of property rights, setting the right level of taxes and royalties, use of market-based instruments, etc. Well-designed forest policies can simultaneously achieve economic and environmental and conservation objectives, at the same time as strengthening compliance and easing the problem of law enforcement (Magrath and Grandalski, 2002). Rational, reasonable and enforceable forest policies are necessary to ensure that other measures to improve governance and reduce corruption have positive and lasting impacts. For example, if there is excessive reliance on command and control type approaches in policy, enforcement is likely to be problematic. Reforming policy in the direction of greater use of market-based instruments will simultaneously make for a better climate of governance. As another illustration, ensuring that property rights to land are clear and non-controversial will enable the rule of law to be implemented effectively and prevent the abuse of forest lands. It is also important to note that the issue is not simply of designing better forest policy but also policy which can be translated into easily enforceable legislation with minimum scope for controversial and discretionary interpretation (Rosenbaum, 2002, Bekhechi, 2001).

Scientific forest management plans must be based on a careful consideration of management objectives, an assessment and inventory of the resource base and its projected trajectory, an estimation of a budget and resource requirements, and provision for evaluation and plan revisions. However, an ITTO study indicated that the extent of tropical forests being managed under

FIGURE 1. Factors Impacting Forests

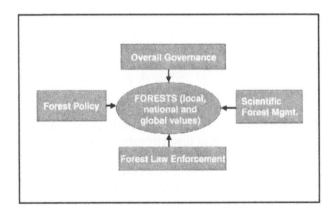

sustainable yield systems was negligible (Poore, 1989). An assessment a decade later found that while significant progress had been made, the challenges of full and coordinated implementation of management plans remained inadequately addressed (ITTO, 2000). It is clear that poor quality of routine forest management will have serious repercussions on SFM. Improving the quality of forest management also reduces the scope for forest crimes such as logging in excess of allowable cuts, logging in vulnerable areas, etc. (bin Buang, 2001, Magrath and Grandalski, 2002). "Win-win" examples of the sort described above need to be identified, and the synergy they offer to counter forest crimes exploited to the full, while developing a strategy to fight corruption and improve governance in forestry to promote SFM.

Improving the *overall quality of governance* requires a system of checks and balances in society that restricts arbitrary actions and bureaucratic harassment, and promotes voice and participation by the population (Thomas et al., 2000). Equally importantly, it reduces the scope for the elite to "capture" the state to serve their own interests, and fosters the rule of law. In theory reforms may be necessary in several areas and a practical approach rests on a thorough examination of the actual situation with respect to: the structure of government, political accountability, a competitive private sector, quality of management of the public sector and the status of civil society with respect to voice and participation. In particular, there is widespread agreement that corruption is a fundamental symptom of public sector malfunction and public sector reforms are a key component in any anticorruption approach (World Bank, 2000).

Finally, SFM will include a specific *forest law enforcement and compliance system*. A recent and important body of work focuses at the impacts of specific monitoring and law enforcement initiatives as applying exclusively to the forest sector. The main message emerging from this literature is that specific steps within forestry to improve monitoring and enforcement aimed at controlling forest crimes do matter. But these are not sufficient in of themselves to ensure compliance and other prerequisite steps are necessary (Magrath and Grandalski, 2002, Melle and Beck, 2001). These (precursive) steps consist of making improvements in forest policy, forest management regimes, and governance. In this context, forest law enforcement is seen as a supplement to poor policies, poor management and weak governance and institutions. The three most important components of a forest law enforcement system are prevention, detection and suppression: (i) *Prevention* includes actions geared to reducing the opportunities for illegal acts such as the formulation of good management plans, reduction in discretionary powers of forest officials, encouraging whistleblowers, etc.; (ii) *Detection* includes monitoring and surveillance to determine if and where crime is occurring. This kind of information is crucial for setting priorities and for evaluating other elements of the enforcement program, and (iii) *Suppression* almost inevitably involves the use of force after unlawful activities have occurred, or while they are underway. Suppression of illegal activity should be the last recourse in a forest law enforcement program, because suppression measures pose risks to agency personnel, the public, and the lawbreaker (Magrath and Grandalski, 2001).

GOVERNANCE DEMYSTIFIED

What Is Governance? And Corruption?

Governance is defined as the manner in which power is exercised in the management of a country's economic and social resources. Good governance is epitomized by predictable, open, and enlightened policymaking (that is, transparent processes); a bureaucracy imbued with a professional ethos; an executive arm of government accountable for its actions; and a strong civil society participating in public affairs; and all behaving under the rule of law (World Bank, 2000). Governance encompasses the capacity to formulate and implement sound policies, and the respect of the citizens and the state for the institutions that govern economic and social interactions among them. It also includes the process of selecting, monitoring and replacing governments (Kaufmann et al., 1999a, 1999b).

It is important to emphasize that governance is not just about the organization, the employees and the policies of the apparatus of government of a country–it extends far beyond that to include the key role and contributions of other stakeholders as well–e.g., the role of civil society in electing a government, the role of the private sector in complying with rules and regulations governing concession management, etc. This point assumes critical importance in any strategy for improving governance. Thus, improvements in governance can and indeed should be the responsibility of all stakeholders–government policymakers, private sector executives, parliamentarians, academics, and civil society at large. Any strategy to improve governance needs to be multi-pronged (Thomas et al., 2000).

Corruption, by contrast, is commonly defined to be the abuse of public office for private gain or for the benefit of a group to which one owes allegiance (Bardhan, 1997). Corruption is most often associated with misuse of power by civil servants and politicians but it can exist in any situation where there is misuse of power (Stapenhurst and Kpundeh, 1997). However, corruption is usually not a cause but a symptom of the weakness of the state and indicates the need for institutional reforms for its control.

Unbundling Governance and Measuring Its Quality

Starting with the broad notion of governance as defined above, governance can be logically subdivided into six main measurable components: Rule of law, Control of corruption, Government effectiveness, Lack of regulatory burden, Voice and accountability, and Political stability and lack of violence (Kaufmann et al., 1999a).

> *Rule of law*–The rule of law is best defined as the opposite of the rule by powerful men or women. Rule of law includes issues such as the protection of property rights, enforceability of contracts, and maintaining the effectiveness and independence of the judiciary, and the perceptions of the incidence of both violent and nonviolent crimes. In overall terms this component refers to that aspect of governance that measures the success of a society in developing an environment in which fair and predictable rules form the basis for economic and social interactions.

> *Control of corruption (or graft)*–The presence of corruption is often a manifestation of a lack of respect, on part of both the one who corrupts and the one who is corrupted, for the societal rules governing the interaction between the two. The existence of corruption itself is often seen as a symptom of weak or poor governance and this component measures the perception of corruption by society.

> *Government effectiveness*–This component focuses on the quality of policymaking and refers to the "inputs" required for the government to be able to formulate and implement good policies and to deliver public services. This combines perceptions of the quality of provision of public services, the quality of the bureaucracy, the competence of civil servants, the independence of the civil service from political pressures, and the credibility of government commitment to its policies.

> *Lack of regulatory burden*–This component focuses on the content of the policies themselves. It includes measures of the incidence of market unfriendly policies such as price controls and inadequate bank supervision, as well as the perceptions of burdens imposed by excessive regulation in areas such as environmental management, foreign trade and business development.

> *Voice and accountability*–This includes indicators of governance dealing with various aspects of the political process, civil liberties, political rights, and the freedom of the press. This component measures the extent to which the citizens of a country are able to participate in the selection and the running of governments. This also includes the independence of the media that plays an important role in monitoring and holding accountable those in authority.

Political stability and lack of violence–This includes indicators that measure perceptions of the likelihood that the government in power will be destabilized or overthrown by possibly unconstitutional and/or violent means. This component captures the idea that the quality of governance in a country may be compromised by the likelihood of "catastrophic" changes in government that have a direct impact on the continuity of policies.

Empirical Measurement of the Quality of Governance

Unbundling "governance" into its key components is useful for several reasons:

i. It enables more precision in policy discussions of which aspects of governance require most improvement and attention, and to identify specific steps through which that might be achieved;
ii. By allowing for a relatively precise measure of its components it enables a benchmarking of the quality of governance in a country, and to track the changes in its quality over time;
iii. It enables a statistical analysis of how the quality of governance impacts upon desired developmental outcomes such as increase in per capita incomes, reduction in infant mortality, increases in literacy, reduction in deforestation and expansion of area under sustainable forest management, control of pollution, etc.;
iv. It allows for a comparison of the development experience across countries. In particular, the linkages between governance and development outcomes will likely yield a useful learning experience from cross-country analysis.

Kaufmann et al. (1999a, 1999b) have analyzed hundreds of cross-country indicators as proxies for various aspects of governance. These are based on more than 150 measures produced by more than 12 different organizations. They are drawn from various published and unpublished sources spanning private forecasting and business risk organizations, think-tanks and other NGOs. They include the results of surveys carried out by multilateral and other organizations such as the Business Environment Risk Intelligence, Standard and Poors, European Bank for Reconstruction and Development, Economist Intelligence Unit, Freedom House, etc. They are based on surveys of experts, firms and citizens and cover a wide range of topics: perceptions of political stability and the business climate, views on the efficiency of public service provision, opinions on respect for the rule of law, and perceptions of corruption.

As an illustration of this approach we present the results for the measurement of the rule-of-law (see Figure 2). Included in this measure are aspects

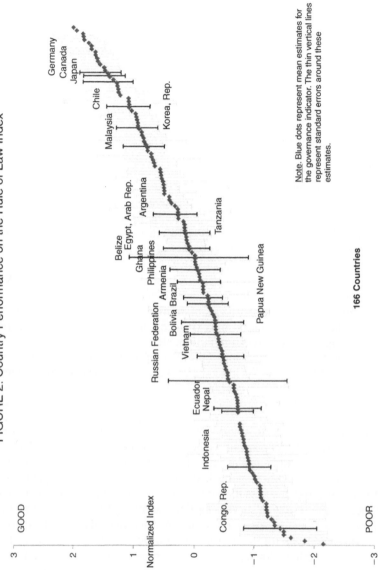

FIGURE 2. Country Performance on the Rule of Law Index

Note. Blue dots represent mean estimates for the governance indicator. The thin vertical lines represent standard errors around these estimates.

166 Countries

Adapted from "Governance Matters" by Daniel Kaufmann, Aart Kraay and Pablo Zoido-Lobaton, May 1999 (Kaufmann et al., 1999b). http://www.imf.org/external/pubs/ft/fandd/2000/06/kauf.htm.

such as the enforceability of contracts, variability in the application of the rule of law applied across groups within the country, police effectiveness in safeguarding personal property, size of a black market, etc. In the figure, countries are ordered along the horizontal axis according to their rankings, while the vertical axis indicates the estimates of the rule of law for each country. The vertical bars represent the country specific confidence intervals (margins of error) for the point estimate for rule of law. Note that this measure (and the other five also) have been scaled to lie between −2.5 and +2.5, and oriented so that higher values correspond to better outcomes.

The margins of error can be considerable and rather than a strict ranking (based on imprecise point estimates) it is deemed more useful to classify countries into three color-coded groups. The red light group consists of those countries that should be considered to be in a crisis of governance characterized in this example by a significant lack of operation of the rule of law (for example, Bangladesh, Ecuador, Guatemala, Indonesia, Myanmar, Russian Federation). The yellow light group characterizes those countries which are vulnerable or at risk of falling into a governance crisis (Argentina, Brazil, Ghana, PNG, Tanzania). And finally, the green light group of countries may be said to have a robust rule of law component of governance and not likely to be at risk of reverting into a crisis situation (Chile, Malaysia, Japan, S. Korea, Canada, Germany, US). The traffic lights approach is useful in identifying the most vulnerable countries that should be high priority for reforms aimed at improving governance (Thomas et al., 2000).

The Development Dividend of Improving the Quality of Governance

Development outcomes such as increases in per capita income, reductions in child mortality, and increases in literacy have been statistically examined with the data on governance measures and development outcomes. The empirical analysis suggests that the direct impact of better governance resulting in better development outcomes is large (Kaufmann et al., 1999a, 1999b, Thomas et al., 2000). As an illustration consider an improvement (of one standard deviation) in the rule of law from the low level in the Russian Federation today to the middling level in the Czech Republic or a similar reduction in corruption from that in Indonesia to that in South Korea. This is associated with an increase in per capita incomes by two to four times, a reduction in infant mortality by a similar magnitude, and an improvement in literacy by 15 to 25 percent points in the long run. Consider also that much larger improvements in governance from the levels in Tajikistan (in the red light group) to those in Chile (green light group) are associated with a near doubling of the development impacts mentioned above.

The relationship between various development outcomes and four measures of governance are illustrated in Figure 3. The heights of the vertical bars show the differences in development outcomes with weak, average and strong governance and illustrate the strong correlation between good outcomes and good governance. The solid lines represent the estimated impact of improving governance on development outcomes–the "development dividend" from governance. This evidence provides strong support for the argument that improving governance is of crucial importance for developing countries.

DATA ANALYSIS

Models of Deforestation

Studies on the causes of deforestation abound in the literature. Given our interest in exploring policy options to control deforestation, however, it is neither our intention nor our task to do an extensive literature review. For this, the interested reader can go to a recent and comprehensive review by Kaimowitz and Angelsen (1999). Instead, we will focus on the literature that deals specifically with the relationship between governance and deforestation.

While extensive concern has been raised regarding the impact of governance on deforestation, few attempts have been made at a careful analysis and measurement of this relationship. Part of the reason may stem from the difficulty in getting objective measures of governance (which we try to address in this paper). However, the study by Deacon is a good attempt (Deacon, 1994). Using cross-country data he tests the hypothesis that insecure ownership exacerbates deforestation. He finds consistent associations between deforestation and political variables (such as constitutional changes, guerilla warfare and revolutions) reflecting insecure ownership. However, the estimates are not very robust and the author suggests the need for better data and better model building (particularly to test for the exogeneity of the political attributes). Shafik (1997), however, concludes that vesting clear property rights *increases* deforestation. Mainardi (1996) concludes that deforestation is likely to be more in countries that are politically unstable. Similarly, Didia (1997) concludes that democracies are characterized by lower deforestation. Palo (2001) uses Transparency International's Corruption Perception Index as an explanatory variable in a cross-country regression equation for deforestation (using the ratio of forest to non-forest area as the dependent variable). For a sample of 36 tropical countries, the coefficient was found to be significant at the 10% level. The coefficient indicates that a 10% improvement in the Corruption Perception Index will increase the ratio of forest to non-forest area by 5%.

The results from these studies generally support the hypothesis that improvements in governance reduce deforestation. However, we build on the

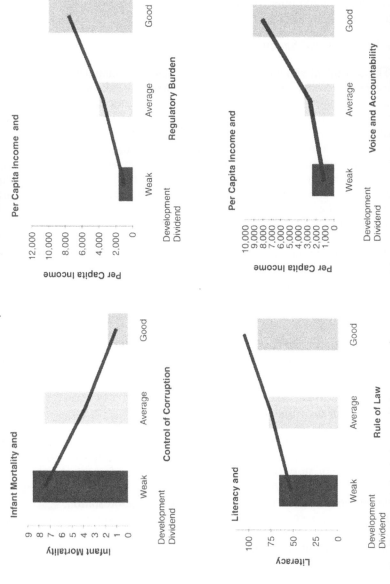

FIGURE 3. The Development Dividend of Good Governance

Note: The bars depict the simple correlation between good governance and development outcomes. The line predicted value when taking into account the causality effects ("Development Dividend") from improved governance to development outcomes (Adapted from Thomas et al., 2000).

previous work by refining the model, particularly through the use of more comprehensive and objectively constructed measures of governance in order to test our hypotheses rigorously.

The Variables and Sample Size

The dependent variable used in the analysis is the annual average rate of deforestation between 1990 and 2000. This variable is taken from the FAO estimates, as reported in the World Development Indicators. We examine 15 basic explanatory variables. The econometric estimates also test the significance of quadratic terms (for income and rural population density), several interaction terms, and dummy variables as explanatory variables. In addition to the six governance measures described in section III above, we also include per capita income, percentage rural population, rural population density, geographical region (1-6), forest type (tropical/others), total roads, globalization, trade openness, and a proxy for government commitment to forest protection. The list of variables, their definition and sources are given in the table in Appendix 1.

The total sample size of 90 countries breaks down as follows: Africa (26), East Asia and the Pacific (12), Europe and Central Asia (30), Latin America (18), South Asia (2), North America (2). Only four countries in the sample (Togo, Albania, Macedonia and Fiji) have forested areas of less than 10,000 hectares. Nineteen countries have a 1995 per capita income of more than US$10,000 per annum. Table 1 below reports the descriptive statistics for the data.

The rate of deforestation ranges from a maximum of 3.11% per annum in the case of Cote d'Ivoire, to a minimum of −1.7% in the case of Portugal. In general, the range of dispersion for most variables indicates that this is likely to be a representative sample of countries.

A First Cut at the Role of Governance

A correlation matrix for the data shows that the simple correlation coefficient between deforestation per annum and governance ranges from −0.22 in the case of regulatory burden to −0.43 in the case of political stability (and rule of law). Thus, at first blush it appears that an improvement in any of the six dimensions of governance is likely to reduce deforestation.

This finding is corroborated when we estimate a simple linear regression for each of the six variables, on deforestation rates. In each case the coefficient for the independent variable is negative and significant at the 5% level (see table in Appendix 2).

The Full Econometric Analysis

To recap, we are interested in testing the hypothesis that our various measures of overall governance impact the forestry sector. Our *a priori* expecta-

TABLE 1. Descriptive Statistics for Model Variables

Variable	Abbreviated Name	Mean	Standard Deviation	Minimum Value	Maximum Value
Percentage Rate of Deforestation per Annum	Dfpa	0.45	0.94	−1.70	3.11
Area Deforested per Annum	Dfarpa	846.82	3642.83	−18063.00	22264.00
Per Capita Income in 1995	Pppcy95	7404.62	7638.15	474.00	28274.00
Rural Population Density	Popdenrur95	324.7345	653.23	5.74	6021.4
Government Commitment for Forest Protection	Forprot	9.44	8.15	0.00	43.09
Total Roads Network	Roads	256662.90	741804.10	3370.00	6296107.00
Rural Populationa as a % of Total Population	Rurpop	46.01	21.12	3.00	87.50
Globalization	Globalness	31.01	24.90	2.01	137.12
Trade Openness	Openess	2.84	1.15	1.00	5.00
Voice and Accountability	Voice	0.27	0.87	−1.62	1.69
Political Stability and Lack of Violence	Polstability	−0.02	0.90	−2.59	1.69
Government Effectiveness	Goveffect	0.07	0.84	−1.77	1.99
Lack of Regulatory Burden	Regburden	0.21	0.66	−2.34	1.21
Control of Corruption and Graft	Corrupgraft	0.03	0.90	−1.56	2.09
Rule of Law	Rulelaw	0.02	0.92	−2.15	2.00

tion is that an improvement in governance will have a positive impact in reducing deforestation. The preliminary analysis suggests that indeed governance does matter for forest management. But this is clearly only a partial picture. Section II of the paper points out that deforestation is likely to be affected by a large number of factors. Moreover, at a conceptual level, these relationships are likely to be complex and two-way, direct and indirect, and strong and diffuse (Kaimowitz and Angelsen, 1999). Thus, we need to incorporate several other explanatory variables in our regression estimates in order to test in a robust manner whether governance matters. To the extent that reliable measures are available, these will be included in the econometric equations being estimated.

As a starting point, deforestation is regressed on all the variables of interest in our sample–including the square of the per capita income–for the entire sample of 90 countries. The results are reported in panel 3.1 of Appendix 3. The overall explanatory power of the model is reasonable ($R^2 = 0.48$). However, *none of the 6 governance coefficients are statistically significant.* Only three coefficients are significant at 10% or less: per capital income, the square of per capita income, and the total road network.

In a regression using the area deforested as the dependent variable, the coefficient of per capita income (PPPCY95) is negative and that for the square of per capita income (PPCY95SQ) is positive. This casts doubt on the "Kuznets curve" hypothesis for this sector and turns out to be an important (and robust) finding of the present analysis (results available upon request). The Kuznets hypothesis states that as incomes rise, deforestation will fall after an initial increase, at a certain threshold level of income. For this to happen, the coefficient of the income variable should be positive and the coefficient of the square of the income term should be negative-just the reverse of our estimates.

The coefficient for the road network is significant, but unexpectedly has a negative sign. *A priori*, one would expect that more roads would lead to greater clearing of forestlands. One explanation for the unexpected sign on the road variable is that because of data unavailability, we measure the total road network in a country rather than the road network in forest areas. The latter variable may still be relevant to deforestation rates.

The apparent contradiction between a reasonable overall R^2 and individual coefficients that are mostly insignificant suggests multicollinearity among the explanatory variables. Multicollinearity occurs when explanatory variables predict each other and do not have individual explanatory power. The simple correlation matrix for this data indicates a very high degree of collinearity among our 6 governance variables with the simple correlation coefficient ranging from 0.71 to 0.94. This suggests that multicollinearity is a likely problem: dropping some of these variables may correct the problem. Changing the sample size is yet another option: this has been explored by regressing on a subset of the sample but the results were not favorable. Because of this problem of multicollinearity, subsequent regressions use only one out of the six governance measures. Various combinations of the governance variables were tried but none were statistically significant when income and the square of income were included as explanatory variables (results available upon request from authors). With the qualification that the results come out much the same no matter which governance measure is tried, the rest of the analysis in this paper will focus on exploring the function of voice and accountability (VOICE) on the rate of deforestation.

The simple correlation coefficients between per capita income (PPPCY95) and the governance measures are also high, ranging from 0.62 to 0.85. They are also almost as highly correlated with the square of the per capita income. This was to be expected and it is consistent with the finding that improvement in governance leads to large increases in incomes, the so-called "development dividend" story discussed in section III above. In this data sample, regressing per capita income on the governance measures individually yields high R^2 (these range from 0.39 to 0.73) and coefficients that are significant at 1%

level, thus corroborating the Kaufman et al. (1999b) results (results available upon request from authors).

These findings suggest that governance variables are likely to have an *indirect* impact on deforestation, through the per capita income variable. But does governance have a *direct* impact on deforestation? In order to test this hypothesis, we first "purged" the influence of income by regressing deforestation on income and the square of income and calculating the residuals. In the second step, these estimated residuals (DFPARES) from the first regression equation were regressed on the other explanatory variables plus one governance measure in turn.

The regression equation for VOICE is reported in Appendix 3.2. The coefficient for VOICE is not statistically significant. The only significant variable is DUM3 (which takes the value 1 if a country is in the Europe or Central Asia region, and 0 otherwise), with a negative sign, indicating that, all else equal, if a country belongs to the ECA region it is likely to experience less deforestation. Qualitatively similar results are obtained when VOICE is replaced by other measures of governance. Thus, in overall terms the message emerging from this data analysis is that governance measures do not have a direct impact on deforestation but a favorable indirect impact cannot be ruled out.

An additional test of whether governance has a direct impact on deforestation was conducted as follows: the most favorable conditions under which this could happen are provided by: (i) not purging the influence of per capita income from DFPA (the dependent variable), and (ii) not including per capita income as an explanatory variable. Appendix 3.3 reports the regression equation using VOICE as one of the explanatory variables. The coefficient of VOICE has a negative sign but is not statistically significant. Only two variables are statistically significant. First, RURPOP (which is the proportion of total population residing in rural areas) has a positive sign–indicating that a higher proportion of population living in the rural areas, other things being equal, is likely to lead to greater deforestation. And second, DUM3, has a negative sign–indicating that if a country belongs to the ECA region, all else equal, it is likely to experience less deforestation.

CONCLUDING REMARKS

To test the basic hypothesis that improving governance has a favorable impact in reducing deforestation, we have performed a variety of statistical analyses on a cross-country dataset for 90 countries. This dataset is unique from others in the literature in that it includes six objectively constructed variables which measure various aspects of governance such as the strength of the rule of law and the extent of civil liberties including the freedom of the press–aspects which are expected to impact deforestation. Our main conclusion is that, with these measures of governance and with the econometric approach adopted, we find little evidence of a *direct* impact of governance on deforestation.

However, two findings from our data analysis are significant. First, income has a statistically significant and negative impact on deforestation, i.e., rising incomes are likely to reduce the rate of deforestation. Second, the governance measures used in this study have a statistically significant and quantitatively large impact in increasing incomes–i.e., improving governance may act as a catalyst to increasing incomes. These two findings taken together imply that *improving governance may have an indirect but strong impact on curbing deforestation.*

The lack of support for a direct impact of governance variables on deforestation is consistent with the notion that governance is a macro-level variable– or an "underlying variable" (Kaimowitz and Angelsen 1999)–and as such, its impact on the forestry sector *a priori* is expected to be diffuse and uncertain. Our findings are also consistent with a "targets and instruments" approach, which states that the best policy levers are those that have (or are thought to have) the most direct impacts on the chosen targets. In our case this approach would suggest that if the main objective is to reduce deforestation in the short run then undertaking steps most specific to the forestry sector are likely to be the most effective both in terms of cost and outcomes. Our review of the literature suggests that specific strategies include reforms in forest policy, scientific forest management, and forest law enforcement and compliance. In the present analysis, variables such as government commitment to forest protection, extent of globalization, trade openness, rural population as a percentage of total population, rural population density were included as proxies for these direct policy options. Only the rural population density variable proved to be statistically significant (equation 3.3 in Appendix 3) and with the expected sign, thus providing support for the targets and instruments approach.

The measures used here are indeed incomplete and imprecise and much more analytical rigor is required. As such, the results in this paper should be treated as preliminary. A key research effort in the future is to collect more and better data on variables measuring sector-specific actions of the type mentioned in section II of the paper (indicators of forest policy, scientific forest management and forest law enforcement). Such data would enable more rigorous testing of the hypotheses proposed in this paper. It is also important to note that, in addition to refining the formal econometric model, it is crucial to cross-check our findings against site specific surveys and country case-studies (Kaimowitz and Angelsen 1999). This also includes an analysis of "outliers" referred to but not pursued in this paper.

Despite the lack of evidence of direct impacts of governance on deforestation, efforts must be made to continually improve the quality of overall governance in any economy, for at least two reasons. First, good governance acts as a catalyst for many desirable social and economic outcomes such as improvements in literacy and child mortality rates, and a sharper rise in growth and incomes. In turn, at least some of these, particularly growth in incomes, have been shown as having beneficial impacts in controlling deforestation. And second we should not forget Gresham's law, which in the context of governance is to be interpreted as stating that an overall environment of poor governance will likely drive out good policies. Thus, in the long run, policy reforms in the forestry sector will likely have a better chance of being implemented and have lasting impacts in a situation of better overall governance in the economy.

REFERENCES

Baird, M. 2001. Forest Crimes as a Constraint to Development in East Asia. Speech delivered at the Forest Law Enforcement and Governance: East Asia Ministerial Conference, Bali, Indonesia, September 11-13, 2001. Available at: *http:// lnweb18.worldbank.org/eap/eap.nsf/2500ec5f1a2d9bad852568a3006f557d/ c19065b26241f0b247256ac30010e5ff?OpenDocument*

Bardhan, P. 1997. Corruption and Development: A Review of Issues. *Journal of Economic Literature* (Sept. 1997) 35, 1320-1346.

Bekhechi, M. 2001. Governance and the Design and Implementation of Forest Legislation. Presentation at the Forest Law Enforcement and Governance: East Asia Ministerial Conference, Bali, Indonesia, September 11-13, 2001. Available at: *http:// lnweb18.worldbank.org/eap/eap.nsf/2500ec5f1a2d9bad852568a3006f557d/ c19065b26241f0b247256ac30010e5ff?OpenDocument*

bin Buang, A. 2001. Forest Management Experiences from East Asia. Paper delivered at the Forest Law Enforcement and Governance: East Asia Ministerial Conference, Bali, Indonesia, September 11-13, 2001. Available at: *http:// lnweb18.worldbank.org/eap/eap.nsf/2500ec5f1a2d9bad852568a3006f557d/ c19065b26241f0b247256ac30010e5ff?OpenDocument*

Callister, D. J. 1999. Corrupt and Illegal Activities in the Forestry Sector: Current Understandings, and Implications for World Bank Forest Policy. Discussion Draft, May 1999. Available at: *http://wbln0018.worldbank.org/essd/forestpol-e.nsf/ HiddenDocView/BCE9D2A90FADBA73852568A3006493E0?OpenDocument*

Contreras-Hermosilla, A. 2002. Law Compliance in the Forestry Sector: An Overview. WBI Working Papers. World Bank Institute, The World Bank 2002. Also available at: *http://lnweb18.worldbank.org/eap/eap.nsf/2500ec5f1a2d9bad852568a3006f557d/ c19065b26241f0b247256ac30010e5ff?OpenDocument*

Deacon, R. 1994. Deforestation and the Rule of Law in a Cross Section of Countries. *Land Economics* 70(4), 1994, pp. 414-430.

Didia, D. O. 1997. Democracy, Political Instability, and Tropical Deforestation. *Global Environmental Change* 7(1), 63-76.

EIA (Environmental Investigation Agency and Telepak Indonesia). 2001. *Timber Trafficking*. September 2001. Available at: *http://www.eia-international.org/*

Glastra, R. (editor) 1999. *Cut and Run: Illegal Logging and Timber Trade in the Trop-ics*. International Development Research Center, IDRC, Ottawa, Canada.

ITTO 2000. Review of Progress towards Year 2000 Objective. ITTO. Yokohama, Japan.

Johnston, M. and A. Doig. 1999. Different Views on Good Government and Sustain-able Anticorruption Strategies. In: *Curbing Corruption. Towards a Model for Building National Integrity* by R. Stapenhurst and S. J. Kpundeh (editors), EDI De-velopment Studies, World Bank, Washington, DC, USA.

Kaimowitz, D. and A. Angelsen. 1999. Rethinking the Causes of Deforestation: Les-sons from Economic Models. *The World Bank Research Observer*,14(1), 73-98. The World Bank, Washington DC.

Kaimowitz, D. 2001. Forest Crimes and Forest Law Enforcement: Issues and Challenges for East Asia. Keynote presentation at the Forest Law Enforcement and Governance: East Asia Ministerial Conference, Bali, Indonesia, Septem-ber 11-13, 2001. Available at: *http://lnweb18.worldbank.org/eap/eap.nsf/ 2500ec5f1a2d9bad852568a3006f557d/c19065b26241f0b247256ac30010e5ff? OpenDocument*

Kaufmann, D., A. Kraay and P. Zoido-Lobatón. 1999a. Aggregating Governance Indi-cators. Policy Research Working Paper 2195, The World Bank, Washington, DC, USA.

Kaufmann, D., A. Kraay and P. Zoido-Lobatón. 1999b. Governance Matters. Policy Research Working Paper 2196, The World Bank, Washington, DC, USA.

Kishor, N. 2002. Sustaining Forest by Improving Governance: An Overview and Ele-ments of a Strategy. Paper prepared for the PREM Anticorruption Thematic Knowl-edge Management Group of the World Bank. Processed.

Magrath, W. and R. Grandalski. 2002. Policies, strategies and technologies for forest resource protection. In T. Enters, P. B. Durst, G. B. Applegate, P. C. S. Kho, and G. Man (eds.). *Applying Reduced Impact Logging to Advance Sustainable Forest Management*. Food and Agriculture Organization of the United Nations. Bangkok, Thailand. Available at *http://www.fao.org/DOCREP/005/AC805E/AC805E00.HTM*.

Mainardi, S. 1996. An Econometric Analysis of Factors Affecting Tropical and Sub-tropical Deforestation, mimeo, Department of Economics, University of Natal, South Africa. (Main findings summarized in Kaimowitz and Angelsen 1999).

Melle, A. and DeAndra Beck. 2001. The U.S. Forest Service Approach to Forest Law Enforcement. Presentation at the Forest Law Enforcement and Governance: East Asia Ministerial Conference, Bali, Indonesia, September 11-13, 2001. Available at: *http://lnweb18.worldbank.org/eap/eap.nsf/2500ec5f1a2d9bad852568a3006f557d/ c19065b26241f0b247256ac30010e5ff?OpenDocument*

Palo, M. 2001. How Does Corruption Drive Deforestation in the Tropics? Unpublished paper presented at "Corruption in Forestry: Roundtable Discussion on Illegal Log-ging and FIN," 10th International Anti Corruption Conference, Prague October 2001. Available at: *http://www.10iacc.org*.

Poore, D. 1989. *No Timber Without Trees: Sustainability in the Tropical Forest*. Earthscan.

Rosenbaum, K. L. 2002. Illegal Actions and the Forest Sector: A Legal Perspective. Paper presented at the ISTF conference on illegal logging at Yale University, March 29-30, 2002.

Shafik, N. 1997. Macroeconomic Causes of Deforestation: Barking up the Wrong Tree? In: Katarina Brown and David Pearce (eds.). *The Causes of Tropical Deforestation: The Economic and Statistical Analysis of Factors Giving Rise to the Loss of Tropical Forests*. University College London Press, London.

Stapenhurst, R. and S.J. Kpundeh (editors).1999. *Curbing Corruption. Towards a Model for Building National Integrity*. EDI Development Studies, World Bank, Washington DC, USA.

Thomas, V., M. Dailami, A. Dhareshwar, D. Kaufmann, N. Kishor, R. Lopez and Y. Wang. 2000. *The Quality of Growth*. Oxford University Press, New York, USA, September 2000.

World Bank. 1997. *Helping Countries to Combat Corruption. The Role of the World Bank*. Washington DC, USA.

World Bank. 2000. *Reforming Public Institutions and Strengthening Governance: A World Bank Strategy*. Public Sector Group PREM Network, The World Bank, November 2000.

APPENDIX 1. Variable Definition and Data Sources

Variable & Acronym	Description and Source
Annual % rate of Deforestation (**DFPA**)	Average annual deforestation refers to the permanent conversion of natural forest area to other uses, including shifting cultivation, permanent agriculture, ranching, settlements, and infrastructure development. Deforested areas do not include areas logged but intended for regeneration or areas degraded by fuelwood gathering, acid precipitation, or forest fires. This covers the period 1990 to 2000. Negative numbers indicate an increase in forest area. Drawn from the World Development Indicators (WDI).
Per Capita Income (**PPPCY95**)	GDP per capita based on purchasing power parity (PPP). PPP GDP is gross domestic product converted to international dollars using purchasing power parity rates. An international dollar has the same purchasing power over GDP as the U.S. dollar has in the United States. GDP is the sum of gross value added by all resident producers in the economy plus any product taxes and minus any subsidies not included in the value of the products. It is calculated without making deductions for depreciation of fabricated assets or for depletion and degradation of natural resources. Data are in current international dollars for 1995.
Rural Population Density (**POPDENRUR95**)	Rural population density is the rural population divided by the arable land area. Rural population is calculated as the difference between the total population and the urban population. Arable land includes land defined by the FAO as land under temporary crops (double-cropped areas are counted once), temporary meadows for mowing or for pasture, land under market or kitchen gardens, and land temporarily fallow. Land abandoned as a result of shifting cultivation is excluded. The information is taken from the WDI for 1995.

APPENDIX 1 (continued)

Variable & Acronym	Description and Source
Total roads network (ROADS)	Total road network includes motorways, highways, and main or national roads, secondary or regional roads, and all other roads in a country, for 1995. For more information, see WDI.
Rural Population as a % of total (RURPOP)	Rural population is calculated as the difference between the total population and the urban population, for 1995. For more information, see WDI.
Globalization (GLOBALNESS)	Trade in goods as a share of PPP GDP is the sum of merchandise exports and imports measured in current U.S. dollars divided by the value of GDP converted to international dollars using purchasing power parity rates (see WDI table 1.1 for a discussion of PPP). For more information, see WDI.
Government commitment to protect forests (FORPROT)	Nationally protected areas are totally or partially protected areas of at least 1,000 hectares that are designated as national parks, natural monuments, nature reserves or wildlife sanctuaries, protected landscapes and seascapes, or scientific reserves with limited public access. The data do not include sites protected under local or provincial law. Total land area is used to calculate the percentage of total area protected. For more information, see WDI.
Trade Openness (OPENESS)	Index of trade openness developed by the Heritage Foundation and the Wall Street Journal. It takes a value between 1 and 5 with higher values representing greater openness. An economy earns "5" if it has an average tariff rate of less than or equal to four percentage points and/or has very few non-tariff barriers; and "1" if the average tariff rate is greater than 19% and there are very high non-tariff barriers that virtually prohibit exports.
Regional Dummy Variables (DUM1, DUM2, DUM3, DUM4, DUM5)	Dum1 = 1 if country is from the Africa region, 0 otherwise; Dum2 =1 if country is from the East Asia and Pacific region, 0 otherwise, Dum3 = 1 if country is from the Europe and Central Asia region, 0 otherwise; Dum4 = 1 if country is from the Latin America and the Caribbean region, 0 otherwise; Dum5 = 1 if country is from the South Asia region, 0 otherwise.
Voice and Accountability (VOICE)	Definition given in main text. Scaled to lie between −2.5 and +2.5 with higher values corresponding to better outcomes (Kaufmann et al. 1999a)
Political Stability and Lack of Violence (POLSTABILITY)	Definition given in main text. Scaled to lie between −2.5 and +2.5 with higher values corresponding to better outcomes (Kaufmann et al. 1999a)
Government Effectiveness (GOVEFFECT)	Definition given in main text. Scaled to lie between −2.5 and +2.5 with higher values corresponding to better outcomes (Kaufmann et al. 1999a)
Lack of Regulatory Burden (REGBURDEN)	Definition given in main text. Scaled to lie between −2.5 and +2.5 with higher values corresponding to better outcomes (Kaufmann et al. 1999a)
Control of Corruption and Graft (CORRUPGRAFT)	Definition given in main text. Scaled to lie between −2.5 and +2.5 with higher values corresponding to better outcomes (Kaufmann et al. 1999a)
Rule of Law (RULELAW)	Definition given in main text. Scaled to lie between −2.5 and +2.5 with higher values corresponding to better outcomes (Kaufmann et al. 1999a)

APPENDIX 2. Deforestation and Governance: Preliminary Regressions

1. regress dfpa voice, robust

dfpa	Coef. (Std. Err.)	t value	P > t	[95% Conf.	Interval]
voice	−.4289583 (.1164568)	−3.68	0	−0.6603917	−0.19752
_cons	.5605176 (.1095465)	5.12	0	0.3428169	0.778218
				R-squared	0.1581

2. regress dfpa polstability, robust

dfpa	Coef. (Std. Err.)	t value	P > t	[95% Conf.	Interval]
polstability	−.4487853 (.0944765)	−4.75	0	−0.6365375	−0.26103
_cons	.4380415 (.089025)	4.92	0	0.2611231	0.61496
				R-squared	0.1872

3. regress dfpa goveffect, robust

dfpa	Coef. (Std. Err.)	t value	P > t	[95% Conf.	Interval]
goveffect	−.434321 (.0900261)	−4.82	0	−0.6132289	−0.25541
_cons	.4780577 (.0950131)	5.03	0	0.2892392	0.666876
				R-squared	0.1512

4. regress dfpa regburden, robust

dfpa	Coef.(Std. Err.)	t value	P > t	[95% Conf.	Interval]
regburden	−.3061839 (.1576361)	−1.94	0.055	−0.6194524	0.007085
_cons	.5091147 (.109868)	4.63	0	0.2907752	0.727454
				R-squared	0.0466

5. regress dfpa corrupgraft, robust

dfpa	Coef. (Std. Err.)	t value	P > t	[95% Conf.	Interval]
corrupgraft	−.4182503 (.0796322)	−5.25	0	−0.5765026	−0.26
_cons	.4590076 (.0922028)	4.98	0	0.2757739	0.642241
				R-squared	0.1608

6. regress dfpa rulelaw, robust

dfpa	Coef. (Std. Err.)	t value	P > t	[95% Conf.	Interval]
rulelaw	−.4403785 (.0792503)	−5.56	0	−0.5978719	−0.28289
_cons	.4562608 (.0905906)	5.04	0	0.2762311	0.636291
				R-squared	0.1859

APPENDIX 3. Estimated Regression Equations

R-squared = 0.4760
Root MSE = .76529

3.1

dfpa	Coef.	Robust Std. Err.	t	P>|t|	[95% Conf.	Interval]
pppcy95	−.0001192	.0000691	−1.72	0.089	−.000257	.0000187
ppcy95sq	4.47e-09	2.34e-09	1.91	0.060	−1.99e-10	9.14e-09
popdenrur95	−.0000717	.0001057	−0.68	0.500	−.0002825	.0001392
Rurpop	.0034669	.00706	0.49	0.625	−.0106139	.0175477
Roads	**−2.70e-07**	**1.43e-07**	**−1.89**	**0.063**	**−5.54e-07**	**1.47e-08**
Forprot	−.0041247	.0084392	−0.49	0.627	−.0209562	.0127068
Globalness	−.0009419	.0049217	−0.19	0.849	−.010758	.0088741
Openess	−.0088526	.1103738	−0.08	0.936	−.2289863	.211281
voice	.0095661	.2970145	0.03	0.974	−.5828106	.6019429
polstability	−.1300711	.2806978	−0.46	0.645	−.6899051	.429763
goveffect	−.1140439	.3581059	−0.32	0.751	−.8282635	.6001758
regburden	.4008919	.3142829	1.28	0.206	−.2259256	1.027709
corrupgraft	.1092471	.309303	0.35	0.725	−.5076382	.7261324
rulelaw	−.1136987	.2987075	−0.38	0.705	−.7094521	.4820546
dum1	.279334	.5497427	0.51	0.613	−.8170932	1.375761
dum2	−.0020431	.4917389	−0.00	0.997	−.9827853	.9786992
dum3	−.6086299	.4543127	−1.34	0.185	−1.514728	.2974681
dum4	.2056796	.4971742	0.41	0.680	−.7859031	1.197262
dum5	.1123848	.6016586	0.19	0.852	−1.087585	1.312355
_cons	.8505096	.9084316	0.94	0.352	−.9613002	2.662319

R-squared = 0.2093
Root MSE = .75325

3.2

dfpares	Coef.	Robust Std. Err.	t	P>|t|	[95% Conf.	Interval]
popdenrur95	−.0000118	.0000933	−0.13	0.899	−.0001976	.0001739
Rurpop	−.0010213	.0052154	−0.20	0.845	−.0114043	.0093617
Forprot	−.0009322	.0070216	−0.13	0.895	−.0149111	.0130468
Globalness	−.0022947	.00633	−0.36	0.718	−.0148966	.0103073
Openess	.0848079	.1072875	0.79	0.432	−.128785	.2984009
voice	.0922575	.1438151	0.64	0.523	−.1940564	.3785714
globaleca	.0110322	.0071316	1.55	0.126	−.0031657	.0252302
dum1	.4594703	.4679036	0.98	0.329	−.4720542	1.390995
dum2	.1481475	.4446566	0.33	0.740	−.7370958	1.033391
dum3	**−.8379219**	**.4543598**	**−1.84**	**0.069**	**−1.742483**	**.0666389**
dum4	.4211026	.4297406	0.98	0.330	−.4344452	1.27665
_cons	−.246896	.6044129	−0.41	0.684	−1.45019	.9563976

| | | R-squared | = | 0.4338 |
| | | Root MSE | = | .74885 |

3.3

----

dfpa	Coef.	Robust Std. Err.	t	P>|t|	[95% Conf.	Interval]
popdenrur95	−8.03e-06	.0000828	−0.10	0.923	−.0001729	.0001568
Rurpop	**.0091163**	**.0050173**	**1.82**	**0.073**	**−.0008703**	**.0191029**
Forprot	.0011052	.0075559	0.15	0.884	−.0139343	.0161448
Globalness	.0018468	.0035676	0.52	0.606	−.0052544	.008948
Openess	.0824247	.097113	0.85	0.399	−.1108739	.2757232
voice	−.1053883	.1426285	−0.74	0.462	−.3892832	.1785066
dum1	.6128054	.3934408	1.56	0.123	−.1703188	1.39593
dum2	−.0479116	.3849347	−0.12	0.901	−.8141049	.7182816
dum3	**−.6064678**	**.3285722**	**−1.85**	**0.069**	**−1.260474**	**.0475388**
dum4	.4937404	.3690511	1.34	0.185	−.2408375	1.228318
_cons	−.3120502	.465309	−0.67	0.504	−1.238225	.6141243

Illegal Logging and Local Democracy: Between Communitarianism and Legal Fetishism

Antonio Azuela

SUMMARY. The paper considers two major views of illegal logging: "communitarian" and "legalistic." The former emphasizes the positive role of local communities and sees law enforcement programs as, at least potentially, counterproductive to environmental policy. While this perception fails to take the rule of law seriously, it shows the importance of local arrangements for sustainable use of forests. On the other hand, there is a view of deforestation that defines it only in juridical terms as 'illegal logging,' without taking into account the variety and complexity of social problems at the local level. The paper reviews some of the ways social sciences help us to overcome the limitations of both views. However, it also points at an issue that has not been sufficiently addressed by social disciplines: the question of local democracy. While most observ-

Antonio Azuela is a member of the *Instituto de Investigaciones Sociales,* at the *Universidad Nacional Autónoma de México.* This paper was written while he was Tinker Visiting Profesor at the University of Texas at Austin.

Comments are welcome at lacueva@servidor.unam.mx. The author wants to express his thanks to the Teresa Lozano Long Institute for Latin American Studies and the School of Law of the University of Texas at Austin, as well as to the Tinker Foundation, for supporting his work in 2002. The author is also grateful to Andrew Mathews, Lorena Flores-Gutierrez and two anonymous reviewers for their helpful comments to a previous draft of this paper.

[Haworth co-indexing entry note]: "Illegal Logging and Local Democracy: Between Communitarianism and Legal Fetishism." Azuela, Antonio. Co-published simultaneously in *Journal of Sustainable Forestry* (The Haworth Press, Inc.) Vol. 19, No. 1/2/3, 2004, pp. 81-96; and: *Illegal Logging in the Tropics: Strategies for Cutting Crime* (ed: Ramsay M. Ravenel, Ilmi M. E. Granoff, and Carrie A. Magee) The Haworth Press, Inc., 2004, pp. 81-96. Single or multiple copies of this article are available for a fee from The Haworth Document Delivery Service [1-800-HAWORTH, 9:00 a.m. - 5:00 p.m. (EST). E-mail address: docdelivery@haworthpress.com].

ers agree on the need for democratic institutions at the local level, the social conditions that make those institutions possible demand further study. This is a challenge for social sciences, due to the growing complexity of rural societies–a complexity that includes *inter alia* conflicts between owners and non-owners of natural resources, as well as the presence of 'external' social actors such as NGOs. *[Article copies available for a fee from The Haworth Document Delivery Service: 1-800-HAWORTH. E-mail address: <docdelivery@haworthpress.com> Website: <http://www. HaworthPress.com> © 2004 by The Haworth Press, Inc. All rights reserved.]*

KEYWORDS. Illegal logging, common property resources, deforestation, Mexico, local government, environmental crime

INTRODUCTION

Social practices that are usually labeled as illegal logging are the subject of two different, and indeed opposing, views. On the one hand, a legalistic view (that at its worst can be characterized as legal fetishism) uses the language of the law to condemn those practices and puts the law above all other consideration. On the other hand, communitarian views put the emphasis on the conditions under which members of local communities engage in logging practices, and tend to avoid any definition of those practices that can imply 'blaming the victims'; the very notion of 'illegal logging' is just one of those definitions. From a legalistic point of view, the law should be enforced by a strong national government, against anyone who misuses natural resources such as forests, regardless of the 'reasons' that they may have at a local level. From a communitarian point of view regulation of natural resources should be left to the local community. National institutions (and especially the legal system) are seen as illegitimate encroachments into the only true space for democracy and the rule of law: the local community. These are indeed extreme views, but it is my impression that too many observers are close to one of them. In this paper, I argue for the need to transcend this dichotomy and I consider some of the contributions and the limitations of social sciences for this task. I take as a reference the case of Mexico, where peasant communities own and control more than two thirds of the country's forests.

My main argument in this paper is that the social sciences can play a major role in assisting our understanding of deforestation and above all, can help us to find a way out of the dilemma between legalism and communitarianism. At the same time, I point to an issue upon which the social sciences are still far from providing us with secure answers: the means by which democratic institutions are created at the local level for the sustainable use of forests.

DIFFERENT ECONOMIC PROCESSES,
ONE SINGLE LEGAL CATEGORY

There is an obvious contribution of social sciences to the understanding of deforestation. It refers to the empirical demonstration that behind the phrase illegal logging one can find a wide variety of economic processes. 'Logging' in the sense of transforming trees into wood for some economic purpose, is by no means the only activity that threatens forests. In the case of Mexico, and this applies to many other countries, deforestation takes pace mainly through three different categories of practices: Unsustainable forestry, the conversion of forested areas into agriculture, and the intensification of certain practices associated with traditional rural society. Although we can find situations in which all three of these categories occur, it is important to recognize their differences because they are part of different economic processes. Let us summarize the context of each of these processes for the case of Mexico.

Unsustainable forestry, as the extraction of timber beyond the rate of renewal, is a well-known cause of deforestation. Misconceived and erratic government policies have played a mayor role in encouraging unsustainable forestry. For most of the twentieth century, the primary way in which the Mexican government tried to avoid the destruction of forests was by establishing *vedas* (i.e., a moratoria upon logging). This of course created a black market for forest products and prevented the development of professional skills among the owners of forests and forest resources. Moreover, when forest projects were approved, permits were given not to the owners of those forests but to private or state owned corporations. Although more than three fourths of Mexico's forests are the common property of agrarian communities (*ejidos* and *comunidades*), the communities were usually only able to collect a small rent for allowing those corporations to use their forests. In the late 1980s, government policies began to promote forestry projects conducted directly by the owners of the forests. Thus, in many regions of the country they have just begun to engage directly in the use of their own forests. The learning process is only beginning and there are still serious problems with community forestry, but most scholars agree that sustainable use of the forest by the owners is a better alternative than just making all logging illegal.

The second category of practices that lead to deforestation is the conversion of forests into agricultural areas. In fact, this has been the major cause of deforestation in many countries like Mexico. Although available data are extremely scarce, there is general agreement that much more tropical forest has been lost to agriculture and cattle raising than to forestry. The causes for this 'change of land use' are multiple, including encouragement by fundamentally misguided public policies. One of the worst mistakes of agricultural policy of the last four decades is the principle of "beneficial use," based upon the belief that land in the tropics would invariably be more productive without its native vegetation. Let us point at some of the most salient features of those policies:

1. Peasants' property rights over forests were protected only if they converted them into agricultural production.
2. Colonization policies populated the tropics with peasants with no previous experience in the management of these fragile ecosystems.
3. Public loans were available for cattle raising, even in natural protected areas.
4. Huge agricultural projects were intended to profit from the supposed fertility of land in the tropics (Tudela, 1988).

Many of those policies were still in effect during the 1980s, so it is not surprising that many rural communities still wish to transform their forested areas to agricultural and/or livestock uses. This happens particularly where there is no timber with significant economic value as in the seasonally dry forest type (the *selva baja caducifólia*) and residents have not developed the ability to make a sustainable use of the resources of the forested areas. In all of these cases, peasants do not take advantage of the forest. They simply remove it because it is an obstacle to short-term profitable activities. In spite of recent efforts (Semarnap, 2000 and Giugale et al., 2001) there is still a long way to go before the inertia of many decades of wrong-headed policies can be offset by new policies.

The third category of practices that lead to deforestation is associated with certain traditions in rural life. The domestic use of wood as fuel, overgrazing in forested areas, and the use of fire in 'swidden' agricultural systems, are some of the traditions that pose a threat to tropical ecosystems. The impact of the last of these may depend not on how carefully or recklessly fire is used, but also on climatic conditions. Mexico had a very dry year in 1998, and fire became the leading cause of forest degradation in that year (Semarnap, 2000).

This brief account does not try to cover all the different processes that lead to deforestation. Talking about local processes does not mean locating 'causes' at that level–regional markets and national policies are freguently the driving forces at work. I only try to illustrate how different practices can fall into the same legal category. Cutting trees without a permit (an illegal action) can be part of very different economic activities. It can be the product of forestry or of agricultural production, of using the forest illegally versus just destroying it. Also, illegal forest activities can be related to consumption practices of poor rural societies (e.g., fuelwood) or to those of affluent urban societies (e.g., the commerce of high value flora and wildlife species). Finally, such activities can be seen as the normal way of doing things (mainly by local communities), or as crimes (through the urban gaze), regardless of the fact that they might be the same from the point of view of the law. In a very important sense, we can be deceived by the categories of law.

Certainly, there are situations in which enforcing the law against those responsible for deforestation is just a matter of 'political will,' and optimal results can be achieved. In fact, some efforts have been made in Mexico to 'map out' the predominant activities leading to deforestation in the different regions where intense deforestation processes occur (Profepa, 1998). Although they are only a preliminary basis for the search for much more robust information, these efforts can help to design different enforcement strategies.

However, once we recognize the variety and the complexity of the social processes behind deforestation, the law itself becomes problematic. Apparently, the 'distance' between law and social reality is so large that we should have to abandon a legal perspective in favor of that of another discipline or a combination of disciplines. It is only through a socio-legal approach that we can fully understand the nature of the problems at hand.

THE ROLE OF LAW IN SOCIETY
AND THE MEANING OF ILLEGALITY

I use the word socio-legal research in a very broad sense, in order to include every analysis that looks at the meaning and the role of law for the organization of society. Rather than a unified discipline, it is a field in which many disciplines and perspectives converge. Even if there is nothing close to an academic community that shares basic tenets, there is a vast literature that documents and reflects upon the relevance of law in society. There is nothing in this literature like a 'general theory of illegality' that we could use in order to understand illegal logging. Therefore, we have to resort to rather general ideas about law and society.

Although the contributions of socio-legal studies are extremely diverse, they help us to see law as performing two basic, and by no means contradictory, functions in society. On the one hand, law brings about what mainstream sociology calls the 'stabilization of expectations'–a function that lends itself to pragmatic analysis. On the other hand, law provides social actors with a series of categories that serve as representations of social relations which in turn inspire interpretive or hermeneutic approaches, i.e., those that try to understand the meaning that social actors ascribe to the situations they confront. There are substantial differences between these two approaches to the law, but taken together they convey what we can learn about the role of law in society from the social sciences–i.e., from that intellectual project that tries to go beyond metaphysics on the one side and common sense on the other. From abstract theories to case studies, socio-legal research teaches us that in modern societies, social relations cannot be maintained if they are not, at the same time, legal relations.

Although there is always the risk of believing that law 'creates' society, it is possible to say that, at least in modernity, society cannot reproduce itself without a legal system.

Pragmatic analyses tend to see social action as guided by some form of rational calculation, based on the knowledge of legal rules and on the expectation that they will also guide the behavior of others. The 'Law and Economics' movement can be considered as one example of this kind of approach (Posner, 1995). However, those analyses take for granted that legal rules will be enforced when violations occur. Since this is not the case when we talk about illegal logging, they are not very useful for understanding it.

There is no doubt that law is made up of categories that provide definitions of the world, its inhabitants, and the relations between them. This is why hermeneutic analyses can be very helpful for our purpose: they look at the way legal categories become part of people's definition of their situation, as well as the situation of others, even when one of them is described as being 'outside' the law. Legal categories transform humans into citizens (and, of course, non-citizens), forests into natural resources, land into property, and so on. In fact, legal categories are so powerful, that it takes some intellectual effort to analyze social relations without using them. Very often, we do not realize that we define a situation through concepts that only make sense within the realm of the law. For example, in most contexts, saying 'I am an Italian' can mean the same as 'I am an Italian citizen.' However, when we come to examine the question of citizenship from a socio-legal perspective, the distinction between identity and citizenship is essential. One has to differentiate between the experience of belonging to a national community, i.e., a complex situation that involves a multiplicity of elements, and even mixed feelings, on the one hand, and citizenship as a legal category, which does not have to correspond to any particular form of experience. It is a way of construing individuals from the perspective of the legal system.

The very idea of 'illegal logging' is a good example of this. At its worst, it is a typical example of 'seeing like a state' (Scott, 1999). There is not one specific form of logging that can be said to be intrinsically illegal, thus it is not difficult to recognize that it is solely in the legal system where certain practices become illegal or legal. However, keeping this in mind in a systematic way is far more difficult.

By the same token, unless one thinks that lawmakers never make mistakes, it should be obvious that not every form of logging that is declared illegal will be a major cause of deforestation, and similarly, that not every form of legal logging is sustainable. The main problem with looking at society through legal categories is the risk of falling into a scheme that separates the law, on the one hand, and social 'reality' on the other. In that scheme, law must be outside reality, it would only exist as our own frustrated expectation of what should be and is not. It is perfectly correct for us to denounce practices that contradict legal norms in which we believe. But it is important to recognize that, in doing this, we are talking within the legal system. Every time we say 'this is illegal' and are listened to by a socially significant actor, even if we are not before a court, we are embracing one version of the law, when in fact there are many of them. And if we want to understand the ways different versions of the law influence the behavior of social actors, we must at least for a moment take some distance from legal categories. If we wish to avoid a scheme in which law is not part of reality, we must bracket, as it were, our own expectations. A recent example of this strategy is the work of Paul Kahn (Kahn, 1999) although it can certainly be traced back to the origins of modern social science. Indeed, since the nineteenth century social thinkers have tried this 'detached' position, although with different implications.

When we stop seeing society through legal categories, and recognize them as (problematic) objects to be analyzed, we can consider two basic questions. First, we can distinguish between the nature of social relations on the one hand (in our case, political and economic relations regarding forests) and, on the other, the legal categories through which they are represented and regulated–a perspective that shows us that ambiguity is an essential feature of law. Second, we can study situations in which illegality appears as a salient feature, in order to recognize the different, and often contradictory, legal categories that social actors use to define the situation. Let us consider these two questions.

When we think about the contrast between social relations and their legal representation it is clear that while social relations are fluid, diverse and in constant change, legal categories tend to be permanent, general and fixed. This generality makes legal categories fundamentally ambivalent. Thus, the same legal category can be used to represent very different social processes, as we saw in the case of illegal logging. Conversely, the same set of social relations can be subject to different legal labels, and we can see this when we think of people and forests. When a social group exerts some form of exclusive control over a forested area, two quite different legal categories can be used: property and sovereignty. When social relations over land are of an economic nature, we use the legal category of 'property' in order to represent it, whereas we use the notion of "sovereignty" when we see social relations of a political nature. In many cases, that distinction is clear enough. A group of people that controls an area as property owners, can be differentiated from a wider group that constitutes a political community, that is, a group that exerts sovereignty over a much larger territory (e.g., a nation-state). Also, the same forest can be subject to two clearly different social relations–e.g., it can 'belong' to the Rotary Club and it can 'belong' to the County of Wartdforshire, *at the same time*. In that case, it is easy to understand the distinction between those two forms of belonging and their correspondence with certain legal categories. However, precisely in the case of forests in developing countries, there are many cases in which things become much more uncertain. Frequently, it is not clear whether a local community is only a group of owners or a political community. Many combinations of economic (property) rights and political (sovereignty) claims may occur and this is why lack of clarity and conflict are so frequent in tropical forests.

Moreover, it is too frequent that national states do not even have a legal category that allows indigenous communities to give any meaning (within the legal system) to their presence in a territory. As long as those territories are not 'theirs' in any legal sense, they will not be able to defend themselves in the realm of the law. In the meantime, deep ecologists in the West can celebrate that those communities belong to the earth, rather than the opposite. It is interesting to observe that such "ecocentric" views are not confined to refined urban societies; they are now also present in many tropical forests around the world, thus bringing additional complexity to the debate on who owns what.

Certainly, we can examine (and even admire) those local communities without the categories of the law; anthropologists have done that successfully for more than a century. But, again, when the issue is 'illegal logging,' we cannot ignore the legal system. Particularly when we want to find a way of including local communities, we have to face the issue of their legal status. And we are forced to do so even if we want to transcend the typically modern public/private dichotomy.

In modern states, sovereignty is frequently expressed at various geographical scales, something that was recognized by the European constitutional legal tradition even in the heyday of the nation-state (Jellineck, 1981). National, sub-national and local authorities claim the political representation of different levels of social groups. Things become even more complicated with the emerging claims for a global citizenship that would entitle members of different national communities to have a voice in the way particular a particular forest is being managed. Federal constitutions have been successful in dealing with this complexity. Far more serious problems arise when it is impossible to make the distinction between political control of a territory, on the one hand, and property rights on the other. The consequences of being a political community, and not only a simple group of property owners, are enormous. It goes without saying that we are talking about the distinction between a private and a public sphere, which is one of the fundamental principles of modern society. Property owners can enjoy the benefits of using a forest, but they are 'subject' to the legal order of a wider political community. They cannot enforce by themselves whatever claim they may have over their forest. However, when the group becomes a political community and creates a local authority, the nature of social relations is completely different. They have to deal with the responsibilities of government–collecting taxes, securing the public order, and so on. In fact, some groups may prefer to be treated as political communities, and even fight for that recognition, whereas others may choose to be seen as simple landowners. External actors, such as government agencies, international organizations or even NGOs may also have different views. Legal definitions are not only varied, they can be the very source of social conflict.

In fact, the lack of a clear distinction between owners and rulers is the main problem for forest management in Mexico at present. Agrarian communities, which are defined as mere owners of forested lands from a legal perspective, have been forced to act as local governments. As I have shown elsewhere (Azuela, 1995) *ejidos* in tropical regions fulfill some of the functions that local authorities are unable to perform: from policing to tax collection and the administration of some basic infrastructures. At the end of this paper, I will consider the problems associated with this lack of clarity. For the moment, the point is that the distinction between social relations and their representation through legal categories is more than an academic abstraction. As long as it is also a field of social conflict, it is important for academic research to maintain that distinction in order to see the way social actors mobilize different legal definitions. Also, from a political point of view, taking that distinction for granted amounts to taking sides *a priori*, for an abstract formula, regardless of the way it works in reality.

There is a second question we can understand better when we see legal categories as external observers, and it is precisely the question of illegality. This can be particularly important for the study of 'illegal logging' where it takes place through practices that are the normal operation of society at the local level and, not surprisingly, are also seen as legitimate by most local actors.

Illegality has given rise to a vast literature from at least three different regions of social sciences. Please take note that, taken as a whole, this phenomenon covers a good part of the human race and its relation to legal systems. Economics, and more recently a sociological version of 'new institutionalism,' have been concerned with the 'informal sector,' which almost by definition operates outside the law (Brinton and Nee, 1998). Anthropological research has produced a huge amount of empirical and theoretical work on 'legal pluralism,' under the idea that rural (pre-modern) societies cannot be ruled out simply as illegal. Instead, they are seen as having a different type of legal system from that of modern societies. Similarly, 'irregular human settlements,' a pervasive element in the urbanization of developing countries, is a field in which the tensions between state legal systems and 'different' local arrangements have attracted the attention of social scientists. I have dealt with this elsewhere (Azuela, 1987) and it is impossible to review all the problems involved here (for an overview, see Fernandes and Varley, 1998). However, it is worth mentioning one version of it as it is closely related with communitarianism in a wider context.

For a good number of scholars in Latin America, the illegality of low-income urban settlements is not a problem, but a solution. Wherever neighbors have managed to avoid the enforcement of state law, and have been able to create their 'alternative' way of settling disputes and create order at the local level, this has fascinated observers. Boaventura de Souza Santos, by far the most prominent figure in this movement, offers a vivid account of the way residents of Rio de Janeiro's *favelas* generate their own legal system (Santos, 1996). By celebrating the existence of an 'alternative' law, a celebration that has an obvious anti-statist component, this literature has succeeded in reaffirming what just one version of state law says: that those settlements are illegal by definition. At the same time, lawyers who are working to defend residents' interests, and who have managed to stop forced evictions and to obtain legal recognition for these settlements, have done so by defining the situation in exactly the opposite terms. Taking as a platform the recognition of housing as a fundamental right in international law, and arguing that forced evictions are violations of that fundamental right, several NGOs have conducted an international campaign against evictions (COHRE, 2001). In many developing countries this has improved the situation of urban settlers, insofar as it has provided some security of tenure. In the view of these lawyers, it is the eviction that can be seen as illegal, not the settlers' possession of the land where they live. The paradox we want to show here is that, by accepting as valid the definition of human settlements as illegal, 'progressive' social research gives a poor service to the cause of those it is supposed to care about. Obviously, the prejudice against all forms of state legality is the origin of this trap. But the analytical error resides in taking legal categories as unproblematic.

There is an obvious danger of repeating this mistake in the case of logging. There can be no doubt that communitarianism has gained momentum in the last two decades. It is the strongest ideology amongst NGOs involved in rural issues and it permeates the work of many academics. In this context, a small dose of legal anthropology and a couple of case studies will apparently be enough to convince ourselves that the very idea of 'illegal logging' is just a demonization of local practices. In that context, all we need is to follow whatever local communities decide by whatever procedure, as if nation-states could be easily 'legislated out.'

The problem with this view is not that it takes seriously local communities, but that it fails to recognize that, apart from local interests, there are wider social interests involved in the use of forests, which can be seen as legitimate by many relevant actors. Unless one believes that sub-national, national and international (both governmental and non-governmental) organizations have no role to play in the definition of the way forests are to be used, one has to recognize the relevance of perceptions and interests that are defined beyond the local level. In fact, one of the main problems in designing institutions for sustainable forestry is precisely the need to balance social interests that are defined at different geographical levels. Before we turn into the area of social science that offers a perspective for this problem, let us summarize the main points in this section. Creating some distance from legal categories is important for three reasons. First, concrete social practices engaged with forest resources can be been seen both as legal or illegal (as can any form of human action), depending on the side one wants to take. Thus, a universal definition of illegal logging means taking sides in a number of cases whose particular conditions remain undetermined. It is only when one wants to influence a decision that it is necessary to embrace the notion of illegal logging as a definition of the situation. A universal fight against illegal logging can be meaningful only for those who want to combat all forms of logging, as well as for those who are ready to place legal arguments over and above any other consideration–i.e., for those who want to defend trees or legal rules at any rate.

Second, taking distance from the law is still more important when it is not only social practices, but also the very nature of social relations that is at stake. This is particularly clear in the tropics, where the legal status of local communities is unclear.

Last, but not least, looking at the law and its categories as external observers is a fundamental methodological requirement in order to see the way social actors represent their positions through normative categories. If law plays any role in social relations, it is because social actors use its categories to give meaning to what they do and what they are.

COMMON POOL RESOURCES AND NEW INSTITUTIONALISM

The main source of optimism for the sustainable management of forests by local communities comes from a movement that, in the last decade, has used the theoretical platform of 'new institutionalism' in order to provide an alternative view from that of the 'tragedy of the commons' (Hardin, 1968). What has also been called the Institutional Development Analysis (IDA) framework, is a growing movement of empirical research that has gathered around the proposals of Elinor Ostrom, Margaret McKean and others. It has established the basis of a theoretical framework that promises a better understanding of the conditions under which local communities can use 'common pool resources' (such as forests or irrigation systems) in a sustainable way. Among the many merits of this movement, is the fact that it recognizes the need to reconcile local institutional arrangements with wider institutional contexts–both at regional and national levels.

One of the concerns that mobilize scholars for studying 'the commons' is that many countries have promoted institutional reforms that suppress communal forms of forest property as part of 'modernization' schemes. As McKean puts it, in many instances common property regimes have been "legislated out of existence" (McKean, 2000, p. 34). Adherents to this movement have shown that in different countries, under certain circumstances, common property can be the best institutional arrangement for forest management. Although the IDA framework has an obvious sympathy for common property regimes, there is also a clear commitment to be more sensitive to empirical findings than to any *a priori* ideological position (Ostrom, 1990).

The case of Mexico is particularly interesting for the IDA framework, because most of its forested land is the common property of agrarian communities, and this is legally recognized. Moreover, the 'modernization' of Mexican agrarian law in the early 1990s did not alter this basic feature. While peasants who are members of an agrarian community are allowed to put their agricultural land into the market as a result of those reforms, the forested areas of communities are to be maintained as common property. Communities are not allowed to subdivide those areas and only in very exceptional cases can forests become private property.

One the most attractive aspects of the IDA framework is that it does not see local communities as self-contained social universes that can succeed isolated from wider political entities. The idea of 'nested institutions' is central to this theory (Ostrom, 1990) as it recognizes that national or sub-national governments have a role to play in the maintenance of common property arrangements at local level. As McKean puts it, " 'where local communities' resource claims go unrecognized by national governments, the best they can hope for is that higher layers of government will overlook them rather than oppose them" (McKean, 2000, p. 43). This implies not only a national legal framework, but also the willingness and the capacity of those higher layers of governments to enforce it. In terms of this theory, this willingness is essential for common property arrangements, because one of the basic conditions for its success is that "the criteria for membership in the group of eligible users of the resource must . . . be clear" (McKean, 2000, p. 44). In consequence, the right of owners to exclude non-owners must also be enforceable. By recognizing the need for a link between local and central institutions, this theory has opened the way to pose problems that most communitarian analysis ignores.

The task of building links between the local dimension and wider spheres meets enormous obstacles. The case of Mexico gives us good examples of at least three major problems. First, national institutions are far from being mere enforcers of local communities' rights. Increasingly, NGOs and public opinion expect to monitor local communities and, if necessary, enforce the law against them whenever they practice unsustainable forestry. In this respect, decentralization is not an option that environmental NGOs support. Moreover, new international actors, both non-governmental and intergovernmental, are bringing ever more pressure upon the national government to act upon whatever is seen as 'illegal logging.' Too often, without knowing what happens at local level, a rock star or a member of the Kennedy family can construct heroes and villains just by calling a press conference.

Second, local communities that embark on conservation or sustainable forestry projects can also have local enemies. Once again, regardless of how we choose to define them (poachers, invaders, defectors, and so on) there can be many sources of local conflict over the use of resources. In a Natural Protected Area like the Lacandona Reserve in Chiapas, the indigenous community that holds common property rights over the area is the victim of other communities' problems. Newcomers penetrate the area in order to settle there. Not surprisingly, the enforcement by national authorities is not an easy task when this happens just twenty miles away from the stronghold of the Zapatista army. In other regions of the country, neighbors can find it difficult to accept that members of a community obtain income from their forests, and may feel tempted to organize different forms of 'civil resistance' in order to stop logging, without waiting to see what forest authorities have to say. Last but not least, communities without strong surveillance mechanisms can be the victims of neighbors who want to make use of their forests.

Third, we can also find the opposite situation when a local community has not found a way of making a sustainable use of its forests. Thus it is frequent that community leaders protect those of its members who embark in something that external observers may see as illegal. When this entails a traditional social practice, like those we described in section 1, federal law enforcers can encounter a hostile environment, which can range from kidnapping to assassination. After two inspectors were killed during a visit to an *ejido* in the State of Campeche in 1996, the Office of the General Attorney for Environmental Protection in Mexico instructed inspectors to avoid 'dangerous places,' a category that could include a large proportion of the forests of the whole country.

In short, the use of forests can give rise to many social conflicts that involve only local actors. That is why they cannot be addressed solely through the intervention of national authorities. To put it in its simplest form: while sometimes communities' property rights are too weak, there are also instances in which they are too strong, i.e., when (non-owners) local actors are completely deprived of the opportunity to express their views about the way resources are used in *their* community. This problem is usually overlooked by communitarian views, but can be very important when rural communities grow and so does the division between owners and non-owners. Citizens in modern societies are allowed to participate in the decisions regarding the use of *other* people's land in the communities where they live, through conventional planning processes. Those places are *their* communities in a fundamental political sense. In many countries like Mexico, 'neighbors' of tropical forests, who are not part of the group that owns them, want to have a say in the way they are used (Azuela, 1995). We are talking about local democratic institutions in which both property rights and citizens' rights are reconciled. How does this sort of institutional arrangement come about?

THE MYSTERY OF LOCAL POLITICAL POWER

We have reached a point in which social sciences do not seem to give us a clear answer. How can we 'build' a local democracy that gives room to such a complex set of social interests? It is true that the origin of state institutions has been the subject of great efforts in the field of historical sociology (Elias, 1994; Mann, 1986). However, it is difficult to find anything specific about the new challenges of the relationship between people and forests at the local level as well as its articulation with larger geographical levels. Indeed, most communitarian analyses of deforestation tend to ignore the complexity of local societies and, in particular, the conflicting nature of relations between owners and non-owners. There are two aspects of this complexity. First, the traditional distinction between the urban and the rural world has been replaced by a complex web of social interests that operate at different geographical levels. Between big urban centers and small rural villages there is a setting of small cities and towns. Within an increasingly complex social geography, at least three groups of actors interact: Regional authorities (that can include elected legislatures), economic agents (that transform or simply transport forest products) and citizens (and citizens' organizations) who want to participate.

Secondly, enforcing the law against 'illegal' loggers is something that cannot be accomplished by national authorities alone. Enforcers in countries like Mexico know that it is not just a question of 'getting there.' Enforcement always implies making decisions about what is legal and what is illegal and, above all, it frequently entails decisions about what to do—what kind of sanctions should be imposed and their severity? What should be done with seized timber? Should permits be temporarily or permanently suspended? and so on. None of these decisions can be successful if they do not have a minimum of legitimacy at the local, if not regional level. To the extent that law is an external rule that is imposed upon a community, that community will not be part of the social basis that supports the rule of law. Some form of participation of local actors in the definition of what is legal and what is illegal seems to be essential to the construction of any legal order. We are not talking here about some utopian process by which *the will* of a community could be easily translated into law. Precisely because the problem is the multiplicity of local and regional wills, perceptions and interests, a strong third party is necessary, i.e., a local state that is both democratic and capable of enforcing what the majority has accepted as legitimate.

Even if we convince *legalists* that local arrangements are important, they would have to drop the view that legal changes are all we need to engender new social arrangements. Law can help in the reproduction of society, but it cannot create it. The consolidation of property (as the operation of markets) depends on what economic agents do, as much as the consolidation of democracy depends on how political actors behave. A legal system can be used to defend interests, but it cannot create them. On the other hand, communitarian views are often associated with prejudices against state institutions in general. This is why the communitarian perspective has difficulty recognizing that local communities must transform themselves into political entities if they are to handle social conflicts successfully. Wherever there are owners and non-owners and wherever there are different views of how the forest should be used, the institutions of political representation become inevitable. Unless we manage to go beyond those two ideologies, we will be unable to think of a sustainable use of forests, which is not only efficient but also legitimate. How can this come about? It is a question that social sciences have yet to answer.

I have tried to show that, in order to understand the relationship between deforestation and the law, it is necessary to transcend prejudices that dominate the two main ways of looking at deforestation practices. When looking at those practices from a legal point of view, it is important to recognize the way society actually works at the local level. Similarly, when looking at the law from the point of view of local communities, it is necessary to recognize that there can be legitimate interests apart from those of the collectivities that own forests. These two steps lead to the construction of alternative legal arrangements that are capable of recognizing the plurality of interests around forests and, at the same time, inhibit practices that lead to deforestation. Strong democratic arrangements at the local level seem indispensable for the attainment of such an ambitious goal. Nevertheless, while it is easy to agree with the need for local democracy, it is difficult to specify the way it comes about. It is my contention that social scientists should make a new effort in order to elucidate the conditions under which this new form of local democracy becomes possible.

REFERENCES

Azuela, Antonio. 1987. Low Income Settlements and the Law in Mexico City. *International Journal of Urban and Regional Research*, Vol. 11 (4).

Azuela, Antonio. 1995. Ciudadanía y gestión urbana en los ejidos de los Tuxtlas. *Estudios Sociológicos 39*, Vol XIII.

Brinton, Mary and Victor Nee. 1998. The New Institutionalism in Sociology. New York: Russell Sage Foundation.

COHRE. 2001. Bibliography on Housing Rights and Evictions. Geneva: Center on Housing Rights and Evictions.

Elias, Norbert. 1994. The Civilizing Process. Oxford: Oxford University Press.

Fernández, Edesio and Ann Varley (Eds.). 1998. Illegal Cities: Law and Urban Change in Developing Countries. London: Sage Books.

Giugale, Marcelo, Olivier Lafourcade and Vinh H. Nguyen (Eds.). 2001. Mexico. A Comprehensive Development Agenda for the New Era. Washington: The World Bank.

Hardin, Garrett. 1968. The Tragedy of the Commons. *Science*, 162:1.

Jellinek, Georg. 1981. Fragmentos de Estado. Madrid: Editorial Civitas.

Kahn, Paul W. 1999. The Cultural Study of Law. Chicago: University of Chicago Press.

Mann, Michael. 1986. The Sources of Social Power. Cambridge, New York: Cambridge University Press.

McKean, Margaret A. 2000. Common Property: What Is It, What Is It Good for, and What Makes It Work? In pp. 27-55 Gibson, Clark, Mckean, M. and Ostrom, E. (Eds.). People and Forests. Communities, Institutions and Governance. Cambridge, Mass: MIT Press.

Ostrom, Elinor. 1990. Governing the Commons: The Evolution of Institutions for Collective Action. New York: Cambridge University Press.

Posner, Richard. 1995. Overcoming Law. Cambridge: Harvard University Press.

Profepa. 1998. Informe Trianual. 1995 B1997. Mexico: Procuraduría Federal de Protección al Ambiente.

Santos, Boaventura de Souza. 1996. Towards a New Common Sense. Law, Science and Politics in the Paradigmatic Transition. New York-London: Routledge.

Semarnap. 2000. La Gestión Ambiental en México. Mexico: Secretaría de Medio Ambiente, Recursos Naturales y Pesca.

Tudela, Fernando. 1988. "Los hijos tontos de la planeación: los grandes planes en el trópico húmedo mexicano" in Garza, Gustavo (Ed.). Una década de planeación urbano-regional en México, 1978-1988. Mexico: El Colegio de México.

You Say Illegal, I Say Legal:
The Relationship Between 'Illegal' Logging and Land Tenure, Poverty, and Forest Use Rights in Vietnam

Pamela McElwee

SUMMARY. Vietnam is estimated to have lost more than half of its forest cover in the past 50 years, with a number of contributing causes. The state nationalization of all forest resources during socialist rule from 1954 to the opening of the economy in the 1980s contributed significantly to illegal logging, as any locally-used forest was considered to be national property. During this period, there was a generalized 'free-for-all' on the forests, contributing to a massive breakdown of local tenure rules and resource allocation. Despite the dissolution of many of the state-owned logging reserves and distribution of this land to local communities, which began in the late 1980s, deforestation has continued. During the 1990s, an export ban on raw logs from Vietnam was enacted to halt the continued deforestation, with generally ineffective results. While the state continues to blame local people for illegal logging and

Pamela McElwee is a PhD candidate at the Yale School of Forestry and Environmental Studies, 210 Prospect Street, New Haven, CT 06511 USA.

The author would also like to sincerely thank the Center for Natural Resources and Environmental Studies, Vietnam National University, for facilitating her research in Vietnam.

Funding for field research in Vietnam was generously provided by the National Science Foundation and the Wenner Gren Foundation for Anthropological Research.

[Haworth co-indexing entry note]: "You Say Illegal, I Say Legal: The Relationship Between 'Illegal' Logging and Land Tenure, Poverty, and Forest Use Rights in Vietnam." McElwee, Pamela. Co-published simultaneously in *Journal of Sustainable Forestry* (The Haworth Press, Inc.) Vol. 19, No. 1/2/3, 2004, pp. 97-135; and: *Illegal Logging in the Tropics: Strategies for Cutting Crime* (ed: Ramsay M. Ravenel, Ilmi M. E. Granoff, and Carrie A. Magee) The Haworth Press, Inc., 2004, pp. 97-135. Single or multiple copies of this article are available for a fee from The Haworth Document Delivery Service [1-800-HAWORTH, 9:00 a.m. - 5:00 p.m. (EST). E-mail address: docdelivery@haworthpress.com].

attendant deforestation, this paper will show that perceived criminal and corrupt actions by the government have significantly contributed to the problem. Furthermore, the paper will explain how local people terms as 'illegal loggers' explain their actions in terms of social justice, poverty alleviation, and local control, rather than in the state's terms of 'illegality.' *[Article copies available for a fee from The Haworth Document Delivery Service: 1-800-HAWORTH. E-mail address: <docdelivery@haworthpress.com> Website: <http://www.HaworthPress.com> © 2004 by The Haworth Press, Inc. All rights reserved.]*

KEYWORDS. Illegal logging, logging bans, corruption, forest management, timber exports, property rights, Vietnam, Cambodia

INTRODUCTION:
PERCEPTIONS OF ILLEGAL LOGGING IN VIETNAM

The Socialist Republic of Vietnam (SRV) has recently become the 'poster child' for perceptions of what is wrong with forest management in Asia. In the past 5 years, many stories have made the international press about Vietnam's rapid deforestation rates, high numbers of illegal loggers, and the role Vietnam plays in creating incentives for illegal logging in neighboring countries such as Laos and Cambodia (Bangkok Post, 1993; Global Witness, 1998; AFP, 2000).

From 1943 to 1976, Vietnam went from 14.3 million hectares (ha) of forest to 11.1 million ha, a yearly deforestation rate of 0.68%. From 1976 to 1995 Vietnam went down to 8.25 million ha of forest, with a yearly deforestation rate of 1.27% (FIPI, 1996). It is believed that Vietnam has been losing about 200,000 ha of natural forest cover yearly since then–a deforestation rate of 2.4% per year (VNS, 1997a). While it is currently estimated that forest continues to cover 28% of the land area of Vietnam, only about 1% of that is undisturbed natural forest (Ogle et al., 1998). Additionally, only 5.5% of Vietnam's remaining natural forests are considered 'rich forests' (with over 120 m^3/ha growing stock), 16.9% are classified as 'medium-quality' (80-120 m^3/ha) and the rest of the natural forests are considered poorly stocked (under 80 m^3/ha) or are replanted (Brown et al., 2001).

The high deforestation rates in Vietnam are usually attributed to one of several causes. The president of Vietnam's Forestry Science and Technical Association has said, "[t]his situation is mainly caused by illegal timber cutting for local use and export, war and natural disaster" (Hoang Hoe, n.d., p. 1). The head of the Forest Science Institute blames "poor people . . . who are highly dependent on forests for food, fuel and income" (VNS, 1997a, p. 1). Other international commentators have blamed high rates of migration to forested highland areas and extensive agricultural expansion in marginal areas (DeKoninck, 1999). Still others blame slash-and-burn agriculture and the felling of trees to make way for plantation crops (Watkin, 1999).

There have been no comprehensive studies in Vietnam assessing causation for deforestation trends. The Forest Inventory and Planning Institute (FIPI), an arm of the national government responsible for land use planning surveys and forest inventories, produces maps that document changes, but does not usually provide comprehensive assessments of the underlying causes for forest trends. A recent study sponsored by the Canadian International Development Agency (CIDA) attempted to draw an analytic picture of deforestation in Vietnam, and bemoaned the difficulties in doing so, including lack of data, lack of cooperation among government agencies, and differing conceptions of deforestation (DeKoninck, 1999).

However, judging by the popular presses in recent years, illegal logging has become one of the most noted culprits in deforestation for the average Vietnamese. Most Vietnamese language newspapers often carry stories of illegal logging, and one paper, *Lao Dong [Labor]*, has devoted one environmental reporter and a special section on its website to uncovering illegal logging schemes. A new term has been coined in Vietnamese to describe these illegal loggers: *lam tac*. (The word is a compound of the root for forest and the verb 'to hijack.') Every week in the papers published in Hanoi, it seems there is a new story about the nefarious deeds of the *lam tac* (see Nguyen Thi Thanh Hai, 1999; Tuan Anh, 2000a; Thai Vu, 2000).

Some estimates have put the total amount of illegal logging occurring each year at 1 million m^3 a year, or 100,000 m^3 more than the allowable cut of approximately 900,000 m^3 (Ogle et al., 1998). According to a recent Asian Development Bank (ADB) review of the forestry sector, the supply of legally obtained, domestically sourced non-plantation forest logs account for only 30% of the total log supply (ADB, 2000). The remainder of the non-plantation hardwood timber in Vietnam is estimated to be made up of timber obtained through illegal logging and through the import of logs from Cambodia and Laos (see Global Witness, 2000).

This paper will attempt to examine the rising phenomenon of 'illegal logging' in Vietnam, by turning a critical lens to both the ecological and social dynamics involved in forest management and extraction. Through interviews with loggers, local villagers, forest rangers, policy planners, and government representatives in Vietnam, and through analysis of newspapers, policy documents and laws, this article will approach the problem of illegal logging in a holistic manner, and in particular, look at the contested dynamics between social actors and among levels–local, regional, national and international.

The use of an 'ethnographic' approach that problematizes differing and conflicting conceptions of environmental issues has been shown in other contexts to be a powerful approach (see Peet and Watts, 1996; Fairhead and Leach, 1996; Brown and Ekoko, 2001; McCarthy, 2002). Instead of taking a term like 'illegality' at face value, the strength of ethnographic approaches to environmental issues is to plumb the contested depths of terminology and concepts. As Peluso notes in the case of illegal teak theft in Indonesia:

> That which is not acceptable is labeled criminal; this labeling is the prerogative of the state. State appropriation of land, water, wood, and other natural resources redefines the rules of access to these and renders unauthorized access a criminal act. Yet what the state defines as criminal often differs substantially from the peasant definition of crime, and the denial of access to vital resources may loom as the most violent crime of the state toward the peasantry. (Peluso, 1992, p. 14)

Rather than gauge the increasing attention to 'illegality' of timber harvesting in apolitical terms and statistics, this paper attempts to question the terms of the debate in one of several ways. First, the article will explore the genesis of increasing attention to 'illegal logging' as a product of debates over land tenure and forest ownership in Vietnam as it undergoes a transition from a centrally planned socialist economy to a market-oriented one. Second, the article will pay attention to multiple scales of 'illegality,' looking not only at so-called illegal loggers but also at the locally contested claims for legal logging by the state. And finally, based on surveys of forest use in villages surrounding protected forests in central Vietnam, the article will attempt to show the role that logging plays in the local economy, and how local conceptions of social justice and rights to trees and land clash directly with national notions of 'forest protection.' All of these factors have led to increasingly rancorous debates within Vietnam among different levels of forest users–the national government, parastatal logging companies, provincial forest policy managers, local protected areas staff, and resident villagers around protected forests.

FOREST RESOURCES IN VIETNAM

How did so much of the forest in Vietnam get logged out so quickly? And when did this boom in 'illegal' logging start? These are difficult questions to answer due to a lack of baseline data on forest resources over time. While production forestry started in earnest under the French colonial regime in the late 19th century, there is evidence that minimally disturbed forest was still significant by the end of the colonial era, perhaps covering around 43% of the total land area (Maurand, 1943). Logging production under the French regime in Indochina was not particularly high, especially when compared with logging in French holdings in Africa. The average production of logs under the French colonial era was generally less than 600,000 m³/yr until 1940 when French control of Indochina started to wane with the start of World War II (Beresford and Fraser, 1992).

The spraying of 72 million liters of chemical defoliants on forests during the Vietnam War from the early 1960s until 1971 is estimated to have resulted in over 1.7 million hectares receiving at least one dousing (Westing, 1975). Including lands damaged by napalm, bombing, and massive "Rome Plows" which flattened whole landscapes like an iron, more than 2.2 million hectares of land in the former South Vietnam were seriously affected by the war's end in 1975 (Kemf, 1988).

Contrary to popular opinion of massive war damage, however, the most serious decline in forest cover in Vietnam appears to have occurred between 1976 and 1990. Starting in the 1970s, the need for wood to supply war material, and after the war, reconstruction material, meant an all-out assault on the nation's forest resources. It has been estimated that 10 million homes needed to be rebuilt after the war, and the demand for timber for these efforts was high (Kemf, 1988). During this time period, production of industrial roundwood peaked in 1987 at 5.4 million m³ (Brown et al., 2001). [Note should be taken that much of the wood exploited during this period for reconstruction may not be reflected in national statistics, as local people often logged with hand saws in the forest.] In addition to the domestic demand, log exports were one of the few ways Vietnam could earn foreign exchange in the postwar years. It was only in 1960 that the production of export logs from Vietnam first exceeded 1 million m³/yr. The number peaked at almost 1.7 million m³/yr during the first '5 year plan' under reunified Vietnam (1976-1980), and declined to about 1.4 million m³/yr during the second '5 year plan' (1981-1985) (Ogle, Blakeney and Hoang Hoe, 1998).

The bulk of this logging was carried out by State Forest Enterprises (SFEs), the logging companies established by the socialist state. The Democratic Republic of Vietnam (DRV), also known as North Vietnam, had nationalized all natural forests starting in the mid-1950s, and this process was extended to the south after 1975 and the reunification of Vietnam. These national forests were then turned over to SFEs to log. This nationalization obviously contributed to the state's perceptions of who was involved in 'illegal logging,' as any locally used forest was considered to be national property. Local populations often continued to manage forests as they traditionally had in many areas, and in areas where SFEs were set up, villages often protested against the nationalization by acts of sabotage. Arson, assaults on SFE personnel, and expansion of agricultural land into SFE forests were all strategies that villagers recalled to the author in 1997 in a village bordering an SFE in northern Lao Cai province. To this day, complaints between villages and SFEs over the borders of land rights and exploitation areas remain hotly contested in many rural areas.

Overall, it can be said that during this period, there was a generalized 'free-for-all' on the forests by both the state and by local communities who felt they had no other choice given extensive state claims to land, thereby contributing to a massive breakdown of local tenure rules and resource allocation. The author was told repeatedly that villages in one province of central Vietnam, Ha Tinh, had severe food shortages as a result of the cooperativization of agriculture during the same time period–starting about 1957–that the forests were nationalized. Given the double whammy of hunger and loss of forests to a logging enterprise (which not only moved into the area but which brought its own personnel from lowland areas and did not hire local people), there seemed no choice but to overexploit forest resources before the SFE did. As one elderly man said during an interview in 2000, "*Khong co gi ma rung* [we had nothing but forests]," particularly in the time of hunger, leading to the conversion of forest land for the expansion of agriculture, as well as the harvesting of forest plants for food to prevent starvation, and for medicine and timber to sell in the provincial market. Several other studies have shown that it was the loss of official access to local forests that contributed to the demise in long-held community tenure rules, particularly in the northwest forests of Vietnam where many non-Vietnamese ethnic minorities lived (Poffenberger, 1998; Sikor and Apel, 1998).

There has been no comprehensive study which has tried to assess the massive impact of logging in the more than 420 SFEs that were set up throughout the country by 1980, but the evidence at the local level combined with state statistics on forest exploitation indicates that vast stretches of Vietnam's highlands were logged by the state, not local people, until they were completely bare (see ANZDEC 1999; ADB 2000). The effects of this large-scale logging began to be seen in the mid-1980s, and there began to arise in some quarters a concern that Vietnam's forests were being depleted at an alarming and irreversible rate. Vietnam adopted a National Conservation Strategy in 1985 and a Biodiversity Conservation Strategy in 1995 in response to these concerns. However, while these documents emphasized the importance of protecting forests throughout the country, they tended to focus on the need to establish a protected areas system, rather than specifically focusing on the out-of-control logging by the state. Even today, with almost universal recognition in Vietnam of the importance of forest protection, there are those who continue to argue that state logging is not the major problem (see Department of Forest Development, 1998).

Current Forest Cover Statistics

Current national statistics about forest cover and forest loss in Vietnam can be misleading because of vast regional differences in forest cover among the major areas of the country. By far the highest forest cover has been and continues to be in an area known as the central highlands, which encompasses the four provinces of Lam Dong, Dak Lak, Gia Lai and Kon Tum in Vietnam's western highland plateau. As of the mid-1990s, the central highlands had 31.6 million hectares (ha) of forest land and approximately 3 million people, while the Mekong delta had only 2.1 million ha of forest and a population of 16 million (see Figure 1).

Along with the regional disparities in forest distribution, there have also been disparities in deforestation rates among regions. It is estimated that during the period of massive state logging from 1976-1995, the central highlands of Vietnam lost 630,000 ha of forest (out of 2.3 million ha), the southeast lost 301,000 ha (out of 467,000 ha), the north central coastal area lost 189,000 ha (out of 1.4 million ha), and the northern central mountains lost 134,500 ha (out of 583,000 ha) (Hoang Hoe, n.d.). A recent report on conservation priorities for Vietnam notes that "[t]he north of the country has lost most of its lowland for-

FIGURE 1. Population and Forest Cover of Vietnam's Regions (in 1993)

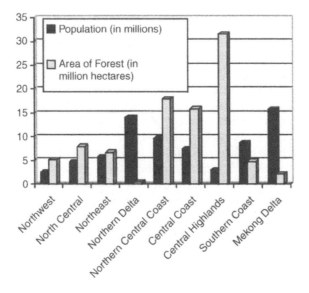

Source: Ban Chi Dao Kiem Ke Rung Tu Nhien Truong Uong [Central Committee for Inventory of Natural Forests], 1993

est, with montane forest now much reduced in extent and highly fragmented. Central Vietnam has lost most of its lowland forest but retains significant blocks of forest in the Annammite Mountains, whereas the south retains extensive forest areas in the Western (or Central) Highlands, with lowland forests again much reduced" (Wege et al., 1999, p. 11). These areas that are depleted are, not coincidentally, some of the areas that harbored large numbers of State Forest Enterprises.

To combat what came to be seen as excessive exploitation of natural forests, a provisional export-logging ban was enacted in 1993, concurrent with export log bans implemented in other Asian countries at that time, such as the Philippines and Thailand. The ban came about as a result of a trip then-Prime Minister Vo Van Kiet took to the central coast of southern Vietnam where a number of wood processing enterprises were located. During his trip, he seemed genuinely concerned that wood processors were exceeding their allocated buying quotas, sometime in excess of 2 to 3 times officially approved quotas. Kiet's directive, effective from 1994, said that the exploitation of any forest must be approved by a central national committee of the Ministry of Forestry together with the State Planning Commission, who would come up with yearly production targets for wood for export for each province, locality and sector. According to the Bangkok Post, the export ban resulted in a major drop in the Prime Minister's support from his home base in the southern and central provinces, as the entire Communist party structure in these areas had come to depend on timber exports as a major source of income (Bangkok Post, 1993).

Though Kiet's provisional ban was later slightly weakened in Decision 286/TTg in 1997, the ban initially tried to be comprehensive. There was to be a logging ban on all 'special-use forests' (i.e., protected areas and nature reserves), and a 30-year logging ban on trees in watershed areas. Commercial logging would be prohibited in the remaining natural forests in all of northern Vietnam and in the southeast and the Mekong delta areas (those areas with the least amount of forest cover remaining). The logging bans codified in 1997 currently cover 4.8 million ha of forest, or 58% of the forest lands (Brown et al., 2001). Figure 2 gives statistics on the diminishing role of production forests in Vietnam in recent years since the logging ban.

Despite Prime Minister Kiet's well-intentioned efforts, the export ban may in fact have been a red herring. The total contribution of the forestry sector to the gross domestic product was estimated at only 3.0% in 1990, which had declined to 1.4% percent by 1995. Furthermore, forest products exports as a share of total exports had declined from about 5% of the total in 1990 to less than 3% percent in 1995 (MARD, 2001). On the other hand, the domestic demand for timber was considerably more important, and the logging export bans did little to curb this domestic need. Rather, the ban simply led to a dra-

FIGURE 2. Decline in Production Forest Land, 1989-2000

Year	Total amount of production forest (in million ha)	Percentage of total national forest cover made up of production forests
1989	12.17	64.4%
1991	12.4	65.3%
1995	10.62	54.8%
1998	6.88	47.2%
2000	4.72	40.7%

Adapted from MARD, 2001.

matic jump in imports from neighboring countries, particularly Laos and Cambodia. Between 1981 and 1997, imports of sawn timber into Vietnam had averaged only 5,000 m^3 per year. The final logging ban resulted in legal sawn timber imports jumping to 100,000 m^3 in 1999 (Brown et al., 2001).

Illegal Imports from Cambodia

In addition to the dramatic rise in legal timber imports, a huge trade in illegal timber imports rumbled to life after the logging ban of 1993. Global Witness, a British-based human rights and environmental organization, has recently highlighted the scope of illegal log imports flowing from Cambodia. They estimate that there is a massive illegal trade in Cambodian logs to Vietnam worth at least US$130 million per year (Global Witness, 1998).

The most damning allegation is that Vietnamese officials, including central ministry officials and the central timber import/export parastatal, Vinafor, have colluded with corrupt Cambodian regional officials to log out the forests of northeastern Cambodia in defiance of Cambodia's own export log ban, enacted in December of 1996 as a result of international pressure. The role of the military and border guards in not only allowing illegal imports through, but also their role in investing in and owning timber-processing facilities on the Vietnamese side of the border, was highlighted in Global Witnesses' reports.

The Vietnamese government has strenuously refuted the allegations by Global Witness. A foreign ministry statement, issued in response to the allegations, said Vietnam had been cooperating with Cambodia since the end of 1996 in efforts to prevent illegal timber shipments from crossing its border (Reuters, 1998). The Vietnamese government highlighted the fact that then-Prime Minister Vo Van Kiet had been actively pushing for a formal regional accord to combat timber smuggling for several years (Reuters, 1996). However, Global Witness continues to allege that the high levels of log smuggling to Vietnam continue. [The author saw ample anecdotal evidence to support this claim during 5 months living in a Vietnamese province bordering Cambodia.]

Reforestation and Plantation Production

As a result of the continuing high demand for lumber, and increasing spotlights on illegal log imports into Vietnam, reforestation has become a concern of the government. There have been several national initiatives since 1992 to reforest large areas of the country and take Vietnam's percentage of total forest coverage back up to 1943 figures. According to the government, since 1992 Vietnam has replanted more than 2.2 million ha of forests. This has resulted in a slight increase in forest cover in some areas. For example, from 1992 to 1998, forests in the northern mountainous province of Son La have increased from 9.7% to 18.5% of the land area; from 27% to 32 % in Quang Ninh Province; from 13.2% to 25.8% in Lang Son Province; and from 27% to 32% in Quang Binh Province (VNS, 1998). The government claims these increases are almost entirely due to deliberate reforestation, not natural regeneration of degraded lands.

With these seeming successes in hand, in 1998 the government embarked on an ambitious nationwide plan to plant 5 million ha of forest, such that by 2010 the total forest area of the country will have reached 14.3 million hectares (equivalent to the figures on forest cover present in 1943). The National Five Million Hectare Reforestation Program (5MHRP), as it is called, has received massive amounts of funding. Current government financial support for the 5MHRP is in the area of $250-300 million US. This is supplemented by approximately $450 million US in foreign direct investment and overseas development aid to the reforestation efforts (MARD, 2001).

The explicit goals of the 5MHRP are not only environmental, but also economic. The government hopes to increase legal wood exploitation to 2 million m^3 a year, of which 85% is supposed to come from plantation forests (Brown et al., 2001). According to a survey of forest resources in 1990, plantations contributed to just 2% of the land area of Vietnam. Currently, there are over 500,000 ha of plantations, and plantation yields have gradually increased to over 480,000 m^3 per year. However, plantations face a number of difficulties, including poor soils, poor management, low prices, and inappropriate choice of exotic species (MARD, 2001).

A foreign-funded review of the 5MHRP objectives has declared that the entire plan is "implicitly based on the questionable assumption that Vietnam needs to be self-sufficient in wood production" (MARD, 2001, p. 9). Given Vietnam's interest in an Asian Free Trade Zone and in joining the World Trade Organization, the hopes for Vietnam's wood processing and forestry sectors to become internationally competitive are slim. Timber harvesting in Vietnam is often carried out with rudimentary machinery (i.e., hand saws) resulting in low harvesting productivity. Transportation costs often account for the largest proportion of timber production costs reflected by the fact that local log prices for plantation wood remain low (ADB, 2000). And the ready supply of good quality hardwoods at low prices (albeit illegally obtained from Cambodia and Laos, which artificially maintains the low cost) keeps the market for plantation softwoods suppressed.

Future Demands for Forest Products

Currently, legal timber harvesting in Vietnam is permitted only under approved forest management plans prepared by FIPI based on forest inventories, which are then approved by the Ministry of Agriculture and Rural Development (MARD), delegated to be carried out by SFEs, and which are enforced by the Forest Protection Department. Each province nationwide is allocated a log quota proportional to the national logging program for the year. For example, the nationwide total quota for logging from natural forests for 1998 was only 460,000 m^3 (Ogle et al., 1998). This official number has been declining in recent years due to the reduction in harvesting in natural forests following the 1997 export ban.

In forecasts of timber and log demand in Vietnam to the year 2010, the estimated total volume that will be required yearly is 11.3 million m^3. After deducting contributions from existing plantations, there would be a need to supply wood from new plantations to the tune of 8.6 million m^3 a year by the year 2010 (MARD, 2001). In addition to natural increases in demand for wood products, the government also has somewhat far reaching and ambitious plans for Vietnam to not only be self-sufficient in wood production, to but develop extensive wood processing industries. MARD has called for 18,000 billion Vietnamese Dong (VND) (at 15,000 VND to the US dollar, this is about US $1.3 billion) in aid from overseas development assistance sources and foreign direct investment to develop the wood processing industry to the year 2010 (VNS, 1999b).

LAWS TOWARD FORESTRY AND FOREST LAND IN VIETNAM

As the preceding discussion makes clear, the limited amount of forest cover and the high demands for timber and wood products play a large role in encouraging illegal logging in Vietnam. However, land tenure and ownership of forests play an equally significant role in contributing to the underlying causes for illegal logging in any country. What may be 'illegal' to the government may in fact be seen as not only 'legal' at the local level, but as an *de facto* right or privilege (see Thompson, 1975; Peluso, 1992).

One of the most difficult problems for forest management in Vietnam is the government definition of 'forestry land.' According to the Forest Resources Protection and Development Act of 1991, "forest land is defined as 'forested land' and 'non-forested lands for which plans have been made for forest plantation,' below referred to as forest plantation land" (p. 7). The law goes on to note:

> The overall management of forests and forest plantations lies with the State. The State will allocate forests and forest plantation land to organizations and individuals-hereinafter referred to as forest users-for protection, development and utilization of forests on a long-term basis in accordance with State planning documents. All organizations and individuals now legally using forests and forest plantation land shall continue to do so following the provisions of this Act. (p.7)

It further emphasizes that "[n]atural forests and state funded forest plantations are the property of the state" (SRV, 1993a, p. 7).

The law established a tripartite classification system for Vietnam's forests: protection forests, special-use forests, and production forests. Special-use forests and protection forests are to be allocated to various departments of the Ministry of Forestry (now subsumed under MARD) and other authorities of the central state. Special-use forests in particular are to have no exploitative activities within them, and into this category fall National Parks and Nature Reserves. Protection forests are less environmentally valuable than the special-use areas, and limited exploitation activities can take place in these (primarily watershed) forest areas. The chairman of each provincial People's Committee is supposed to confer with the central government to decide upon the classification and allocation of protection forests within each province. And finally, production forests are to be allocated to SFEs when necessary. The chairman of the district People's Committee (a sub-provincial level) is given the power to decide the allocation of production forests to the state, to cooperatives, or to individuals. Each individual forest, whether protection, special-use or production, is to have a Management Board, which is to prepare a plan for the management and utilization of the forest to be submitted to national authorities for approval (SRV, 1999b).

In other words, the state has control over most forest areas in Vietnam, despite the move in recent years away from state socialist planning in other sectors of the economy (see Fforde, 1993). This has led to a number of constraints on community forestry, including most prominently the stipulation that natural forests are the property of the state and state-approved entities only. During fieldwork in Ha Tinh, the author heard several shouting matches between managers of the state forests and villages and community organizations (such as the local veteran's organization) who wanted to be allocated a plot of forest to be managed locally and collectively. The local state representatives were absolutely opposed to the idea that forest land with stocked forest should be managed by anyone other than the state. Another obstacle to local forest management is the fact that these simplified classification systems assigned to forests do not acknowledge the fluidity in local areas between agricultural zones and forest areas, particularly in the highlands where swidden agriculture is practiced (for examples, see McElwee, 1999).

Shifting Land Ownership

The role of forests and ownership over forests is intimately tied up with the changing roles of land ownership in Vietnam. After 1954 in the North and after 1975 in the South, the government declared that all productive land was under the ownership of the socialist state, and peasants were organized into cooperative farms (Vickerman, 1986). Slowly, in the early 1980s, cooperatives began contracting out individual production tasks to households (Pillot, 1995). On the basis of these experiments, in 1988 a Land Law was passed that 'entrusted' (in Vietnamese: *giao*) land to organizations for long-term use. At this time, however, no rights of transfer or sale were given to households for land, particularly forest land (Gayfer and Shanks, 1991).

Once these tentative land rights had been established, Vietnam began experimenting with a program of allocating not only agricultural lands, but also forest lands to the management of individual households. These lands were allocated to households with the stipulation that they agreed to properly reforest and manage the lands and to eventually share the proceeds from any timber sales from the land with the government. Households wanting to receive forest lands for protection were to make applications to local authorities, and upon approval the household would receive a 50-year lease for the forest land. The share of timber revenues that was supposed go back to the government after allocation was 20 percent where trees were newly planted by the household after the land allocation, and 50 percent where trees were already planted before by others (Hayami, 1993). Most households allocated land in this way accepted some government support in the form of credit to buy seedlings, mainly of fast growing exotic trees such as *Acacia* spp. and *Eucalyptus* spp. Given the constraints on management of these lands and the continuing government role in collecting profits from them, these land allocations are more properly classified as 'joint forest management' activities, rather than the 'privatization' of property as they have been called elsewhere (MARD/FAO, 1997).

True forest land allocation began to speed up when the Land Law was revised in 1993, when a package of 'five rights' of land use were given to individual households: the right to use, inherit, mortgage, rent, and transfer lands. These leasehold rights were given for 20 years for agricultural land and 50 years for forestry land. Land holdings were still limited, however, to 2 ha of annual plants and 10 ha of perennials per household (SRV, 1993b).

The major goal of forest land allocation was to create incentives for better protection of forest resources by villagers and to promote investment in forestry at the household level. So far, however, mostly bare lands have been distributed to households under this decree, despite the fact that these lands are called 'forestry land' (MARD, 1998). These are lands the government would like to see as forest, or were formerly forest, but almost no lands containing well-stocked trees have been allocated to households. As a result of the newness of this program and the poor quality of land that has been allocated thus far, success has been limited. Another major question remains about what institutional arrangements should be used to manage forest lands—continued allocation to individual households, or to groups of households and communities. Given the economies of scale required for many types of successful forest management, allocating forest land to groups and communities makes more sense than allocating to individual households. However, community management of forests remains a politically sensitive topic in Vietnam and has only been officially implemented in a few experimental pilot projects (Anon., 2000).

By 1998, the Government had given out management responsibilities for a total of 7.2 million ha of forest land (38% of the total forest land area) to the following stakeholders:

- 5.4 million ha of forest lands (primarily production and protection forests) managed by 356 units of State Forest Enterprises (SFEs) and Forest Management Boards (FMBs),
- 0.9 million ha of forest lands managed by 143,723 households ,
- 0.9 million ha of forest lands managed by 326 non-state units (localities and cooperatives),
- The remaining 11.8 million ha of forest lands (both non-forested and forested) are yet to be allocated (National Assembly, 1998).

In comparison with previous years, the transfer of ownership of forests is dramatic (see Figure 3). What is remarkable, however, is that the government is primarily shifting responsibility for forests not from the state to households, but from the state to the state. For example, vast stretches of State Forest Enterprise lands have been officially 'allocated' back to the SFE. (Some SFEs, however, do in turn allocate these lands to their workers or other individuals to manage.) Other SFEs are being turned into Forest Management Boards and the area of forest formerly slated for production turned into 'protection forest.' In this case, the government has merely transferred land from oversight in one domain of the state (SFEs) to another (MARD's Forest Protection Department).

FIGURE 3. Shifting Land Management and Ownership in Vietnam, 1990-2000

Management of Forests in 1990

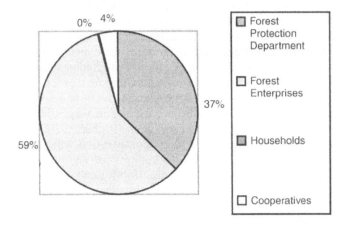

Management of Forests in 2000

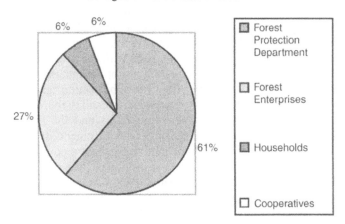

Adapted from MARD, 2001

The overall effects of land allocation remain to be seen. The massive increase in reforestation that was the expressed goal of forest land allocation has not yet happened in most areas. And in those areas that have seen increases in forest cover, this has mostly been linked to increased productivity of wet-rice agriculture and the shift away from a need for forest land for swiddening

(Sikor, 2001). Given that other studies of land tenure in forested areas have not definitively concluded that security of tenure for households leads to forest promotion and conservation, it is not surprising that Vietnam's program has had ambiguous results. For example, a recent study in Ghana concluded that "in the Ghanaian setting there is no correlation between land registration and sustainable forestry practices . . . These findings caution generalizations about security of tenure and suggest clarification of the contexts within which the term is used" (Owubah et al., 2001, p. 261). These findings have been echoed by others who emphasize that it is by no means certain that secure tenure rights will automatically result in sustainable forestry. Nothing prevents land users who have secure property rights from behaving in ways that impose social costs on others, such as through deforestation (Richards, 2000). Secure land tenure may be less significant for forest conservation when pressures from outside intervene, such as increased market pressure and population pressure (Henrich, 1997; Angelsen and Kaimowitz, 1999). Land titling programs have also in some cases led to more deforestation by creating consumer demand for land that was previously ignored by the state and other powerful actors (Christensen and Rabibhadana, 1994; Angelsen, 1999).

As a result of these fundamental uncertainties about the link between land tenure and sustainable forestry, and the general ineffectiveness of land titling programs in Vietnam thus far, there remain serious questions regarding to what degree the state ought to control forest management through its Forest Protection Department, and to what degree other individuals and communities ought to have a full 'bundle' of use rights to forests. One possible solution to the dilemma is to institute land tenure regimes with a strong 'social incentive' for forest management, through a combination of community use rights and inter-household land tenure regimes (i.e., forest protection 'groups' of multiple households). Other nations have been more attuned to the value of combinations of use rights between local communities and the state, variously known as 'joint forest management' or 'co-management' (see Appasamy, 1993; Sundar, 2000) or devolution of land use rights to communities in the form of recognition of common property regimes (see Richards, 1997; Arnold, 1998). However, in contrast to these situations, Vietnam remains a country mostly divided between the extremes of total state management of forests and individual household management of other lands.

THE ROLE OF STATE LOGGING COMPANIES

No exploration of the role of forests in Vietnam can be complete without an examination of the role of State Forest Enterprises (SFEs, known in Vietnam as *Lam Truong*). According to MARD, SFEs are the "semi-autonomous state-owned entities which were initially created and responsible for management of all production forest land" (ANDZEC, 1997, p. 5). Some SFEs were directly operated by the central government through MARD, while others reported to provincial People's Committees. All operated as parastatal companies, which did not pay the state for logging concessions, but rather returned income to the state through taxes on the produce harvested. At the height of logging in Vietnam in the 1970s and 1980s, there were over 420 SFEs, most all charged with extensive exploitation of the lands allocated to them. According to MARD, SFEs were not only logging operations, but they were supposed to see to it that forests were replanted and nurtured. In reality, though management plans looked good on paper, repeated cycles of exploitation gradually reduced the quality of the remaining stands, ultimately leaving very little of value and huge swaths of Vietnam denuded of significant forest cover (MARD, 1996).

A clear example of the great changes wrought in highland forested areas by SFEs is seen in a 10-year review of one of the SFEs set up by the state after the reunification of Vietnam. In 1979, the Easup Forestry-Agriculture-Industry Union was created in Dak Lak province of the central highlands. Prior to the government interest in the area, there had been approximately 10,000 local people living there, primarily ethnic minorities practicing shifting rotational agriculture. After the establishment of the Easup SFE, more than 20,000 people were moved from lowland areas to take jobs at the logging company. The SFE controlled more than 313,000 ha of forest, mostly deciduous forest dominated by teak (*Tectona grandis* L.f.) and *Hopea* spp. Despite several major river systems flowing through the area, only 15% of this land area was under protection. The rest was under production for large diameter timber, pulpwood and sawn veneer wood. The forest was logged under 20-year cutting rotations, and reforestation was carried out in 10-year cycles. It was imagined that this enterprise would eventually support 200,000 people organized into communities around the production lands in towns with names such as "Green Forest Town" (Pham Huu Van, 1989). Another explicit goal of the enterprise was to stop itinerant shifting cultivators in the area from burning the economically valuable forest and instead turn to settled wet-rice agriculture.

Over the first 10 years of the enterprise, over 863,000 m^3 of large diameter timber was logged and 200,000 m^3 of roundwood was processed into veneer. This was approximately half the standing volume of wood originally surveyed in 1979 that ended up being cut in the first 10 years. The export value of the wood was estimated to bring to the union about \$2-3 million US per year. About a third of this went to the state as duty, and the rest went to run the enterprise and pay the workers (Pham Huu Van, 1989).

The SFE has resulted in not only a huge decline in standing volume of timber in just 20 years, but has caused massive changes throughout the province in which it was located–high rates of lowland-to-upland migration, loss of indigenous minorities' land rights, and expansion of wet-rice agriculture in areas more suited for shifting cultivation given low irrigation levels and highly variable soils and inclines. This was not an uncommon scenario. SFEs have had a variety of social, economic and environmental impacts in the communities in which they were located (ANZDEC, 1997). On a purely economic basis alone, most were actually complete failures. As a recent review of the sector noted:

> In many SFEs, forest management prescriptions were altered in order to achieve the high production targets imposed on them. Selection felling cycles, originally set at 20-30 years, were reduced to 5-10 years; minimum residual standing volume targets of 200-300 m³/ha was reduced to 100 m³/ha; and 50 cm minimum felling diameter was reduced to 30 cm . . . Consequently, although sound in theory, in practice the selection forest management system upon which most of Vietnam's forest management is based has not led to sustainable forest management. (Ogle et al., 1998, p. 5)

More often than not, whole areas were completely logged-over, and the land converted to agriculture or cash crops to feed the large numbers of SFE employees who moved to the hills and who could not live on the low salaries of the state (Vu Huu Tuynh, 2000).

Changing Times for SFEs

The government ban on large-scale logging of natural forests since 1997 has meant that SFEs' *raison d'etre* was substantially changed. As a result, SFEs have had to try to diversify into other income generation activities or expand their provision of community services in order to survive. Many SFEs have been dissolved, and those that remain have been forced to compete in a market system under the new open-economy policies in Vietnam. Under a decision of the prime minister on "Renovation of State Forest Enterprises' Organization and Management Mechanism" issued in 1999, the SFE system needed to be revamped to encourage "economic and commercial effectiveness" of the SFEs (SRV, 1999b, p. 1). Those SFEs that have been retained in the market-oriented system are primarily those that are: (1) profitable; (2) on a large scale or have other reason to require the State's direct management and investment; or (3) in very remote or isolated regions where SFEs can play a role in general rural development (SRV, 1999b). Only 120 SFEs will still continue to operate under these new rules, and another 130 will continue to run, but only as land managers of watershed and protection forests, not as forest exploiters (see below). The rest of the SFEs will have been dissolved (Vu Huu Tuynh, 2000).

SFEs covering more than 5,000 ha and with more than 70% of their forest land belonging to the critical/very critical protection forest category will be transformed into Forest Protection Management Boards (FPMBs) (Ogle et al., 1998). However, asking SFEs to take on the tasks of protecting forests with limited budgets and staff is difficult. For example, the Minh Hoa SFE in Quang Binh Province has only 5 staff members for over 83,000 ha of forest (Vu Huu Tuynh, 2000). The other problem with this system of simply shifting SFEs into protected forest managers with FPMBs is that it creates an incentive *not* to protect forests at all. As one observer has noted, "if the forests are still allowed for wood exploitation, the province proposes to keep the SFE as a business unit. When the wood is over-exploited, the SFE will be proposed to reform to a FPMB in order to get a subsidy from the Government. It is absolutely not a sustainable way to manage forests" (Vu Huu Tuynh, 2000, p. 3).

The main tasks of the remaining SFEs with licenses to log will be engaging in "forest plantation, protection, enrichment, forest products exploitation and processing in addition with the provision of inputs to industrial processing units to respond to consumption demands of the national economy" (SRV, 1999b, p. 1). Fortunately for the SFEs, there are a number of government subsidy programs for which they are still eligible, such as the reforestation programs mentioned earlier like the 5 Million Hectare Program. SFEs can apply for government funds for seedlings and so forth from 5MHRP to "mitigate their descent into insolvency," as one report put it (ADB, 2000, p. 28). Others have noted that SFEs, though inefficiently operating and completely unprepared for funding from 5MHRP or other sources, still want to continue to exist for the sole reason that staff can have enough time working for the Government to earn a retirement pension (Vu Huu Tuynh, 2000).

Unspoken in the prime minister's recent review of the sector (see SRV, 1999b) is the role SFEs have historically played, and continue to play, in illegal logging in Vietnam. SFEs have to pay taxes on sales of their forest products, estimated at 37%-50% of roadside market value. As a result, there is an embedded incentive in the system for SFEs to log outside their approved plans for tax evasion purposes (ANZDEC, 1997). Furthermore, the structure of wood processing industries—generally separate from SFEs and owned by local governments—means the incentives to skirt the system at all levels are high.

THE FOREST 'PROTECTION' DEPARTMENT

Running parallel to the SFE system has been the government's Forest Protection Department (or *Kiem Lam*) under the Ministry of Forestry (now part of MARD). The forest administration system under the *Kiem Lam* has been set up essentially to prevent abuse by the SFEs, and to investigate cases of illegal logging by SFEs. What has happened in reality is the *Kiem Lam* is often in collusion with SFEs to illegally log, or in the cases where the *Kiem Lam* does try to catch illegal loggers, usually finds itself outmanned and outgunned.

Ho Chi Minh established the first forest protection unit in 1946, with the objective of "achieving justice regarding any cases of forest crimes" (Kiem Lam, 1998, p. 1). In 1972, the objectives were broadened by a new law to include the general work of "protecting the forest" and the modern *Kiem Lam* was born. In 1993, Resolution 39/CP codified the organization, duties, and rights of *Kiem Lam* (SRV, 1993a). *Kiem Lam* was then established as a national branch of the central government with a unified system of authority over localities.

Currently, *Kiem Lam* is located under the rubric of MARD and promotes public awareness of forest protection, engages in management audits of SFEs and their approved cutting plans, and manages the oversight of most protected forests through a series of field offices located in every province in Vietnam (Ogle et al., 1998). The local *Kiem Lam* branches report to the provincial and district People's Committees and are responsible for forest protection activities on all forest lands. The management boards of special-use forests–Nature Reserves and National Parks–report to *Kiem Lam* for all major management decisions. On lands allocated to SFEs, *Kiem Lam* officers are responsible for taking enforcement actions on specific problems identified by the SFE's own forest guards. In other words, *Kiem Lam* retains a significant 'bundle' of rights usually associated with land tenure rights: the right to monitor and inspect all forests; to tax and regulate trade in forest products; and to determine violations of any national forest laws, even on non-state lands.

Recently, the deputy minister of MARD proudly proclaimed that *Kiem Lam* has in the last five years been responsible for catching "179,280 cases of forest destruction, illegal exploitation, trade, transport and processing of forest products. Of the figure, 1,110 cases have been brought to trial. More than 190,600 m³ of rare timber, 190,854 kg of wild animals, 38,380 animals such as tiger, leopard and cobra, 18 guns, and 106 automobiles have been confiscated" (VNS, 2000d, p. 6). In a recent self-review, the head of *Kiem Lam* declared that "[t]ens of thousands of people who used to live by slash-and-burn farming and illegal exploitation of the forest have been brought out of the forest. Forest destruction has fallen by 40 percent compared with 1995" (Kiem Lam, 1998, p. 10).

Problems in Forest Protection

Although the *Kiem Lam* appears on paper to have broad authority to combat illegal logging, the department faces a number of obstacles. First is the lack of personnel. In 1979, there were 9,700 *Kiem Lam* officers. In 1994, there were only 8,400 *Kiem Lam* and the total population of Vietnam had risen to over 70 million. The head of *Kiem Lam*, Nguyen Ba Thu, has written, "Our force is very thin, missing people, and weak (8,000 *Kiem Lam* have to protect 11 million ha of forest) and our equipment is poor" (Kiem Lam, 1998, p. 3).

The other major problem is the administrative structure of *Kiem Lam*. Currently, *Kiem Lam* is part of MARD. But local *Kiem Lam* stations remain under the control of the local provincial People's Committee, leaving unclear who has ultimate authority to arrest loggers. As a local *Kiem Lam* officer said to a newspaper recently:

> In the decision of the law, *Kiem Lam* has the function to carry out the laws of protection and developing the forests. But the rights that are delegated to *Kiem Lam* are rather limited. For example, the scope of the forest law is that we can settle administratively any fines up to 100 million VND, but the lawyers only gave to us at *Kiem Lam* the right to fine up to 5 million VND. The rights of *Kiem Lam* are the 'rights of straw'. (Xuan Quang, 1999b, p. 6)

In order to combat these weak 'rights of straw,' several policy orders have been issued in the last few years, including Decision 286/TTg on 'strengthening urgent measures to protect and develop forests' and 287/TTg on 'checking and arresting individuals and organizations caught destroying forest.' There have also been requests from MARD for an official police force under the security ministry, which would be charged with enforcing forest laws more effectively and prosecuting them under the legal system.

The final problem is the sheer number of people violating the forest laws since the extensive logging bans were introduced in 1993 and codified in 1997. As the national Vietnam News Service admits,

> Serious illegal exploitation stems from the ban on forest exploitation. The ban has led to a reduction of timber output in many provinces. Therefore, demand far outweighs supply and the price for timber increases, particularly for rare species. In many provinces, illegal traders provide themselves with weapons and remote controlled machines and build camps deep in the forests for people they hire to cut down trees. They organize illegal exploitation in bordering areas between districts and provinces so the perpetrators can easily flee . . . Illegal traders often use forged documents, false forest rangers, and set up trading spots along rivers and roads. (VNS, 2000c, p. 4)

In addition to the rising numbers in reports of illegal logging, there also appear to be increased incidences of violence directed again *Kiem Lam* officers.

As the official news agency states, "The central forest department is alarmed over the risks rangers now face from poachers who are increasingly turning to violence when caught plundering the country's richest reserves of wood and wildlife . . . A shortage of forest rangers and equipment made the task of protecting the nation's natural resources almost impossible and those poachers who were caught often got off lightly, allowing them to continue in their illegal trade" (VNS, 2000b, p. 21). It is not uncommon to hear of stories about groups of 50 or more people with knives, crowbars and hand saws that were uncovered by one or two *Kiem Lam* officers, who were unable to act against such large crowds (Tuan Anh, 2000b). According to a study by *Kiem Lam* in 2000, from 1996-2000, 12 *Kiem Lam* officers were killed and 490 injured in clashes with illegal loggers (VNS, 2000b).

As a result, *Kiem Lam* now finds itself more often than not the object of sympathy in press reports, which portray the noble Forest Protection Officer defending his green turf from the onslaught of hordes of loggers. This patriotic image is commonplace among *Kiem Lam* officers themselves. In an interview with a *Kiem Lam* officer, the officer noted that the recent spates of violence directed against them by illegal loggers "despairs them—but also inspires them, in their memory, to pick up the fight. And it's a fight for all, he stresses. For with the country despoiled, then there's no country for tomorrow. For anybody" (Thanh Long, 2000, p. 5). The officer went on to explain the motivation for many of his fellow officers: "If you do not possess a great love for the green of the forest, it would be very hard for you to accept our job for a living." There is even a "Love Song for the Forest Protection Officer" printed up in the monthly *Kiem Lam* newsletter to inspire the rangers: "Crossing over how many streams, through how many deep jungles, his tired feet toiling over the mountains and blind from fog, I hope you remember him . . . Planting trees on the mountains, stopping the swidden farmers, so the forest is always green, so the streams run clearly . . . Our soldierly Forest Protection officer!" (Kiem Lam, 1998, p. 27).

Corruption and Collusion in Illegal Logging

What is less commonly heard from the leaders of *Kiem Lam* is the rampant problem of corruption of local officers. Most rangers earn salaries somewhere between US$30-50 per month, and one boatload of illegal timber may be worth the entire yearly salary of *Kiem Lam* officers. *Kiem Lam* rangers can easily be bribed to turn an eye to small-scale local extraction. A bribe of 300,000 VND (approximately US$15) is the usual amount to extract timber worth about 1 million VND on the market (about US$66), as noted by the author in many interviews with illegal loggers in central Vietnam in 2000-2001. Note should be taken, however, that these bribes are often the only contact between local villagers and *Kiem Lam* officers, as the rangers often do not live in local villages or originate from the local areas; hence, maintaining a relationship with the *Kiem Lam* is not necessary for most villagers, who may only encounter officers once or twice every few years. Thus this situation differs from those outlined in other parts of Southeast Asia in which forest rangers or state managers act as 'patrons' to local 'clients' and provide a range of services and access to resources–at a price–to villagers (for example, see Kahn, 1999 and McCarthy, 2002).

In addition to bribes, *Kiem Lam* officers can confiscate illegal lumber cut by others, and then turn around and sell it, hereby saving themselves the trouble of cutting it themselves. If the ranger does report confiscated lumber to superiors, it is usually used by the station or else sold, and the money goes straight into the local station budget or other state budget. The 'donations' the state receives from confiscating illegal timber–rather than stopping it before it is cut–is equal to about US$450,000 per month (VNS, 2000a). *Kiem Lam* yearly collects about 100,000 m^3 of wood, which is estimated to be between 1/5 and 1/10 of the total amount of illegally logged timber (Hoang Hoe, n.d.).

The official figures on enforcement of forest laws clearly indicate that confiscation of timber after the fact is the major job of *Kiem Lam* (see Figure 4). Overall in the country, from 1992-1997, *Kiem Lam* uncovered 404,648 cases of violations of the forest laws, or about 80,929 cases per year. The majority of these are cases where someone was stopped trying to transport or sell timber illegally. The actual number of people caught in the act of logging is relatively low (9% of the violations). This seems to point to an incentive for *Kiem Lam* to discover violations only after they have happened (i.e., when the logs have al-

FIGURE 4. Number of Violations of Forestry Law 1992-1997 Discovered by the Forest Protection Department, by Violation

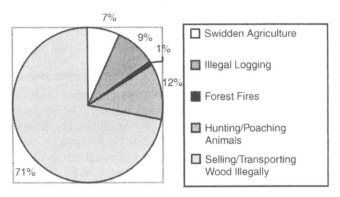

Adapted from Kiem Lam, 1998.

ready been cut, rather than before) so these products can be confiscated to add to the budget of the department. This was the conclusion that many local villagers who had had their timber confiscated would express to the author on various occasions. The confiscation raises serious questions in local community members' minds about the slippery definitions of 'illegal' and 'legal' timber extraction and sales.

The evidence does seem to support the claim of local loggers that *Kiem Lam* stops them only after they have already cut the logs, and rarely imposes more significant fines than simply confiscation of the wood. Of the over 400,000 violations of the forest code discovered by the *Kiem Lam* from 1992-1997, only 1,300 cases were referred to the justice system (i.e., solved outside of the *Kiem Lam* system). The rest were 'administratively solved' with confiscation of 678,795 m^3 of wood–equaling a donation to the national budget of 640 billion VND, or almost US$425 million (Lam Nghiep, 1998).

Perhaps the most troubling dilemma, however, is the collusion between some *Kiem Lam* offices and other government officials and provincial departments to profit individually by forging paperwork and allowing massive cases of deforestation to take place virtually undetected. The author was able to witness some evening spot-checks at a remote *Kiem Lam* station on the main road between Cambodia and the Vietnamese coastal port of Nha Trang. Logging trucks with local government license plates could barrel through the checkpoint carrying one or two enormous logs (clearly harvested illegally in Cambodia), but they always had enough forged paperwork to satisfy the local *Kiem Lam* officers. In the meantime, several local farmers with three-wheeled tractors coming down the same road with firewood and small amounts of lumber for housing were repeatedly stopped and admonished and told to pay fines, as they had no paperwork.

In recent years a more active and independent press in Hanoi has begun to uncover some of the government collusion in large-scale deforestation cases. A review of several recent cases published in national newspapers provides a snapshot of some of the difficulties in catching illegal loggers and poachers, especially when they are government officials. [The news reports below are taken verbatim from local newspapers]:

- *Duc Linh District, Binh Thuan Province:* The police discovered a deforestation case when a SFE made a road illegally. The road not only cost 3 billion VND to build, but the company building it colluded with the SFE to cut down 12,663 m^3 of timber while 'building the road' (which was actually unneeded). The police started criminal proceedings against the director of the SFE, who asked them to delay, as he needed to be admitted to a hospital in Ho Chi Minh City. He then fled the country and is now believed to be living in Singapore (Huynh Phan, 2000).

- *Khanh Hoa Province:* 2 *Kiem Lam* officers got news that a car illegally carrying wood was coming through their road. They erected a barrier, but the car drove through. One *Kiem Lam* officer fired 6 shots into the car's tires to get it to stop. One bullet hit the driver and killed him. The car was later found to belong to the Ministry of Defense. The court of Khanh Hoa Province forced the *Kiem Lam* officer to serve 30 months in jail and pay almost US$1000 to the family of the deceased, as well as pay US$4/month as long as the deceased's children were young, and US$2/month to his elderly father until death. The provincial head of Khanh Hoa said afterwards, disgusted at the verdict, "When we solve cases this way, how on earth can we stop illegal logging?" (Xuan Quang, 1999a).
- *Gia Lai Province:* The head of MARD in Gia Lai province held a meeting in January 2001 to assess work for the year 2000, and said he was angered by district leadership cadres who decided on their own to grant individuals the right to go into the forests to log in order to make "official houses" for the government. In reality these logs were sold in the market for individual gain. The estimated amount cut was more than 20,000 m^3, almost half the amount of wood allowed to be cut legally in logging plans by the province [the MARD director later took his statement back under pressure and said it was only around 300 m^3 they could document for sure that was stolen this way] (Nguyen Thinh, 2001).
- *Tuyen Hoa District, Quang Binh Province:* Tran Dinh Long, head of the *Kiem Lam* office in Quang Binh from 1993-1998 stood trial with several other defendants accused of deforestation. They, in collusion with the SFE in Tuyen Hoa, removed several thousand cubic meters of ironwood [*Erythrophloeum fordii* Oliver] from the forest. The head of *Kiem Lam* was dismissed, and his underlings received suspended sentences (Tung Lam, 2000).
- *Dong Xuan District, Phu Yen Province:* Twenty-eight cadres of the new economic zones program [of MARD] were accused of deforesting 73 hectares of watershed forest to clear agricultural land. Included in the number was the head of *Kiem Lam* for the district of Dong Xuan. All were given suspended sentences (Yhan Eban, 2000).
- *Ia Grai District, Gia Lai Province:* After several years of unsolved arson cases in a SFE, the SFE requested to cut down 111 hectares of pine forest because of these 'difficulties in management' and because the SFE needed 'agricultural land' for its workers. While they were waiting on a decision from the province to cut, the land was cleared anyway and planted with coffee by a private company that had contracted with the SFE (Nguyen Thinh, 1999).
- *Ba Ren District, Quang Binh:* The SFE Ba Ren in Quang Binh province was discovered to have cut 54 m^3 of rare hardwoods illegally. An inves-

tigation by the provincial People's Committee investigated these charges, found them to be true, and asked for reparations only equal to 0.96 m^3. The *Lao Dong* newspaper printed an expose of this and as a result the province called a meeting with all interested parties to reassess the issue. The SFE was fined 20 million VND and the province confiscated 13.2 m^3 of wood (Lao Dong, 2000). [However, if we assume these not-named valuable hardwoods were worth approximately US$500/m^3 (an average figure used by the state forestry corporation) then the total value of the logs was US$27,000. The fine the SFE incurred was US$1,300 plus the value of the confiscated wood (US$6,600), resulting in a net profit for the SFE of more than US$19,000.]

- *Duc Linh District, Binh Thuan Province:* The head of the provincial people's committee and the vice head of the provincial MARD signed contracts allowing the People's Committee of the district of Duc Linh permission to cut trees that were located on paddy bunds as part of an effort "to clean up rice fields." The total amount of timber cut was 3,381 m^3 and 1,027 stere of firewood (Trung Phuong, 1999). [Note that most paddy fields in Vietnam rarely have standing trees on the bunds, and finding over 3,000 m^3 of timber in a paddy field would be virtually impossible. The wood clearly came from other nearby forests.]

Two provinces in particular, Quang Binh and Binh Thuan, are notorious in Vietnam for having been run by corrupt officials for years and having significantly reduced forest cover as a result. As the Vietnam News Agency reported in the year 2000: "A number of forest stations and management boards of protection forests have made false plans for timber exploitation. Some are allowed to use timber from one area but exploit two. The largest tracts of forest destruction over the past years are at Tuyen Hoa forest ranging station in the central province of Quang Binh and Tanh Linh forest ranger station in the southern central province of Binh Thuan" (VNS, 2000c, p. 4).

The Tanh Linh case in particular captured the nation's attention in 1999 for the sheer enormity of corruption and the mendacity of officials involved. It is estimated that these corrupt government officials logged 53,429 m^3 of trees in various wildlife sanctuaries and protected forests between 1993 and 1995 with the cooperation of local forestry officials, for a total value of more than US$1.5 million. The case only reached the court system after 74 written reports were submitted to Hanoi on the problem by a local retired official-turned-whistleblower. The ensuing investigation resulted in the trial of 36 defendants, who were accused of "violating forest protection rules, irresponsibility, corruption and illegally stockpiling military weapons" (VNS, 1999a, p. 5). Of the 36 defendants, 29 were district and provincial officials. The highest forestry official sentenced was Le Thi Phuong, former deputy director of the Binh

Thuan Department of Agriculture and Forestry, who was charged with issuing illegal logging licenses. In addition, the provincial People's Committee chairman was removed by his underlings due to "gross neglect" of his charge to both manage and protect the province's forests during his tenure as chairman between 1992 and 1998 (AFP, 2000). In the end, more than 50 people associated with the case received jail terms, including all the officials on trial. However, these jail sentences for officials are certainly an exceptional case in Vietnam overall.

THE LITTLE GUY VERSUS THE BIG GUY: WHO IS TO BLAME?

Cases like the Tanh Linh case bring up an important question: how much of Vietnam's illegal logging is perpetrated by big operations covered up by the government, and how much is attributable to the small actions of hundreds of thousands of local loggers desperate for timber? In other words, how do you catch the big guy (in Vietnamese: '*ong trum*') who is out to get rich from illegal exploitation, while still maintaining access to forests for the little guy who just wants to build a sturdy house for his family?

The major problem today is that the private market for logs has created incentives for both large-scale and small-scale illegal logging operations throughout the country. The market for processing of large logs remains mainly in the hands of the private sector and supply and demand creates the incentive for illegal logging at all levels. "If we buy the timber from the state, how can we make a profit?" asked one sawmill owner of a reporter covering illegal log trade (Lam Giang, 2000, p. 5). In fact, the enormous profits to be gained in wood trade have led to vulnerability of the system at all levels to corruption, tax evasion, and systematic cover-ups of deforestation.

The fact that actual local log prices remain relatively low should not cause observers to conclude that logging in Vietnam is not profitable; rather, the relative amounts that can be made through large-scale illegal logging compared with other economic activities provide the incentive to log. The fact that timber prices remain relatively low here seems to support the theory advanced by Kummer (1991) that the actual price of timber can often play a smaller role in illegal logging than other political factors. As he writes, "it seems clear that government control of logging concessions and tolerance, if not encouragement, of illegal logging and smuggling have been more influential in determining the rate of logging than the price of logs" (Kummer, 1991, p. 139). This is certainly the case in Vietnam, where the lack of a comprehensive private concession system thwarts transparency in large-scale logging and leads to immense profits, as there is no stumpage cost for illegally harvested trees–only the cost of transporting them.

Who Is the Big Guy?

The scope of illegal logging operations by '*ong trum*' organizations is sometimes unbelievable. In one case in Yen Bai in northern Vietnam, illegal loggers were so sure they would not get caught by *Kiem Lam* that they had invested significantly in developing logging operations, such as building permanent roads with dynamite and going through and marking clearly the trees they wanted. Over 1,000 ha of Yen Bai have been deforested by that particular gang already (Nguyen Thi Thanh Hai, 1999).

Illegal loggers with connections and power have various strategies for getting through the weak forest law enforcement systems. Through interviews at various administrative levels, the author found that illegally cut wood is almost always cut into smaller lengths and transported at night. Many small sawmills operate at night deep in forested areas, mixing legal and illegal timber. Oftentimes, relatives of *Kiem Lam* officers are involved in smuggling and will report on the work habits and patrol areas of *Kiem Lam* rangers. Other loggers will call up and make false reports to *Kiem Lam* stations to spread resources thin and move *Kiem Lam* into other territories away from their own logging sites. Some logging groups even try to report on each other, calling *Kiem Lam* on other people they encounter in the forest in the hopes that *Kiem Lam* will reward the callers by allowing them to take their own illegally cut timber.

In order to transport illegally-cut wood long distances, some sort of vehicle is needed. It has been estimated that in 1997 about 60% of all the vehicles traveling south out of Quang Binh Province were likely carrying some smuggled forest products (VNS, 1997b). Generally speaking, these traffickers are not poor peasants. Most residents of Vietnam cannot afford a motorcycle, let alone a four-wheeled vehicle. Therefore, illegal wood transport is usually not carried out by the poorest people, but by those with connections. Trucks are the most common forms of transport. Wood is often carried under the space in floorboards, hollows in the body of the car, or under the carriage. Other items such as pesticide are transported at the same time to cover up the smells of freshly cut wood. There are also many recorded cases of the national railroad system being used to transport illegal timber, and even of valuable timber and rare poached animals being transported regularly on Vietnam Airlines (VNS, 1996). These types of connections indicate that the people involved are not small-scale operators, but rather powerful figures with wide webs of smuggling networks.

Who Is the Little Guy?

In addition to large scale logging gangs, much of the so-called 'illegal logging' in Vietnam is carried out by millions of small-scale loggers. But who are these local people who also cut wood illegally? During a year's fieldwork in one province in north central Vietnam (Ha Tinh) with high rates of illegal logging, the author conducted interviews and surveys in 5 main study villages that lay on the outskirts of the Ke Go Nature Reserve (a special-use forest) and the Cam Xuyen SFE (a production forest). During the course of fieldwork, a household survey to measure dependence on forest resources and household income derived from forests and other productive activities was administered to 104 households total in the 5 villages (a 20% random sample of each village).

Of the 104 households surveyed, 13 households were engaged in illegal commercial logging activities (12% of the villagers). Including those engaged in logging, exactly half the households (52) derived some commercial income from forest-based goods (selling fuelwood, charcoal, rattan, medicinal plants, etc.). Timber on average contributed just 2% of the total income sources in the villages; however, for those households that logged, they received 15% of their income from timber. Comparing the logging households (13) versus the non-logging households (91), the average contribution of all forest-based goods to the household income was 28% and 10% for each group, respectively. This is likely due to the fact that logging households knew the forest better and could harvest other forest products more easily, and that loggers often collected other forest items (rattans, fruit, honey, etc.) at the same time they cut timber. Clearly, although logging was not particularly widespread in this area, it did contribute fairly significantly to the incomes of those who participated.

What was most interesting was the fact that there did not seem to be overall income differences between logging families and non-logging families. While 38% of the logging families were classified as poor by the government versus 31% poor in the non-logging families, this difference was not borne out by looking at the details of income levels in each household. In terms of overall household income, logging families had an average household income of 5,904,154 VND (US$393) and the non-logging families has an average household income of 5,544,418 VND (US$369), not a statistically significant difference. The evidence from this case in Vietnam supports the idea that logging from state or common lands is a forest activity pursued by a range of income classes, including richer ones, while non-wood forest product collection is more commonly dominated by the poor (Beck and Nesmith, 2000; Southworth and Tucker, 2001; Wunder, 2001).

The consistent use of timber and wood products across income classes is not surprising, given the common needs for wood for house construction and other basic household needs such as furniture. The average household reported needing about 1 m^3 of timber per year for various construction projects. Approximately every 10 years, households would need to rebuild their homes, in which case timber needs would rise to about 4-5 m^3 per year. While some households could meet their needs through home gardens and private forests (25% of households had informal or formal tenure on private forest lands), most households had no such legal access to forest lands (75% of households had no forest lands), and were forced to rely on wood from either the nearby SFE or the Ke Go Nature Reserve (a former SFE-turned-nature-reserve).

The increasing enforcement of laws preventing illegal logging by locals did seem to be having some effect, as several families who used to log now no longer did so. The typical punishment in this area for illegal logging was confiscation of timber and any equipment used to cut it, and a monetary fine up to 500,000 VND. During a field trip to Ke Go Nature Reserve the author herself discovered a group of 8 illegal loggers who were coming out of the Nature Reserve with 4 bicycles full of sawn timber and handsaws (see Figure 5). They had about 20 planks, which they were planning to sell for 50,000 VND (approximately US$3.30) each. By that estimate, each person would have made around US75 cents per day for this logging trip, a very small amount. However, during the slack agricultural season when food supplies are running low and other off-farm employment opportunities are practically non-existent, even the meager 75 cents a day contributed by logging could buy rice rations for the families involved.

Food security remains one pressing problem of households in this field study area. Gradually, as food needs are met and surpluses generated by some of the wealthier families, these households have been turning to planting trees

FIGURE 5. Small-Scale Illegal Loggers in Ke Go Nature Reserve, Ha Tinh, Vietnam (February 2001)

on private lands to supply their timber needs. However, even with this move away from the use of state forest lands for timber, problems with definitions of illegal logging remain. One of the major problems with forestry law for local villagers is the bureaucratic paperwork necessary to declare timber legally logged. The wood needs to have with it a bill of sale; a history of the timber's origin attached to it; and must be marked with a ranger stamp on the wood. Certain 'precious' hardwoods require a 'special transportation permit' issued by the local *Kiem Lam* station. This onerous law also applies to people who would like to harvest wood on their own lands, in which case "the forest owner needs only to notify the nearest rangership agency of the People's Committee of the local commune or township so that within 10 days he can be issued with a certificate that the products are legal and can be freely circulated in the market" (SRV, 1999a, pp. 25-26). In other words, even private landowners need official permission to use timber on their own lands. Laws like this make it too difficult for the average person who might want to engage in legal logging to bother to do so. Therefore, most small-scale logging by villagers in village forests and unclaimed forest lands would be classified by the state as illegal, even though it takes place on non-state lands.

Social Justice Claims at the Local Level

Besides the seeming necessity of skirting the law due to lengthy and inconvenient paperwork, there are also notions of social justice that underlie local claims to trees in many areas. In Ha Tinh, there is a sense of entitlement and social equity underlying some people's responses to why they use lands that they do not have official permission to use. This is not uncommon; as Peluso notes for the case of teak theft in Java, Indonesia, "Whether or not they are intentionally breaking the state's law (by stealing wood or appropriating game), peasants have their own notions of morality, rights, criminality, and subversion. Frequently, these differ from the assumptions embedded in state ideology" (Peluso, 1992, p. 12).

For example, it is well known that large boats are used to transfer wood from the far side of Ke Go Nature Reserve by wealthy illegal loggers; therefore, why shouldn't families be able to cut a few trees now and then, especially if it is just to serve their own household's needs? If illegal loggers can get away with it, why shouldn't small farmers? Again and again, interviews revealed people saying, "It's just for my family, and I didn't deliberately mean to take from the Nature Reserve land–it was just there!" Villagers see their small collection of wood products as having an almost insignificant impact on forest lands in comparison with the large-scale loggers they see flouting the same regulations, but on a much larger scale. As several people said to the author,

"What's wrong with taking a tree now and then? It's just to build a house. Everyone needs a house."

Furthermore, local people's perceptions toward *Kiem Lam* guards were almost uniformly negative. Most *Kiem Lam* officers are sent to areas far from their home provinces, so they will not be tempted to help out relatives and friends in procuring illegal wood. However, this policy usually only ends up making local people in their new area distrust the *Kiem Lam* as 'outsiders.' The *Kiem Lam* in Ha Tinh did not help the situation by almost entirely focusing their enforcement efforts at the market where people went to sell fuelwood they had collected and charcoal they had made. One woman at the market asked, "Do you agree with the policy of *Kiem Lam* to take away things I've harvested? Of course they do it at the market–they're too darn lazy to go into the forest and stop me then."

The situation in this local area was further compounded by the inability of *Kiem Lam* to stop army and border police from harvesting large logs in the surrounding forests and hauling them off in army trucks in full view of everyone (Figure 6). When villagers saw these large trucks roaring off from the forest, they felt they had just as much of a right to the wood as anyone else, and perhaps more so. People would often say how poor they were–much poorer than

FIGURE 6. Army Truck Carrying Logs from the Ke Go Nature Reserve, Ha Tinh, Vietnam (July 2001)

the army and border guards who at least had guaranteed government salaries and pensions–and that as a result of their poverty, any illegal actions they took were entirely justified. This is not a feeling reserved to Vietnamese; it is often a common feeling among those who are removed from resources in other parts of the world that 'if I can't have it, neither can you' (c.f. Thompson, 1975). As Peluso notes, "Sometimes the rage felt by people deprived of resource access derives not from the denial itself but from the reassignment of access to others whose claims are considered invalid" (Peluso, 1992, p. 14).

To make matters worse, the government had started a radio campaign to tell people about the purpose of the Nature Reserve, and to explain that it was government property on which people were not to trespass. These campaigns were having the opposite effect of inducing people to stay out of the forests; rather, they created complete apathy among the local villages who now had no incentive to protect forests for the village's own benefit. Outsiders from coastal villages were now coming through the forest villages to use the Nature Reserve's wood resources, and the local villagers, who might have been given incentives to stop the outsiders, were completely uninterested in protecting the government's lands. As one village headman said, "Why don't we stop people from going into the nature reserve? We could–they pass through our village every day. But the government tells us again and again, *that is the government's property*. Then the government ought to stop people from going in, not us." Another elderly man noted, "'No one cries for the father of everyone'–do you know that phrase? It means these forests–no one takes care of them. How can we be expected to when the government does not care themselves?"

CONCLUSIONS

This article has attempted to enlarge the boundaries of debate over illegal logging in Vietnam by expanding the definitions of illegal and legal to include not only official perceptions of law but also claims of social justice and use rights at various spatial levels. Additionally, this article points out the need to study illegal logging not just in terms of statistics of forest cover decline, but through a combination of quantitative and qualitative approaches to the economic, political and social factors which result in forest cover loss (see McCarthy, 2002). As another author has noted for other countries in Southeast Asia, "Detailed case studies of tropical deforestation are needed to capture the uniqueness of each nation's deforestation process . . . Generalizations based on cross-national studies may be of little relevance for formulating specific programs to control deforestation" (Kummer, 1991, p. 138). This paper has outlined the variety of problems–land tenure insecurity, poverty, ineffective state management, poorly enforced logging bans, and endemic corruption–which have all contributed to the specific cases of illegal logging in Vietnam.

Given the enormity of the problems with illegal logging, are there any steps that can be taken to solve this dilemma and reduce the scale and scope of illegal logging? Other observers have suggested cracking down on middlemen, particularly wood processors and sawmills that buy their wood from illegal loggers, rather than trying to target the loggers themselves who are both powerful and diffuse (Ogle et al., 1998). Other steps to solve the problem of the middlemen would include increasing monitoring and enforcement of major transportation routes, as there are a limited number of highways and train routes on which the majority of stolen timber is transported by the large mafia-type logging networks.

Continuing plans for allocating forest lands to households and individuals, and for group management in communities, will also likely create personal incentives for forest conservation and management, although it is not clear to what degree this will be successful given the long-term breakdown of community tenure rules. The current efforts to allocate only poorly stocked or bare forest lands are also not going to be as effective as giving out lands of high quality. This type of individual and community ownership of forest land seems the only solution to the wide-scale illegal logging and the difficulties of state patrols over more than 10 million ha of forest land. Extending the protected areas network in Vietnam and trying to patrol areas with forest guards, as has been advocated by some international conservation NGOs (see Wege et al., 1999) is simply not feasible given low salaries for rangers, limited personnel to patrol large areas, and the historical animosity between *Kiem Lam* rangers and local populations in many areas.

The most fatal problem for forest lands in Vietnam appears to be what one *Kiem Lam* officer explained as "the forest of the 3 no's": forest areas with no people living close enough to protect it; areas where no land had been given out to individuals or communities to protect as part of land allocation policies; and areas with no *Kiem Lam* forest guards to check forest quality regularly. In these situations, forest destruction through illegal logging of all types is almost guaranteed to continue to occur. Only by changing the 3 no's land to 3 yes's land–local people with incentives to conserve, local land use rights codified in law, and the support of local *Kiem Lam* officers with incentives to truly protect forest lands–can this situation of illegal logging be reversed.

REFERENCES

ADB [Asian Development Bank]. 2000. Interim report: Key issues affecting the forestry resource sector. Project TA 3255-VIE: Study on the policy and institutional framework for forest resource management. Asian Development Bank, Hanoi, Vietnam.

AFP [Agence France Press]. 2000. Top Vietnam official sacked over logging scandal. *Wire story*, August 24.

Angelsen, A. 1999. Agricultural expansion and deforestation: Modeling the impact of population, market forces and property rights. *Journal of Development Economics* 58:185-218.

Angelsen, A. and D. Kaimowitz. 1999. Rethinking the causes of deforestation: Lessons from economic models. *World Bank Research Observer* 14: 73-98.

Anon. 2000. Workshop proceedings: Experiences and potential for community forest management in Vietnam. National Working Group on Community Forest Management, Hanoi, Vietnam.

ANZDEC. 1997. Restructuring of state forest enterprises. ANZDEC Ltd. Consultants, Hanoi, Vietnam.

Appasamy, P. P. 1993. Role of non-timber forest products in a subsistence economy: The case of a joint forestry project in India. *Economic Botany* 47(3): 258-267.

Arnold, M. 1998. Managing forests as common property. UN Food and Agriculture Organization, Rome.

Bangkok Post. 1993. Forest reserves being depleted despite logging and export bans. *Bangkok Post*, Bangkok, May 18: p. 3.

Beck, T. and C. Nesmith. 2000. Building on poor people's capacities: The case of common property resources in India and West Africa. *World Development* 29(1): 119-133.

Beresford, M. and L. Fraser. 1992. Political economy of the environment in Vietnam. *Journal of Contemporary Asia* 22(1): 3-18.

Brown, C., P. Durst and T. Enters. 2001. Forests out of bounds: Impacts and effectiveness of logging bans in natural forests in the Asia-Pacific. UN Food and Agriculture Organization Regional Office for Asia and the Pacific, Bangkok.

Brown, K. and F. Ekoko. 2001. Forest encounters: Synergy among agents of forest change in southern Cameroon. *Society and Natural Resources* 14 (4): 269-290.

Christensen, S. R. and A. Rabibhadana. 1994. Exit, voice, and the depletion of open access resources: The political bases of property rights in Thailand. *Law and Society Review* 28 (3): 639-655.

DeKoninck, R. 1999. Deforestation in Vietnam. International Development Research Centre, Ottawa.

Department of Forest Development. 1998. Proceedings of the national seminar on sustainable forest management and forest certification. Department of Forest Development, Hanoi, Vietnam.

Fairhead, J. and M. Leach. 1996. Misreading the African landscape: Society and ecology in the forest-savanna landscape. University of Cambridge Press, Cambridge.

Fforde, A. 1993. The political economy of reform in Vietnam-some reflections. In: pp. 293-325. B. Ljunggren (ed.). The challenge of reform in Indochina. Harvard University Press, Cambridge.

FIPI [Forest Inventory and Planning Institute]. 1996. Final report on forest resource changes (1991-1995). Forest Inventory and Planning Institute, Hanoi, Vietnam.

Gayfer, J. and E. Shanks. 1991. Northern Vietnam: Farmers, collectives and the rehabilitation of recently reallocated hill land. *Social Forestry Network Paper 12a* Summer/Winter 1991.

Global Witness. 1998. Vietnamese government attempts to cover up collusion with Cambodia over illegal timber trade–Global Witness releases the evidence. Global Witness, London.

_____. 2000. Chainsaws speak louder than words. Global Witness, London.

Hayami, Y. 1993. Strategies for the reform of land property relations in Vietnam. UN Food and Agriculture Organization, Hanoi, Vietnam.

Henrich, J. 1997. Market incorporation, agricultural change, and sustainability among the Machiguenga Indians of the Peruvian Amazon. *Human Ecology* 25: 319-351.

Hoang Hoe. n.d. Deforestation and reforestation in Vietnam. Vietnam Forestry Science and Technical Association, Hanoi, Vietnam.

Huynh Phan. 2000. Mot giam doc-can pham pha rung o Binh Thuan tron ra nuoc ngoai [A director, accused of deforestation in Binh Thuan, flees the country]. *Lao Dong [Labor Newspaper]*, Hanoi, May 29: p. 1.

Kahn, J. S. 1999. Culturalising the Indonesian uplands. In: 79-103. T. M. Li (ed.). Transforming the Indonesian uplands: Marginality, power and production. Harwood Academic Publishers, Amsterdam.

Kemf, E. 1988. The re-greening of Vietnam. *New Scientist* June 23: p 53-57.

Kiem Lam [Forest Protection Department]. 1998. Ban tin kiem lam [News from the forest protection department]. Kiem Lam, Hanoi, Vietnam.

Kummer, D. 1991. Deforestation in the postwar Philippines. University of Chicago Press, Chicago.

Lam Giang. 2000. Crafty timber poachers foil Quang Binh forest protection officers. *Vietnam News*, Hanoi, June 9: p. 5.

Lam Nghiep [Forestry Journal]. 1998. Qua 5 nam thuc hien 'luat bao ve va phat trien rung' (1992-1997) [Through 5 years of implementing the 'law on protecting and developing forests' (1992-1997)]. *Lam Nghiep [Forestry Journal]* 37: 2-10.

Lao Dong [Labor Newspaper]. 2000. Mot vu pha rung tra hinh [A case of disguised deforestation]. *Lao Dong [Labor Newspaper]*, Hanoi, Nov. 23: p. 3.

MARD [Ministry of Agriculture and Rural Development]. 1996. Vietnam case study: Development of the forests in a brief historical perspective. Ministry of Agriculture and Rural Development, Hanoi, Vietnam.

_____. 1998. Plan for implementation of the 5 million hectare reforestation national programme, 1998-2010. Ministry of Agriculture and Rural Development, Hanoi, Vietnam.

_____. 2001. Five million hectare reforestation program partnership synthesis report. Ministry of Agriculture and Rural Development, Hanoi, Vietnam.

MARD/FAO [Ministry of Agriculture and Rural Development/Food and Agriculture Organization]. 1997. Proceedings of the national workshop on participatory land use planning and forest land allocation. Agricultural Publishing House, Hanoi, Vietnam.

Maurand, P. 1943. L' Indochine forestiere [The forests of Indochina]. Imprimerie d'Extreme Orient, Hanoi, Vietnam.

McCarthy, J. F. 2002. Power and interest on Sumatra's rainforest frontier: Clientelist coalitions, illegal logging and conservation in the Alas valley. *Journal of Southeast Asian Studies* 33(1): 77-106.

McElwee, P. 1999. Policies of prejudice: Ethnicity and shifting cultivation in Vietnam. *Watershed* 5(2):30-38.

National Assembly, SRV. 1998. Report of the National Council of the General Assembly on "Review of the implementation of the forest law." Quoc Hoi [National Assembly], Hanoi, Vietnam.

Nguyen Thi Thanh Hai. 1999. Lam tac lam chu rung Yen Tu [Loggers own the forests of Yen Tu]. *Lao Dong [Labor Newspaper]*, Hanoi, July 24: p. 1.

Nguyen Thinh. 1999. Pha rung theo hop dong [Deforesting according to contract]. *Lao Dong [Labor Newspaper]*, Hanoi, May 7: pp. 1, 7.

_____. 2001. Cua rung, thich thi mo? [The forest gate, if you like you can just open it?]. *Lao Dong [Labor Newspaper]*, Hanoi, Feb. 7: p. 3.

Ogle, A., K. J. Blakeney and Hoang Hoe. 1998. Natural forest management practices. Asian Development Bank, Hanoi, Vietnam.

Owubah, C. E., D. C. Le Master, J. M. Bowker and J. C. Lee. 2001. Forest tenure systems and sustainable forest management: The case of Ghana. *Forest Ecology and Management* 149: 253-264.

Peet, R. and M. Watts, eds. 1996. Liberation ecologies: Environment, development, social movements. Routledge, London.

Peluso, N. 1992. Rich forests, poor people: Resource control and resistance in Java. University of California Press, Berkeley.

Pham Huu Van. 1989. Easup Forestry-Agricultural-Industry Union: Forest management plan and results of 10 years of the implementation. Easup Forestry-Agricultural-Industry Union, Ea Sup, Vietnam.

Pillot, D. 1995. La fin des cooperatives: La decollectivisation agricole au Nord Vietnam [The end of cooperatives: Agricultural decollectivization in north Vietnam.]. *Cahiers d'Outre Mer* 48(190): 107-130.

Poffenberger, M., ed. 1998. Stewards of Vietnam's upland forests. Asia Forest Network, Berkeley.

Reuters. 1996. Vietnam says Indochina's forests need protection. *Wire story*, Sept. 25.

_____. 1998. Hanoi denies collusion in illegal Cambodia logging. *Wire story*, Feb. 27.

Richards, M. 1997. Common property resource institutions and forest management in Latin America. *Development and Change* 28(1): 95-117.

_____. 2000. Can sustainable tropical forestry be made profitable? The potential and limitations of innovative incentive mechanisms. *World Development* 28: 1001-1016.

Sikor, T. 2001. The allocation of forestry land in Vietnam: Did it cause the expansion of forests in the northwest? *Forest Policy and Economics* 2(1): 1-11.

Sikor, T. and U. Apel. 1998. The possibilities for community forestry in Vietnam. Asia Forest Network Working Paper Series, Berkeley.

Southworth, J. and C. Tucker (2001). The influence of accessibility, local institutions, and socioeconomic factors on forest cover change in the mountains of western Honduras. *Mountain Research and Development* 21(3): 276-283.

SRV [Socialist Republic of Vietnam]. 1993a. Luat bao ve va phat trien rung va nghi dinh huong dan thi hanh [Law on protection and development of forests and declarations for guiding implementation]. SRV, Hanoi, Vietnam.

_____. 1993b. Vietnam land law (12th Revision). SRV, Hanoi, Vietnam.

_____. 1999a. Decision No. 47/1999/QD-BNN-KL issuing the regulation on inspection of the transportation, production and business of timber and forest products. *Cong Bao [Official Gazette]* 26 (July 15): 23-29.

_____. 1999b. Decision of the government No. 1187 QD-TTg on the renovation of organization and mechanisms for management of State Forest Enterprises. Ministry of Agriculture and Rural Development, Hanoi, Vietnam.

Sundar, N. 2000. Unpacking the 'joint' in joint forest management. *Development and Change* 31: 255-279.

Thai Vu. 2000. Hang ngay doi mat voi lam tac [Every day meeting the faces of the illegal loggers]. *Lao Dong [Labor Newspaper]*, Hanoi, Oct. 3: p. 3.

Thanh Long. 2000. Forest protectors dodging bullets on the frontline. *Vietnam News*, Hanoi, Sept. 11: p. 5.

Thompson, E. P. 1975. Whigs and hunters: The origins of the Black Act. Pantheon, New York.

Trung Phuong. 1999. Nhung 'sang kien' doc chieu de lay go rung cam: Bai II: Con duong lam tac [Malicious 'initiatives' to get wood from prohibited forests: Part II: The path of the loggers]. *Lao Dong [Labor Newspaper]*, Hanoi, May 15: pp. 1, 7.

Tuan Anh. 2000a. Ba can bo kiem lam Thanh Hoa bi lam tac hanh hung [3 cadres of Thanh Hoa's Forest Protection Department are attacked by illegal loggers]. *Lao Dong [Labor Newspaper]*, Hanoi, Aug. 29: p. 1.

_____. 2000b. Mot lam tac con do hung han bi ban chet o Lam Dong [A cruel hooligan caught illegal logging is shot dead in Lam Dong]. *Lao Dong [Labor Newspaper]*, Hanoi, Sept. 22: p. 1.

Tung Lam. 2000. Xu nhu the, rung con bi pha [Solving cases this way, it's no wonder the forests are still being destroyed]. *Lao Dong [Labor Newspaper]*, Hanoi, Sept. 11: p. 1.

Vickerman, A. 1986. The fate of the peasantry: Premature transition to socialism in the Democratic Republic of Vietnam. Yale University Southeast Asian Studies Monograph, New Haven.

VNS [Vietnam News Service]. 1996. Centre needed for smuggled wildlife. *Viet Nam News*, Hanoi, Aug. 14: p. 4.

_____. 1997a. Deforestation is still 'alarming' says FAO. *Vietnam News*, Hanoi, March 27: p. 1.

_____. 1997b. Timber theft proves difficult to stop. *Vietnam News*, Hanoi, Apr. 1: p. 2.

_____. 1998. Viet Nam's forest plan in figures. *Vietnam News*, Hanoi, Dec. 25: p. 6.

_____. 1999a. Lone crusader gets his day in court. *Vietnam News*, Hanoi, Apr. 10: p. 5.

_____. 1999b. Wood processing industry needs VND 18,000 billion. *Vietnam News*, Hanoi, Jan. 28: p. 5.

_____. 2000a. Nearly 3,740 cases of forest destruction uncovered in January. *Vietnam News*, Hanoi, Feb. 18: p. 6.

_____. 2000b. Poachers ready to kill for nature's bounty. *Vietnam News*, Hanoi, Aug. 19: p. 21.

_____. 2000c. Rangers tackle forest threat with crackdown on loggers. *Vietnam News*, Hanoi, Aug. 14: p. 4.

_____. 2000d. Safeguarding Viet Nam's forests is a job as dangerous as it is crucial. *Vietnam News*, Hanoi, Sept. 9: p. 6.

Vu Huu Tuynh. 2000. Reform of organization and management mechanism of state forest enterprises according to the Decision 187-1999/QD-Ttg. Paper presented at Workshop on the 5 Million Hectare Program, Hanoi, Vietnam.

Watkin, H. 1999. Farming and logging cut forests by third in 15 years. *South China Morning Post*, Hong Kong, Aug. 6: p. 21.

Wege, D. C., A. J. Long and Mai Ky Vinh. 1999. Expanding the protected areas network in Vietnam for the 21st century: An analysis of the current system with recommendations for equitable expansion. Birdlife International, Hanoi, Vietnam.

Westing, A. H. 1975. Environmental consequences of the 2nd Indochina war: A case study. *Ambio* 4(5-6): 216-222.

Wunder, S. (2001). Poverty alleviation and tropical forests–What scope for synergies? *World Development* 29(11): 1817-1833.

Xuan Quang. 1999a. Dieu tra vu kiem lam ban chet lam tac o Khanh Hoa: Mot khang nghi hop tinh hop ly [Checking the case of a Forest Protection Officer who shot an illegal logger in Khanh Hoa: An appropriate and sentimental protest.] *Lao Dong [Labor Newspaper]*, Hanoi, Oct. 10: p. 1.

_____. 1999b. Nguoi dung tung, ke quyen rom [People are allowing [deforestation], people have limited power]. *Lao Dong [Labor Newspaper]*, Hanoi, Aug. 9: p. 6.

Yhan Eban. 2000. Phu Yen: Pha hon 73 ha rung, chi bi an . . . treo?! [In Phu Yen: Deforestation of over 73 hectares, and only given . . . a suspended sentence?!]. *Tuoi Tre [Youth]*, Hanoi, Sept. 7: p. 5.

Can 'Legalization' of Illegal Forest Activities Reduce Illegal Logging? Lessons from East Kalimantan

Luca Tacconi
Krystof Obidzinski
Joyotee Smith
Subarudi
Iman Suramenggala

Luca Tacconi is Senior Economist at the Center for International Forestry Research, Indonesia (E-mail: l.tacconi@cgiar.org).

Krystof Obidzinski is Research Fellow at the Center for International Forestry Research, Indonesia.

Joyotee Smith was Senior Economist at the Center for International Forestry Research, Indonesia, at the time of writing.

Subarudi is Scientist at the Forestry Research and Development Agency of Ministry of Forestry, Indonesia.

Iman Suramenggala was Executive Director of Yayasan Pioner Bulungan, Indonesia, at the time of writing.

Paper prepared for 'Illegal Logging in the Tropics: Ecology, Economics, and Politics of Resource Misuse,' March 29-30, 2002, Yale University. Organized by the International Society of Tropical Foresters, Yale Student Chapter, and the Yale School of Forestry and Environmental Studies.

The work reported in this paper was partly funded by a grant from the U.K. Department for International Development (DfID). The views expressed in this publication are those of the authors and not necessarily those of the organizations associated with this research.

[Haworth co-indexing entry note]: "Can 'Legalization' of Illegal Forest Activities Reduce Illegal Logging? Lessons from East Kalimantan." Tacconi, Luca et al. Co-published simultaneously in *Journal of Sustainable Forestry* (The Haworth Press, Inc.) Vol. 19, No. 1/2/3, 2004, pp. 137-151; and: *Illegal Logging in the Tropics: Strategies for Cutting Crime* (ed: Ramsay M. Ravenel, Ilmi M. E. Granoff, and Carrie A. Magee) The Haworth Press, Inc., 2004, pp. 137-151. Single or multiple copies of this article are available for a fee from The Haworth Document Delivery Service [1-800-HAWORTH, 9:00 a.m. - 5:00 p.m. (EST). E-mail address: docdelivery@haworthpress.com].

SUMMARY. Illegal activities are one of the most pressing problems facing the Indonesian forest sector today. The debate on illegal forest activities has focused primarily on legal and governance issues. Economic forces, however, are increasingly recognized as fundamental drivers of illegal forest activities. We ask the question whether the legalization of small logging concessions and their development can teach us anything about how to address the illegal logging problem. We find that legalization alone–when a legal timber concession is granted to a previously illegal operator–does not necessarily result in a significant reduction in illegal activities. When illegal activities are profitable, they can be expected to continue. Changing the regulatory framework to increase monitoring and enforcement can affect the profitability of these illegal activities. By changing the underlying economic incentives for logging, such interventions hold greater promise of success. In the medium to long term, however, legalization may help reduce illegal logging when it entrusts local people with ownership and control of forest resources and maintains a monitoring role for government agencies. *[Article copies available for a fee from The Haworth Document Delivery Service: 1-800-HAWORTH. E-mail address: <docdelivery@haworthpress.com> Website: <http://www. HaworthPress.com> © 2004 by The Haworth Press, Inc. All rights reserved.]*

KEYWORDS. Corruption, Indonesia, Malaysia, economics, timber, law enforcement, monitoring

INTRODUCTION

Illegal activities are one of the most pressing problems facing the Indonesian forest sector. Illegal logging is one of five declared priorities for the Ministry of Forestry of Indonesia, and it has been one of the main forest sector topics discussed with at recent Consultative Group on Indonesia (CGI) meetings. Illegal logging–defined as the illegal harvesting of timber–is one of many illegal forest activities (FAO, 2001). In the Indonesian context, the term illegal logging typically refers to illegal forest activities.

This paper follows this definition of illegal logging, focusing on a range of extractive activities that include those entirely outside of the legal framework, e.g., without any license, to operations that have some degree of legal recognition, e.g., small-scale concessions. As noted later, some logging activities may be considered to be illegal on the basis of central government regulations, but legal according to district government regulations.

Annual estimates of the amount of logs harvested illegally illustrate the magnitude of illegal activities in the Indonesian forest sector. These estimates are derived by comparing supply data–i.e., log production from State forests-with data on industrial production of value-added products and log exports less imports. Estimates of annual log harvest range from 32.6 million cubic meters for 1997/98 to 49 million cubic meters in 1997 and 64.6 million cubic meters in 1998, and between 55% and 75% of timber production *could* be illegal (Palmer, 2001; Scotland et al., 1999). At the current average aggregate tax rate of US$30/m^3 (ITTO, 2001), this figure suggests that the Indonesian Government loses between US$1 billion and US$1.9 billion annually. It is much more difficult to assess the environmental impacts but it is likely that millions of hectares of primary (e.g., protected areas) and secondary forest are being affected (FWI/GFW, 2002).

In Indonesia, the government has stated that law enforcement is the most important approach to the problem. Also, a log export ban was introduced in October 2001 to stem the flow of illegal logs to other countries. Despite these and other public commitments to addressing the problem, such as the 2001 Ministerial Declaration at the East Asia Forest Law Enforcement and Governance conference in Bali, the evidence available indicates that no significant reduction in illegal logging activities has taken place.

The debate on illegal forest activities has focused particularly on the legal and governance aspects (Contreras-Hermosilla, 2001). Economic forces, however, are increasingly recognized as fundamental drivers of illegal forest activities. The incentives and disincentives, including the cost of 'being caught,' faced both by illegal and legal operators must be understood in order to identify potential institutional and economic policies that might curtail illegal forest activities.

The objective of this paper is to consider these institutional and economic issues by deriving policy-relevant lessons from recent regulatory changes regarding small-scale logging activities in Indonesia. We ask the question whether the legalization of small-scale logging through community-based forest concessions could result in a reduction of illegal logging activities.

It is important to stress at the outset that the focus on small-scale activities does not imply that they are most responsible for illegal log extraction in Indonesia. Rather, we focus on small-activities because they have been the most greatly affected by recent regulatory changes. As such, they help us understand the effectiveness and distributional implications of regulatory strategies that legalize illegal activities in order to bring them under control.

To be clear, we do not advocate that all illegal logging activities should be legalized. We analyze legalization because of its potential merits in (a) regulating otherwise uncontrollable illegal activities, and/or (b) harmonizing legislation with current social norms vis-à-vis "problematic" activities. Legalization of small-scale logging activities could also legitimize the income of rural people involved in those activities as well as their control over forest resources.

This kind of legalization could have the effect of redistributing rent capture from forests from companies to local small-scale operators. According to the literature on narcotics legislation, legalization reduces illegal activities by decreasing their profitability. As the risk involved in supplying the products in question decreases upon legalization, production would be expected to increase. The price of the products would then decrease and so too would the profitability of their illegal trade. Increasing legal access to forests, then, could reduce log prices and returns to illegal logging. Finally, the legalization of illegal activities would allow the government to recoup lost tax revenues.

The price effect described above may be substantial when demand for the product is inelastic–i.e., if demand does not increase substantially in response to price declines. Also, it should be noted that the analogy between illegal drug and timber markets is complicated by the fact that narcotics are widely recognized as illegal products. Timber, however, is generally a legal product that may or may not be produced and/or traded illegally.

We recognize that legalization may not be always an appropriate strategy, especially in those situations where the total harvest exceeds the sustainable one. A mix of approaches and instruments would be required, including reduction of the harvest level through law enforcement as well as legalization of appropriate activities.

The paper proceeds by presenting the methods and considering recent policy changes and small-scale logging activities. It then looks at the management, distribution of revenues, and illegal activities in small-scale logging operations. Finally, the recent changes in timber prices, exports, and bribes, and how these factors affect illegal activities are reviewed.

We would like to stress that given the limited and inaccessible nature of information on these subjects, this paper represents a preliminary attempt to consider some of the institutional-economic relationships that drive illegal logging activities. Understanding these relationships will be key to designing successful interventions to reduce illegal logging activities.

METHODS

This paper draws on literature reviews and a series of interviews conducted in East Kalimantan (Indonesia) and Sabah (Malaysia) by a team that included Joyotee Smith, Subarudi, Krystof Obidzinski, and Iman Suramenggala during a period of three weeks in August 2001. Field research focused on interviews with key stakeholders in government, timber extraction and trade activities in the Districts of Bulungan, Malinau, and Nunukan, as well as in the municipality of Tarakan. Government officials interviewed in each of the three districts and Tarakan included staff from District Forestry Office, Trade and Industry,

Tax and Finance, and Customs Office. Following discussions with government officials, village meetings (that included village leadership and open to other interested villagers) and field visits to IPPK logging activities were conducted in three villages with IPPK licenses in Bulungan. In Malinau, IPPK logging activities were investigated in one village. One IPPK operator (individual managing IPPK logging activities on behalf of Malaysian investors) was interviewed in each of the Districts of Bulungan, Malinau and Nunukan. Finally, in Sabah interviews were conducted with Officials of the Department of Forestry, timber sector associations, and two traders importing timber from East Kalimantan. This field research was carried out as a part of a grant from the Department for International Development (United Kingdom).

RECENT FOREST POLICY AND SMALL-SCALE LOGGING

In the late 1960s, control over forest resources in Indonesia had a certain degree of decentralization. Provincial Governors could issue forest concessions of up to 10,000 ha in size and District Heads (*Bupati*) could grant concessions of up to 5,000 ha. Timber was abundant and accessible at the time and few concessionaires used heavy machinery. Instead, many small firms and families used manual harvesting methods and a traditional extraction method known locally as *banjir kap*. This logging method relies on monsoon floods to float timber to market down-river. Large-scale concessionaires, however, perceived *banjir kap* logging as a threat to their access to cheap labor and control of timber markets. These large concessionaires found sympathetic ears among central government officers, including the armed forces, who saw *banjir kap* logging as a threat to their control of forest resources and revenues. In 1971 the central government's Ministry of Forestry took control of forest concessions, increased the minimum concession size to 50,000 ha, and banned *banjir kap* logging. Small and medium sized concessionaires merged and soon abandoned labor-intensive cutting methods in favor of mechanized logging (Manning, 1971; Dixon, 1974; Daroesman, 1979; Ruzicka, 1979; Peluso, 1983; Magenda, 1991). *Banjir kap* style logging continued illegally after the 1971 ban and eventually adopted chainsaws for felling. These small independent operations often worked inside timber concessions and sold their wood back to the concessionaires or to outside brokers.

In 1999, the Indonesian Parliament passed new legislation on regional governance and on fiscal balance between the central government and the regional governments (Law No. 22/1999 and Law No. 25/1999). This was partly due to an attempt to contain the risk of national disintegration resulting from the unhappiness of resource rich regions that sought increased authority to manage their resources after the fall of the Suharto regime. These laws gave greater financial and decision-making powers to local government, particularly at the district level. The fall of Suharto also ushered in a new era of reduced capacity and willingness to enforce a forestry regulatory system that was widely perceived to have benefited only the central government, the presidential family, and a small cadre of timber tycoons (Barr, 1999; *Bisnis Indonesia*, 1999; Brown, 1999, 2001; *Jakarta Post*, 1999). It is not surprising, therefore, that small-scale logging activities–illegal according to the regulatory system at the time–quickly resurfaced.

As a result of the political pressures generated by the fall of Suharto, the new government gave the authority to governors (head of province) and regents (head of district) to issue small forest concessions permits, respectively up to 10,000 hectares and up to 100 hectares (Government Regulation No. 6 1999, Decision of the Minister of Forestry No. 310/Kpts-II/1999 and Decision of the Minister of Forestry No. 317/KPTS II/1999). The concessions granted by the regents came to be known as IPPK (*Izin Pemungutan dan Pemanfaatan Kayu*, license to extract and use timber). IPPKs were to be issued to communities residing in or near forest areas to conduct extractive forest activities (including timber) through village cooperatives and associations (PP No. 62/1998 and SK Menhutbun No. 677/1998). IPPKs, however, have not brought about the return of *banjir kap* logging. IPPK permits became highly sought after as they were easy to obtain, formally required comparatively little investment, generated windfall profits and provided a venue towards legality for those who were operating without a license. In addition, IPPK logging now features heavy machinery such as tractors, skidders, and logging trucks. This expensive equipment as well as the skilled labor, managers and investors required to operate it typically come from the large Suharto-era concessions. IPPKs, which ostensibly make community-based forest management possible, have effectively been co-opted by those with access to capital and heavy machinery (Obidzinski, 2002).

Patlis (2002) notes that the introduction of Government Regulation 34 in 2002 (which provides the details for the implementation of Law 41/1999) has in effect brought about a re-centralization of decision making over forests by clearly stating that the central government has sole authority over them, and the permits issued at the regional level, such as IPPKs, are no longer valid. The time between the introduction of Laws 22/1999, 25/1999, and 41/1999, and the introduction of Law 34/2002 was a period of great uncertainty about the distribution of authority and responsibilities over forest management among national, provincial, and district governments. In the midst of this uncertainty, district governments attempted to assume *de facto* control over forest resources (Rhee, 2001), as they were anxious to capitalize on them both economically and politically. Presently, however, the legal framework clearly states decision-making authority over forests. Obviously, this does not imply that regional governments will not attempt to challenge the central government in its attempt to recentralize decision-making authority.

The introduction of IPPK licenses may have resulted in some illegal operations becoming legal as a result of the acquisition of a license. However, the introduction of IPPKs was not an intentional act aiming to legalize existing illegal logging activities. Rather, as already noted, the introduction of IPPKs was in response to political pressures to decentralize the forestry management systems and to increase the participation and benefits derived by local populations.

This helps to put into context the key question addressed in this paper–i.e., whether the legalization of illegal logging operations by awarding licenses such as IPPKs would result in a reduction of illegal logging. To do this we learn from the recent experience with IPPK concessions.

MANAGEMENT OF IPPK CONCESSIONS, DISTRIBUTION OF REVENUES, AND ILLEGAL ACTIVITIES

The process of obtaining an IPPK concession license began with local timber contractors searching for village forest areas with potential for logging. This was the most common way of initiating a search for IPPK concessions, although village communities became increasingly pro-active in seeking out venture partners on their own.

The timber contractors were usually local entrepreneurs who previously had been involved in small-scale logging. Since the fall of Suharto, they vigorously sought to expand their timber business operations, particularly through IPPK concessions. They did so by working with local 'strong men' (key timber barons in the area) who help facilitate financial, technical and logistical support from neighboring Malaysia.

If the village forest areas had potential for commercial log production, contractors drew up IPPK joint venture agreements offering communities timber fees, infrastructure projects and/or village plantations as compensation from logging. Once the community agreed to a contract, the next step was to establish a corporate body in the village that could serve as the basis for the IPPK application. Most often, this entailed the formation of a village cooperative. As the community moved to establish the required cooperative, the contractor finalized preparation of the concession map and timber production plans, which were then submitted to the District Forestry Office for technical evaluation and, if approved, presented to the regent for signature.

Interviews with IPPK operators in East Kalimantan revealed that Malaysian investors were funding many of the operations through Indonesian contractors. Both locals and migrants were employed as laborers and the contractor sold the timber to the Malaysian investor. The logs were exported and processed in Malaysia or they were re-exported for processing in China or elsewhere. Despite the legalization of this logging network through the granting of IPPKs, exports to Malaysia were illegal because of Indonesia's log export ban.

Our interview data also suggest that IPPK logging operations returned a lower share of direct financial benefits to the local people compared to *banjir kap* style activities (see Table 1). Incomes earned by local community members were lower in the IPPK system than in the *banjir kap* system. This is not surprising given that the timber was harvested mechanically and the sale controlled by the contractors and investors. That local people earned less from IPPKs than from *banjir kap* activities does not necessarily imply that they were worse off. They spent less time working manually to extract the timber and could use their labor to earn alternative income–when alternatives were available.

Table 2 presents the distribution of economic rents under the IPPK system. Rents captured by contractors and investors were calculated by deducting their estimates of raw material, production and transport costs, informal payments and taxes from the timber price they receive at their stage of the production/distribution chain. Rents captured by police/military and government officials consist of informal payments, while villagers' share of rent is the stumpage fee. The distribution of rents in the IPPK system shows that various government officials may have captured a significant share (up to 24%) and the police, the military, and customs both in Indonesia and Malaysia received about 20%. Villagers, meanwhile, obtained a share of about 7% of revenues. These data are based on very small samples of two or three respondents per category. The sensitive nature of the subject matter also made cross checking the information difficult. Results should therefore be interpreted with caution. Nevertheless, we can conclude that police, military, and customs officials appropriated a large share of logging rents and that they did so illegally.

Why were they being paid? Police, military and customs officials were paid to close their eyes to the illegal export of logs from East Kalimantan to Malay-

TABLE 1. Incomes earned by villagers in different logging systems (per m³).

	Manual extraction[1]	IPPK[2]
Extract and sell logs	$8.0-14.0	0.0
Stumpage fee	0.0	About $3-3.5
Wage rate	0.0	About $0.3
Total	$8.0-14.0	$3.3-3.8

[1](Obidzinski et al., 2001)
[2]Visits to four IPPK areas, and data from government offices (see Methodology)

TABLE 2. Distribution of revenues in the IPPK system (per m^3).

Stakeholders	Revenues	Share
Villagers	$3.25	7%
Contractor (Indonesian)	$12.0	25%
Investors (often foreign)	$23.0	48%
Local government officials	$2.2	5%
Police/Military/Customs (Indonesian & Malaysian)	$7.5	16%
Total	$47.95	100%

Source: Interviews in four villages, IPPK operators, district government officials, Malaysian timber buyers (see Methodology)

sia. District government officials received payments to award IPPK permits and to close their eyes to the illegal forest activities perpetrated by the concessionaries and contractors. The latter involved at least:

- Contracted areas under the IPPK permits exceed the maximum allowed area of 100 ha;
- The area actually harvested may be up to 2.5 times greater than authorized area.

A review of 10% of the IPPK permits in Bulungan and Malinau districts showed that the average holding area of IPPKs was 1,793 ha–an order of magnitude greater than their 100 ha permit. This review found that contractors actually harvested only 20 m^3/ha while they had permits to harvest 49 m^3/ha. Contractors often bribed district forestry officials to overstate inventories in order to secure permits for timber volumes larger than those actually existing in the IPPK concession areas. Having gotten permits to harvest more wood than they had in their concession, they then illegally logged in other areas.

Another important aspect of IPPK activities was the lack of tax collection at the district level (see Tables 3, 4). This failure to collect taxes may result from a lack of capacity in the district administration and/or purposeful lack of collection by government officials. The increase in the rate of tax collection from 2000 to 2001–to an impressive 88% in Malinau in 2001–may indicate that capacity is the key factor.

'Informal payments' to various government officials (including local police, etc.) were substantial when compared to tax receipts. This pattern suggests that officials face perverse incentives: the less they collect in taxes, the more they may receive through informal payments.

The findings presented above suggest that legalizing small-scale logging activities does not necessarily reduce illegal activities in other stages of the

TABLE 3. Tax collected by the districts.

District	Tax receipts as % of tax due, 2000	Tax receipts as % of tax due, 2001
Bulungan	8	21
Malinau	51	88
Nunukan	No tax levied	4

Source: Tax and Finance, Customs Offices in each district

TABLE 4. Informal payments and tax revenues from IPPK logging in East Kalimantan in 2000 and 2001.

District	Estimated informal payment ($,000)		Tax revenue as % of informal payments	
	2000	2001	2000	2001
Bulungan	142	311	11	16
Malinau	451	684	73	43
Nunukan	52	124	No tax	3

Source: Interviews in four villages, IPPK operators, district government officials, Malaysian timber buyers (see Methodology)

logging process. The large-scale logging concessions in Indonesia are notorious for ignoring forestry regulations (Kartodihardjo, 2002), often with the consent of corrupt officials. For government officials–now at the district level rather than central level–small-scale logging activities are new sources of rents. In some cases, these rents are taken for private benefit. In other cases, they are taken for the benefit of the district–as demonstrated by the Bupatis who 'legalize' illegally harvested logs simply by charging district fees (Casson, 2001).

TIMBER MARKETS AND THE ECONOMICS OF ILLEGAL LOGGING

According to interviews with logging contractors and timber brokers, the price of industry-standard Red Meranti in East Kalimantan declined about 50% between 2000 and 2001. The price of Red Meranti logs from Indonesia imported in Tawau, Malaysia declined by about 36% over the same period. Possible reasons for these declines are an increased supply of timber (in Indonesia) due to easier access to forest resources, and/or the reduced demand resulting from the effects of the Asian crisis, and/or competition from cheaper and lower cost plywood production in China.

Over the 2000-01 period, the unaccounted export of logs from East Kalimantan to Sabah increased from 1.8-2 million m^3 in 2000 to 1.6-1.7 million m^3 in only the first seven months of 2001–i.e., over 3 million m^3 on an annualized basis. These export estimates are based on the difference between log production in Sabah and official log supply (domestic and imported). According to the officials at the Department of Forestry in Sabah, the difference between the official log supply and demand in the state is almost entirely filled with undocumented, or unaccounted for, Indonesian imports (interviews at the Sabah Department of Forestry, Sandakan, April 2001). We use the term 'unaccounted' rather than 'illegal' because before the export ban went into effect in October 2001 many log exports–regardless of the legality of their harvest–went unreported in order to avoid customs dues.

This increase in unaccounted log exports in a period of falling timber prices seems counterintuitive. What might explain this trend? Unlike long-term timberland investors in developed economies–who face incentives to reduce production when prices drop–speculative liquidation investors have a short window of time to recoup their investment and earn their profits. For them, reducing production would lower returns to sunk investments such as machinery and networks. Confidential interviews with logging contractors and investors revealed that entrepreneurs instead cut operation and distribution costs in order to maintain returns on their investments. Cutting costs allowed them to maintain production levels–and even increase them in East Kalimantan–despite falling prices.

In a period of decreasing prices, by employing illegal strategies that range from production and export under-reporting to tax evasion, IPPK operators are more likely to be able to maintain production levels than legal operators. The former are more competitive as they can be expected to have lower production costs given, for example, that they do not have to pay taxes (or pay a small amount) and are able to access resources quickly by paying bribes.

Bribes are a component of the cost structure of illegal activities. While they may not pay taxes, these activities have to pay bribes and if these were high enough the competitiveness of these activities would be reduced. Therefore, it is worthwhile considering whether there are ways to force up the level of the bribes to be paid.

Loggers pay bribes to gain access to resources and to carry out activities that would otherwise be illegal. These bribes are similar to insurance policies taken out to avoid paying penalties for carrying out those activities. Theoretically, (see also Akella et al., 2002; Oh, 1995), the value of the bribe is the expected cost of the penalty:

Expected cost of penalty (value of bribe) =
penalty ×
probability of detection ×
probability of arrest ×
probability of prosecution given arrest ×
probability of conviction given prosecution.

This formula implies that even if penalties for illegal logging are very high, with limited monitoring and enforcement the actual expected cost of the penalty can be quite low, or equal to zero if there is no monitoring and/or enforcement. In such cases, the value of a bribe would also be low. This situation occurs in Indonesia, where the forestry department detects very few cases of illegal forest activities, the police follow up on even fewer, and almost none are brought to court at all, let alone found guilty. Increasing the penalty for illegal logging will achieve nothing when the monitoring of forest activities is limited, law enforcement by police is lacking, and the judiciary is not prepared for the task, or even worse, is not 'willing.'

It is also important to note that there may be situations in which bringing the level of the actual cost of penalty to the maximum possible would not alter the willingness to undertake illegal activities: this is when the expected cost of the penalty (all probabilities 100%) still allows the entrepreneur to the earn a satisfactory return on the investment. In this situation the penalty would have to be increased to a level that would bring the return to the illegal activity to a level lower than the average return on capital. Making the illegal activity unprofitable would be even more effective.

This analytical approach suggests that increasing the level of the penalty may not be effective in situations in which monitoring is lacking. It also suggests that an increase in monitoring may not be effective if there is not appropriate law enforcement, and a similar argument applies to the work of the judiciary. Does this mean that all improvements in the equation need to be undertaken simultaneously, making this an almost impossible task? Not necessarily.

Setting the penalty at an extremely high level–e.g., complete closure of a business and possibly imprisonment of the person responsible for the business–could create a perceived risk (i.e., cost) that exceeds the potential return on capital. Even in an environment where the cumulative probability of being found guilty is low, stiff penalties may disproportionately increase the perception of downside risk. One risk with this approach is that high penalties are applied only to the 'small fish' in the business–e.g., rural people who may harvest few logs to make a living–to show that the government or its agencies are doing something about the problem. High penalties should be set for 'significant' infringements; this raises, however, the problem of setting differentiated levels of infringements and proving which infringement applies to the specific case. Finally, it should be noted that the higher the penalty is the greater the incentive to pay the bribe and the higher the amount of the bribe to be paid. This implies that small businesses unable to afford the bribes will be the first ones to go out of business (unless a differentiated penalty system is applied) and that grand corruption prevails.

CONCLUSIONS

Legalizing the operations of illegal timber harvesters may not necessarily result in the cessation or significant reduction in illegal activities. When illegal activities are profitable, it can be expected that they will continue.

Successful approaches to illegal forest activities should target their profit-ability by changing the regulatory framework and strengthening its monitoring and enforcement. While effective law enforcement in required, in an environment where corruption is rife–as it is in Indonesia (Partnership for Governance Reform in Indonesia, 2001)–the government's capacity to effectively monitor and enforce forest laws is currently very limited. Simply allocating resources to law enforcement agencies will not be effective. Reducing corruption in the responsible agencies may seem the best approach, but it is unlikely to take place in the short term. Forest monitoring and law enforcement functions are, at the very least, in dire need of reform and may well require reassignment to other parties–governmental or otherwise. Without drastic change in the *status quo*, illegal activities will continue.

Improving forest monitoring is possible–even in an environment where corruption is still rife–by assigning responsibility to a third party, such as a specialized international company. Corrupt domestic institutions, however, would still be responsible for the regulatory framework so this improved monitoring may have a limited effect. This application of the "expected cost of penalty" framework should not leave us utterly without hope. Increased monitoring would bring more cases of illegal activities to the light of the day. Appropriately publicized, this exposure would increase public pressure on the government and its agencies to take action, not only in relation to illegal logging, but also to the corruption affecting those agencies. The question of how to get governmental agencies to agree to a genuine initial step towards improving forest monitoring remains.

The Jakarta-based Partnership for Governance Reform (2001) suggests that improving the regulatory framework by clarifying rules and reducing significant discretionary powers has been more effective than disciplinary actions and higher salaries. Nevertheless, if low salaries in the public sector persist in the long run, the incentive to seek bribes will continue, the cost of being caught (for the public servants) will remain low, and corruption will remain endemic to the system. It is also important to stress that it is not just corruption in the forest sector that needs to be addressed, but also that in the army and navy, the police, the customs department, and the judiciary (Partnership for Governance Reform in Indonesia, 2001). In the context of constant economic deficiency, corruption associated with illegal logging plays a significant role as an alternative source of income both for individual as well as institutional use. As a result, it is not a foregone conclusion that illegal logging is generally viewed in negative terms. For the stakeholders who are directly involved in illegal logging practices or playing part in corruption associated with it, illegal logging is beneficial as it generates a range of economic and political benefits.

In conclusion, we suggest that an approach to legalization that relies simply on the allocation of small-scale timber concessions without considering the issue of permanent control over forest resources might further marginalize local people. Local people often have little capital and know-how to invest in such operations, as well as little bargaining power. In the medium to long term, legalization strategies should entrust local people with the ownership and control of forest resources, and should maintain a monitoring role for government agencies. Restructuring the interests of various forest stakeholders would thereby break the current 'vicious alliance' in which most, if not all, forest stakeholders have an interest in perpetuating the *status quo*.

REFERENCES

Akella, A.S., J.B. Cannon, and H. Orlando. 2002. Enforcement economics and the fight against forest crime: Lessons learned from the Atlantic forest of Brazil. Paper presented at 'Illegal Logging in the Tropics: Ecology, Economy and Politics of Resource Misuse,' March 29-30, 2002, Yale School of Forestry and Environmental Studies, New Haven, CT, USA.

Barr, C. 1999. Discipline and Accumulate: State Practice and Elite Consolidation in Indonesia's Timber Sector, 1967-1998. MS Thesis, Cornell University.

Bisnis Indonesia. 1999c. "Koperasi Diprioritaskan Kelola HPH Jatuh Tempo." 12 November.

Brown, D. 1999. Addicted to Rent–Corporate and Spatial Distribution of Forest Resources in Indonesia; Implications for Forest Sustainability and Government Policy. Report No: PFM/EC/99/06, Indonesia-UK Tropical Forest Management Programme.

Casson, A. 2001. Decentralization of Policy Making and Administration of Policies Affecting Forests in Kotawaringin Timur. Case Studies on Decentralisation and Forests in Indonesia. Case Study 5. Center for International Forestry Research, Bogor.

Contreras-Hermosilla, A. (2001). Forest Law Compliance: An Overview. World Bank. *http://lnweb18.worldbank.org/eap/eapnsf/Attachments/FLEG/_OB3/$FILE/OB+ 3+Overview+Paper+-+Arnoldo+Contreras-Hermosilla.pdf.*

Daroesman, R. 1979. "An Economic Survey of East Kalimantan." Bulletin of Indonesian Economic Studies vol. 15, pp.43-82.

Dixon, G. 1974. Some questions regarding timber exploitation in East Kalimantan. Centre of Southeast Asian Studies, Monash University, Melbourne.

FWI/GFW. 2002. The State of the Forest: Indonesia. Forest Watch Indonesia, Global Forest Watch, Bogor, Washington DC.

ITTO. 2001. Achieving sustainable forest management in Indonesia. International Timber Trade Organization, Yokohama.

Jakarta Post. 1999. "Soeharto's Family Owns 4.3 m Hectares of Forest." 2 January.

Kartodihardjo, H. 2002. Structural problems in implementing new forestry policies. In: C. J. P. Colfer and I. A. P. Resosudarmo eds. Which Way Forward? People, Forests, and Policymaking in Indonesia, 144-160. Resources for the Future, Washington, D.C.

Magenda, B. 1991. East Kalimantan: The Decline of a Commercial Aristocracy. Cornell Modern Indonesia Project, Southeast Asia Program, Cornell University, Ithaca, New York.

Manning, C. 1971. The Timber Boom with Special Reference to East Kalimantan. Bulletin of Indonesian Economic Studies 7 (3): 30-60.

Obidzinski, K., I. Suramenggala, and P. Levang. 2001. L'exploitation forestière illégale en Indonésie: Un inquiétant processus de legalisation. Bois et Forêts des Tropiques 270 (4): 85-97.

Oh, Y. 1995. Surveillance or punishment? A second best theory of pollution regulation. International Economic Journal 9 (3): 89-101.

Palmer, C.E. 2001. The extent and causes of illegal logging: An analysis of a major cause of tropical deforestation in Indonesia. Centre for Social and Economic Research on the Global Environment, London.

Partnership for Governance Reform in Indonesia. 2001. A National Survey of Corruption in Indonesia. Partnership for Governance Reform in Indonesia, Jakarta.

Patlis, J. 2002. Mapping Indonesia's Forest Estate from the Lawyer's Perspective: Laws, Legal Fiction, Illegal Activities, and the Gray Area. Draft October 2002. Prepared for The World Bank/Wildlife Fund for Nature Alliance, Jakarta.

Peluso, N.L. 1983. The Markets and Merchants: The Forest Products Trade of East Kalimantan in Historical Perspective. Cornell University Press, Ithaca, New York.

Rhee, S. 2000. De facto decentralization and the management of natural resources in East Kalimantan during a period of transition. Asia-Pacific Community Forestry Newsletter 13(2): 34-40.

Ruzicka, I. 1979. Economic aspects of Indonesian timber concessions 1967-1976. University of London, London.

Scotland, N., A. Fraser and N. Jewell. 1999. Roundwood Supply and Demand in the Forest Sector in Indonesia (draft prepared 8 December 1999). Indonesia-United Kingdom Tropical Forest Management Programme.

CASE STUDIES

Social and Environmental Costs
of Illegal Logging
in a Forest Management Unit
in Eastern Cameroon

Philippe Auzel
Fousseni Feteke
Timothée Fomete
Samuel Nguiffo
Robinson Djeukam

Philippe Auzel, Faculté Universitaire des Sciences Agronomiques de Gembloux (FUSAGx), Forestry Unit, Gembloux, Belgium.

Fousseni Feteke is affiliated with PAPEL NGO and EU (DG VIII) funded project "Community forest in the Periphery of Dja Faunal Reserve," Yaounde, Cameroon.

Timothée Fomete is affiliated with Dschang University, Agricultural and Agronomy Science Department, Dschang, Cameroon.

Samuel Nguiffo, and Robinson Djeukam are both affiliated with the Center for Environment and Development (CED), Yaounde, Cameroon.

Address correspondence to: Philippe Auzel at the above address (E-mail: AuzelP@ aol.com).

[Haworth co-indexing entry note]: "Social and Environmental Costs of Illegal Logging in a Forest Management Unit in Eastern Cameroon." Auzel, Philippe et al. Co-published simultaneously in *Journal of Sustainable Forestry* (The Haworth Press, Inc.) Vol. 19, No. 1/2/3, 2004, pp. 153-180; and: *Illegal Logging in the Tropics: Strategies for Cutting Crime* (ed: Ramsay M. Ravenel, Ilmi M. E. Granoff, and Carrie A. Magee) The Haworth Press, Inc., 2004, pp. 153-180. Single or multiple copies of this article are available for a fee from The Haworth Document Delivery Service [1-800-HAWORTH, 9:00 a.m. - 5:00 p.m. (EST). E-mail address: docdelivery@haworthpress.com].

SUMMARY. Cameroon's forests have experienced the development of a flourishing informal forestry sector. This case study was designed to provide detailed information on illegal logging in the East Province of Cameroon, Messok district and focuses on Forest Management Unit (FMU) 10 030. Between 1999 and 2000, 21,750 ha of the unit's 79,757 ha (27%) was logged illegally. In total, 162,255 m^3 (7.46 m^3 per ha) of timber worth approximately US$ 26.5 million were removed from this area. The area logged is equivalent to 8.2 annual cut blocks which should have generated US$ 7.5 million in tax revenues. Locally, no long-term investments were made and the employment generated by the logging operations lasted only 2 years. Imposing sanctions against companies responsible for illegal logging activities remains an untested approach in Cameroon. The company recently challenged the administration's decision at the administrative court, on the grounds of the lack of evidence, and won the case. There is a risk that the company will escape all legal charges because of a four-year statute of limitations on tax infractions. The FMU 10 030 case shows the extent of financial and environmental damages from illegal logging and gives an idea of the level of losses incurred in Cameroon and other countries in the Congo Basin. *[Article copies available for a fee from The Haworth Document Delivery Service: 1-800-HAWORTH. E-mail address: <docdelivery@haworthpress.com> Website: <http://www.HaworthPress.com>* © *2004 by The Haworth Press, Inc. All rights reserved.]*

KEYWORDS. Illegal logging, sustainable forest management, forest policy, road detection, remote sensing, Cameroon, Congo Basin

INTRODUCTION

Thanks to easy access to valuable resources and corruption, most forested areas in Cameroon have experienced some form of illicit extraction of forest products as part of a flourishing informal forestry sector (Faure, 2000). The development of this wide base of illegal activities in the forest sector stemmed from the convergence of a variety of forces: a new 1994 forest law, the ongoing attempt to implement the various aspects covered by the application articles, and the intense industrialization of the forest sector following the 1994 devaluation of F CFA, the national currency (Carret, 1999).

The State Forest Policy objectives are seriously threatened by the increase in the informal sector (Durrieu de Madron, 2000), in terms of:

   i. The reduction of economic sustainability for industrial logging
  ii. The prevention of sustainable management plans for allocated forest management units (FMUs)

iii. A decline in the contribution of forest sector revenues to local and national economies in an era where poverty alleviation is necessarily a high priority.

The negative impact of illegal logging cannot be denied; however, it is still difficult to derive accurate figures showing the damage to the forest or the financial loss to the country due to this activity. We conducted studies that were designed to provide detailed information on illegal logging. We chose well-identified cases of illegal logging and decided to evaluate the effect of these practices on tax collection, the national economy at various levels, and the sustainable management potential in forest management units. The results and further debate that arise from these cases are the basis upon which we propose modalities by which the forest sector can try to meet the objective of sustainable forest management (SFM).

After a brief background presentation of the forest sector in Cameroon, we describe the study sites and methodology used to assess the situation in FMU 10 030 and finally, present our findings and conclusions. To date, mid-2002, the FMU 10 030 case has still not come to a satisfying end and the World Bank, employing a consulting firm with international repute, has formally requested an assessment of the damages.

This case study illustrates the necessity of developing practical and legal tools to fight both illegal logging specifically and the informal forest sector in general, both of which adversely affect the local and national economies and hinder the ability of Cameroon to achieve the objective of sustainable forest management in a reformed forest sector.

BACKGROUND

Permanent and Non-Permanent Forest

In Cameroon, forests feature layered appropriation and utilization rights that are often very complex. In the wake of the 1994 forestry law, the forests have been fractioned into two categories with an arbitrary assignment of spaces and rights to resources by the state. With the implementation of recent laws that alter the country's forest zoning plan, the southern Cameroonian forests are now divided, at least in theory, into a multiple-zone management area. The new regime turns on one central distinction between permanent forests and non-permanent forests. Permanent forests may be divided into forest management units or (FMUs) which, under the plan, become private property of the State as permanent timber production areas. Non-permanent forests, also called "the agroforestry zone," are areas where local populations may legally clear fields. This second zone is most often located along roads or following human settlements, and generally does not exceed, even in the best cases, more than 2 or 3 kilometers on each side of the road.

Traditional rights and contemporary commercial rights must now overlap and are negotiated in a context where intrusion by outsiders, attracted by forest resources that they might sell in the swelling markets of the country's urban centers, is increasingly frequent and intense. The stakeholders are constantly shifting and this makes consensus for sustainable management difficult to imagine, let alone achieve. Most use of forests by local populations extends well beyond the geographical and legal limits of use that is sanctioned under the new regime, and all sorts of agreements are reached beyond the sight of the state, between longer-term residents of forested areas, international corporations that manage the areas for timber, and nationals who, through actual or social capital, are able to facilitate timber extraction and other lucrative activities despite the complexity of recent laws on the matter.

Illegal logging constitutes one of the biggest threats to forest management in Cameroon. It is estimated that about 20% of national timber production is from illegal practices.

Allocation of Forest Resources

National forest inventories were carried out in 1984 and 1991. These inventories were used to prepare a zoning map of the forest zone, and allocate forest resources to logging companies mandated to log in accordance with a management plan. Following the new law and the new procedures, FMUs are allocated through an auction process where both technical and financial criteria of offers are examined. This process also determines the amount of tax per hectare to be paid by the concession owner each year. An inter-ministerial committee examines technical criteria and establishes a short list of bidders that are given a technical quote. The second step consists of the opening of financial offers during an open ceremony and the attribution of a financial quote according to the highest bidder. The result of the allocation process is based on the combination of these two quotes. Financial bids account for 70% of the final decision while technical consideration represents only 30% of the final quote and ranking of bids. Allocation is then confirmed, or not, by the Prime Minister. Some major changes occurred in the final ranking after the 1997 auction (Brunner, 2000).

Small permits of 2,500 ha, mostly located in the agroforestry zone, often behind villages, are also allocated through the auction process. Communities can be granted community forests after a long application process and a significant investment of time and money. Many logging companies believe they can pursue their unsustainable logging operations through these community forests and ensure the timber they needed for their industries at reduced cost because of an advantageous tax regime for community forests.

Decreased Timber Resource Availability

Since the introduction of the new forest law, logging has increased tremendously both inside and outside the permanent forest estate. Consequently, forest resources are decreasing both quantitatively and qualitatively. Quantitatively, the decrease in forest areas opened for logging has led to lower volumes per hectare that may be harvested. Logging is taking place on lands where former title-holders have already logged once or even several times. Inside FMUs only one logging right, equaling the right to log 2,500 ha or 1/30 of the FMU total surface, is issued annually in the three-year preliminary allocation. During this time, logging companies are supposed to develop management plans. The new rule of following implementation of sustainable forest management (SFM) has reduced the total area opened for logging each year.

Qualitatively, species composition has changed due to the overexploitation of high value timber species. With the transition from a forest management system with no FMU allocations between 1997 and 2000 to the current system of FMUs, logging companies' access to their desired supply of timber species has significantly changed and illegal logging has developed as a result of these difficulties.

The Level of Financial Resources Allocated to the Forestry Sector

Financial resources for the forestry sector are channeled through the state budget, the special fund for forest development, the special fund for wildlife, and private sector investments and cooperation. Investments in the forest sector have increased significantly since 1994 (Carret, 2002). These investments are mainly in wood processing capacities and are a reaction to log export restrictions imposed by the forestry law as well as to the devaluation of the national currency (F CFA) that reduced the industry's running costs and substantially increased profit margins.

According to the fiscal and economic review of the forest sector, the State collects roughly 35 billion F CFA in forest taxes annually; this amount will reach approximately 40 billion F CFA once all FMUs are allocated (CIRAD-Institutions & Development, 2000).

In 1999, the global turnover of Cameroon's forest sector was estimated at about US$ 450 million. The total share of the forestry sector to the overall exports of the country soared from 15% in 1996/97, to 25% in 1998/1999. This sector employs 90,000 people directly and indirectly.

Forestry Policy and Legislation

The current forest policy was adopted in 1995 and the revised forestry law was enacted in January 1994. The main principle underlying the new forest policy is to ensure the sustainability and development of the economic, ecological and social functions of forests through an integrated management system that ensures the sustained and lasting conservation of resources and forest ecosystems (MINEF, 1993). The new forestry law further mandates greater transparency in concession allocation than its 1981 predecessor. One of the major objectives of the 1994 law is to increase State income from logging revenues and to secure this revenue through improved control of forest activities.

Monitoring and Control of Activity in the Forest Sector

The monitoring of forest activities is legislated under the 1994 forestry law N° 94-01 and its application decree N° 95/531/PM of August 1995. On March 19, 1999, a program was developed to ensure complete collection of taxes from companies logging in Cameroon's forests. This program is administered by the Tax Department in collaboration with the Forest Department and was put in place by decree N° 99/370/PM. Monitoring of forest activities at the FMU, regional and national level is the responsibility of the State Forest Service.

In the Ministry of Environment and Forest, monitoring of forest activities is done by the Central Control Unit (UCC), which is directly linked to the Minister's office. The UCC is comprised of field units based at the provincial level. These are called Provincial Control Units. There are 6 units operating in regions where timber is being exploited. The UCC's duties include:

- Collecting and centralizing data from its provincial officers.
- Undertaking Control Missions.
- Comparing field data with the National Forestry Database System (SIGIF). The SIGIF database is a computerized system of forest information management. Through SIGIF yearly logging permits are issued based on cross verifications of previous rule compliance by logging companies (e.g., taxes being paid), log volumes are recorded, logging companies are taxed, and statistical data on forest exploitation can be produced at any moment.
- Proposing sanctions against offenders.

Monitoring is also carried out by all the provincial and departmental Forestry delegations across the country. Monitoring of forest activities, including logging, requires the knowledge of legal procedures and other management tools such as exploitation titles and their list of engagements, the SIGIF database, management plans in the permanent forest, intervention norms pre-scribed by the law, as well as forest mapping units and updated forest maps following inventory and logging.

The monitoring process consists of tasks carried out by Forest Service per-sonnel to ensure that regulatory norms are respected and management data are accurate. This monitoring system requires precise implementation of forest management systems. Forest activities are monitored in the non-permanent and permanent forests where management plans are being implemented.

In principle, the monitoring system requires field visits. Field visits are un-announced, spontaneous and frequent. However, the enforcement practices are not based on field visits, but are usually initiated after a denunciation or third party reporting of an alleged infraction against forestry regulation or dis-respectful behavior toward local communities.

This paper examines the current gap between the responsibilities and the actions of the government agencies in charge of the environment, forests, and finance. We hope this study will provide a platform on which particular argu-ments can be built and to strengthen the debate on illegal logging issues, in-cluding the long-term impact on tree species diversity and the social impact of lost tax revenue on forest communities' potential for development.

STUDY SITE AND METHODS

Study Site

Prior to the establishment of an independent observer position to support the Ministry of Environment and Forests (MINEF) in monitoring logging ac-tivities in Cameroon, Global Witness–a British based non-governmental orga-nization with a mandate to highlight links between natural resource exploitation and human rights–assisted Cameroon's Forestry Minister's efforts to control forest activities by logging companies. Global Witness discovered illegal log-ging at FMU 10 030 during a scoping mission in July 2000.

Located east of the remote village of Lomié, bordering the Dja reserve and the heart of Cameroon's rain forest, FMU 10 030 was part of one of the last in-tact forest blocks in the region and had yet to be allocated for use when a Cam-eroon-based and Lebanese owned logging company, was found exploiting the FMU through its neighboring FMU, 10 029, allocated in 1998 (Figure 1).

It should be noted that a few "Vente de coupe" or cutting permits of 2,500 ha each had been allocated in the agroforestry zone (Figure 2). Some small permits were allocated partially or even totally in FMUs. These small permits were often used to log beyond legal limits and constituted entry points for ille-gal loggers. To date numerous cases are known and some were documented, as in the case of FMU 10 030 where the entry point has been a logging opera-

FIGURE 1. Location of the study site (FMU 10 030) in south-eastern Cameroon, east of the Dja Reserve, near the Messok town.

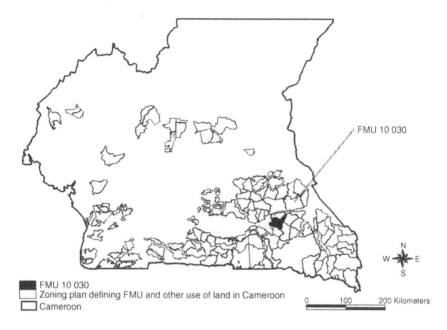

FMU 10 030

FMU 10 030
Zoning plan defining FMU and other use of land in Cameroon
Cameroon

0   100   200 Kilometers

tion in a small title adjacent to a FMU. Illegal logging took place before the FMU was allocated to another company (R. Pallisco) during the 2001 bidding process.

Detecting Illegal Logging by Mapping Logging Roads

Logging roads that were first discovered in July 2000 were mapped using GPS. Assistants drove motorbikes, rode bicycles and walked along the logging trails. Recordings take into account only the trails that were used by trucks and not skidding trails (Figure 3). GPS recordings were then transferred into ARCVIEW GIS for mapping purposes. Data from Auzel (2002b) were also used to detect roads and confirm the extent of the road network in the FMU. Landsat 7 ETM images taken in early 2001 were provided by GTZ Profornat and used to detect logging roads (Figure 4).

Assessing the Spatial Extent of Illegal Logging

The extent of logging was estimated using the road network. The first river not crossed would be the limit of the logged area. Where no natural limit was

FIGURE 2. Small permits allocated in the agro-forestry zone, the white band between FMUs, in the Messok area, near FMU 10 030, East Province.

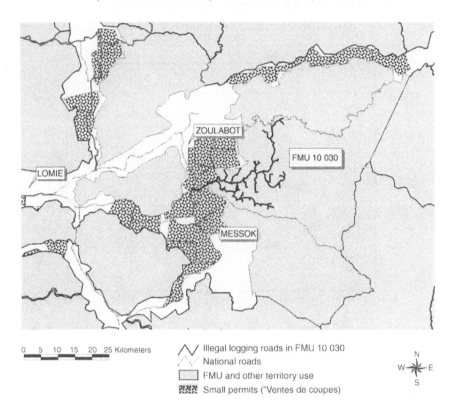

available we considered a distance from the closest road that could have been the start of a skidder trail. Skidding distances were found to be abnormally long, sometimes extending to 1.5 km. After discussion with logging companies' employees and managers we estimated that the average skidding distance normally ranges from 450 m to 850 m maximum. In this case, D7 Caterpillars were mainly used rather than the usual Caterpillar 528. This was probably due to the long skidding distance.

To evaluate the areas that have been logged, we applied buffers of 750 m and 1.5 km using ArcView GIS to the extracted roads detected inside the FMU (Figure 5). We also used remote sensing data in order to have another estimate of the areas that had been logged. The two methods were compared for accuracy in detection of the extent of logged areas.

FIGURE 3. Mapping of logging roads using GPS was done to map the extent of logging as a first step and before planning a more complete assessment that we conducted later with the realization of a forest inventory.

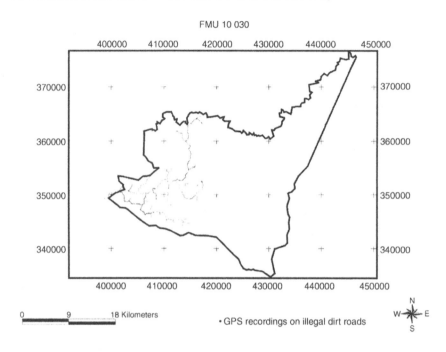

Assessing the Value of the Damage

In FMU 10 030, we opened 5 transects oriented North-South, on each side of a 10 km access transect, oriented East-West (Figure 6). The inventory transects, starting from the access transect, were 2.5 km each for a total length of 89.2 km, and a total of 189.5 ha inventoried. This included 361 parcels of 0.5 ha each located along the transects.

A systematic inventory was conducted on the 22,300 ha area (0.81% sampling rate), and included logged and un-logged areas that could be compared. The estimate of logging intensity was performed by counting and measuring stumps along the transects. A full inventory following management inventory protocol was conducted to assess the logging potential of the forest.

We compared these data with the inventory data that was collected by an accredited company prior to logging by the current owner and with the exact quantity of timber that was extracted by this company in the annual permit N° 12 and declared to the MINEF administration (Figure 7).

FIGURE 4. Detecting logging roads using remote sensing with Landsat 7 images.

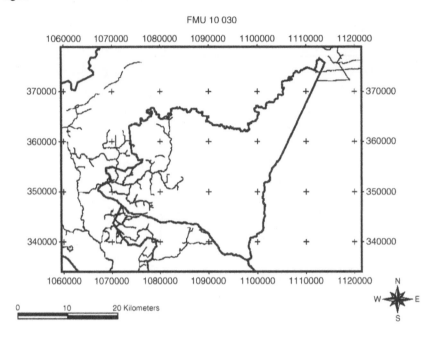

FMU 10 030

Assessing the Impact of Illegal Logging

Calculating the legal provision for forest taxation against the scale of illegal logging and considering the absence of a forest management plan, we also examined the impact of illegal logging on tax loss, the local economy and sustainable development, as well as sustainable forest management.

RESULTS

Detection and Spatial Extent of Illegal Logging

The FMU 10 030 case study allowed the team to test and investigate techniques to assess the level and extent of forest damage through field investigations, use of remote sensing and conducting forest inventories. Prior to any calculation, 21,750 ha, representing more than 27% of the 79,757 ha of the FMU, were estimated to have been logged between 1999 and 2000.

FIGURE 5. Estimating logging extent applying buffers to logging roads detected using remote sensing images of Landsat 7 ETM.

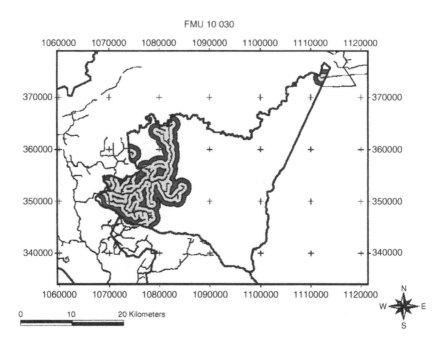

We conducted this assessment in FMU 10 030 at the same time that we conducted a study on the state of the forest sector using Landsat 7 imagery. In the FMU 10 030 case, the roads were below the appropriate size of openings (~100 meters) to facilitate the drying process of the road after periods of extreme rain. In some cases, several parallel roads existed because original roads were abandoned after being destroyed by heavy rains. The cost of opening new roads is much lower than having to bring numerous trucks of materials to repair damage to existing roads. This was particularly significant for the company that was logging illegally because it had not invested in long-term infrastructure development. The company preferred using numerous caterpillars that were permanently working on the road and pulling logging trucks heavily loaded to get them out of the FMU even if there was no investment in a proper road network allowing forest exploitation all year round.

It appears that in at least 2 other parts of FMU 10 030 there might have been some other logging incursions from neighboring permits. Further research is necessary to assess the extent of these activities.

FIGURE 6. The FMU 10 030, East Cameroon Province, Cameroon, and an overview of the transect put in place to study the incidence of illegal logging that occurred between 1999 and 2000.

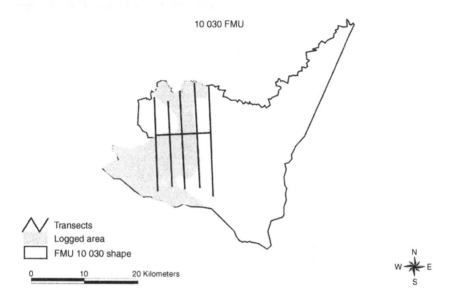

It is almost certain that 750 m buffers underestimate the extent of logging within the 12,045 ha logged-over area, representing only 55% of what has been estimated in the field. In this case, skidding distances were known to have been very long. Logging companies' employees reported skidding trails of more than 1.5 km on occasion. The 1,500 m buffer applied on the detected roads shows an estimated extent of 18,102 ha logged and fits more closely (83%) to the field estimate of 21,750 ha (Table 1). If we only consider the 1,500 m buffer estimate, the other possible incursions would have an extent of 533 ha in the north and 626 ha in the west of the FMU.

Estimating the Significance of the Illegal Harvest

Estimating Damages–The inventory we conducted in the logged area is the basis for our calculations assessing financial damages resulting from illegal logging. We found that 4% of the area was impacted at ground level, which is consistent with selective logging operations that were found to have removed 0.64 tree per ha on average according to the inventories we conducted. Logging intensity recorded shows an extraction rate of 7.46 m^3/ha for 8 species:

FIGURE 7. Annual permit for year 2001-2002 also named AAC n°12–Surface of 2612 ha. The FMU 10 030 was allocated after illegal logging occurred in the FMU. The company that won the biding process did some forest inventory to position its first annual permit outside the illegally logged area (AAC 12 in gray on the Figure).

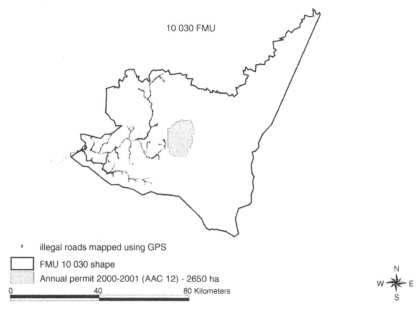

10 030 FMU

*   illegal roads mapped using GPS

☐   FMU 10 030 shape

▨   Annual permit 2000-2001 (AAC 12) - 2650 ha

0   40   80 Kilometers

N
W ✳ E
S

TABLE 1. Logged surface estimates using GPS recording and remote sensing of roads.

	GPS	Remote Sensing Detection	
	Recording	Buffer 750 m	Buffer 1500 m
Logging extent estimates	21,750 ha	12,045 ha	18,102 ha
Western border of the FMU		242 ha	626 ha
Northern border of the FMU		218 ha	533 ha

Sapelli (*Entendophragma cylindricum*), Assamela (*Pericopsis elata*), Kossipo (*Entandrophragma candollei*), Tali (*Erythrophleum ivorense*), Iroko (*Milicia excelsa*), Ayous (*Triplochiton scleroxylon*) and Pao Rosa (*Swartzia fistuloides*) with the total removal extrapolated to be 162,255 m^3 (Table 2). This rate of extraction is low when compared to the rate allowed as stated on the annual cutting right certificate (AAC) delivered by MINEF to the actual owner of the

TABLE 2. Species and logged volumes.

Species	Volume per ha in m³	Total volume logged (m³) in the 21750 ha logged illegally
Entendophragma cylindricum	6.04	131,370
Pericopsis elata	0.36	7,830
Entandrophragma candollei	0.58	12,615
Erythrophleum ivorense	0.12	2,610
Milicia excelsa	0.12	2,610
Triplochiton scleroxylon	0.18	3,915
Swartzia fistuloides	0.06	1,305
Total	7.46	162,255

FMU. This was after conducting an exploitation inventory of 100% on 2,650 ha which represents 1/30 of the total FMU surface as requested in a 30 year rotation system designed to ensure sustainable forest management. The certificate showed a potential harvest of 37,488 m³ for 31 species to be logged. For the area logged illegally, this represents the extraction of more than 307,000 m³. Some of these species, however, often have either no market value at all or are not profitable to log, according to company criteria.

Data from Official Harvest by Owner of the FMU–The certificate for the annual permit delivered by MINEF administration through provincial delegations shows the possibility of 37,488 m³ of timber to be logged from the 2,650 ha of AAC 12. This represents 14.15 m³ of timber per ha, and most likely underestimates the real volume due to a poorly conducted inventory by the accredited firm. The valuable trees like Sapelli (*Entendophragma cylindricum*), Iroko (*Milicia excelsa*) and Kossipo (*Entandrophragma candollei*) were removed during illegal logging operations. Furthermore, other species such as Assamela (*Pericopsis elata*), one of the most valuable timber species, were probably cut under their official diameter. For less valuable species like Tali (*Erythrophleum ivorense*) or Ayous (*Triplochiton scleroxylon*), only the most merchantable logs were removed leaving behind the majority of the population of these species. The figures coming out of the forest inventory we conducted are consistent with other inventories conducted by specialized firms in the FMU, and they also highlight the strategy of illegal logging operations which is to log as quickly as possible the most valuable trees and trees that best conform to market standards in terms of diameter, length, and other desirable characteristics (Table 3).

TABLE 3. Comparative examination of inventory results, AAC certificate and DF 10 records for logged timber.

Tree species found logged during the inventory	AAC 12 certificate 2650 ha	DF 10 record of logged volume for AAC 12	Illegal logging in an area equivalent to one AAC (equivalent to the 2650 ha)
Entendophragma cylindricum	17,117	14,823	16,006
Pericopsis elata	872	112	954
Entandrophragma candollei	689	857	1,537
Erythrophleum ivorense	8,596	10,061	318
Milicia excelsa	424	173	318
Triplochiton scleroxylon	2,160	2,432	477
Swartzia fistuloides	10	44	159
Total	29,868	28,502	19,769
Average m³ per ha	11. 27	10.75	7.46

Estimating the Value of Illegally Extracted Logs

We estimated the value of the illegal harvest by considering that 55% of the timber logged was sawmill quality and the remaining 45% was export quality that would likely be exported directly without processing, as was in the case of most of the logging operations we visited at the time. This was also confirmed by an interview with the General Manager of the logging company currently owning this FMU following a regular biding process. The total value of the volume logged was calculated to be US$ 26.5 million, based on the Free On Board (FOB) value for each species. Most of the volume and the value are represented by *Entendophragma cylindricum* (Table 4).

Effects of Illegal Harvest on the Local Economy

Tax loss was estimated using fiscal data available for FMU 10 030, which was allocated in 2001. The owner will pay an annual surface area payment of $US 11.5 per ha per year on the total FMU surface of 79,757 ha (Box 1). This tax must be paid to have the right to log the annual permit of 2,650 ha. The owner thus pays US$ 900,000 in taxes, which constitutes a charge to apply to the 31,101 m^3 that were logged in the annual permit (AAC 12) allocated for 2001-2002. This represents a tax load of US$ 29.5 per cubic meter for this year.

Up to 21,750 ha of the area was logged illegally. If we consider an annual permit right of 2,650 ha, this represents 8.2 annual permits (AAC) which, if

TABLE 4. Estimation of the value of the illegal timber.

Timber species harvested	Species value in US$ per m^3		Total value in US$ (US$1 = 700 F CFA)
	Sawmill quality log (55%) price entry sawmill	Export quality logs (45%) price FOB Douala	
Entendophragma cylindricum	128.6	214.28	21,957,557
Pericopsis elata	200	285.71	1,868,014
Entandrophragma candollei	107.14	121.42	1,432,703
Erythrophleum ivorense	85.71	107.14	248,882
Milicia excelsa	171.43	200	480,986
Triplochiton scleroxylon	85.71	121.43	398,491
Swartzia fistuloides	185.71	242.86	134,135
Total			26,520,768

BOX 1. The annual area tax system in Cameroon.

In Cameroon, 30 years is understood to be a sustainable rotation length. The FMU is logged by annual permit equivalent to 1/30 of the concession surface:

Annual cutting right = Surface of the FMU/ 30 (recommended rotation time)

An exception is made for concession under 75,000 ha wherein logging companies may request an annual permit for 2,500 ha during the three-year preliminary agreement with the State, during which time they are supposed to write a management plan.

Taxation is calculated on the entire concession even if the companies have the right to log 1/30 of the area each year.

Annual Forest Surface Tax = Surface of the FMU × (biding price + minimum price).

This allows for a competitive biding process and avoids the concentration of the forest resources in the hand of people that would not be able to use it properly. Even if the taxes are paid on the entire area, the amount of the bid reflects the potential in one annual permit (AAC) or 1/30 of the concession surface. The concessions with the highest bid support an area tax of US$ 30.6 per cubic meter extracted. Some other taxes are added such as the felling tax (2.5% of the Free on Board value for the felled timber).

taxed, would result in a payment of 8.2 times the current annual surface area tax, for a total of US$ 7.5 million.

We have no information on the payment of felling taxes, which represent 2.5% of the Free On Board value for all species logged. In this case, the amount to be paid reached US$ 0.66 million. Often, illegal timber is declared on a valid title and becomes legal when the administration ignores or does not pay attention to the fact that the volumes harvested do not correspond with the inventory done prior to the deliverance of the (AAC). There are also problems with declarations that lead to reduced felling tax payments. For example, it is common to see a declaration of only 10% of the logged volume or even no declaration at all. It is therefore difficult to assess the felling taxes left un-recovered from illegal logging operations.

Illegal logging activities and the non-recovery of taxes mean that there is a drastic reduction in tax revenue that would otherwise benefit the State, local councils and neighboring communities of the FMU. According to the previous calculations, we estimate the following loss for FMU 10 030:

- An estimated US$ 7.5 million fiscal loss for the State considering the US$ 0.9 million annual tax payment as the 21,750 hectares logged illegally is worth 8.2 years in annual surface allowed each year under sustainable management and fiscal schema.
- An estimated US$ 66,000 lost in felling taxes (2.5% of FOB value).
- US$ 2.8 million lost for local development of the Messok district is and,
- US$ 700,000 lost for the development of neighboring communities.

Locally, no long-term investments were made and the employment generated from the logging operations lasted no more than the two years that were necessary to log 25% of this FMU.

What About Sustainable Forest Management?

Taking these logged areas into account when developing current management plans presents a difficult problem. Sustainability in forest management requires that logged tree populations recover from the off-take. In other words, as forests require time to reconstitute via natural regeneration, the quality and quantity of a second rotation of timber is affected by the short rotation periods when logging companies start logging previously logged areas before the end of the rotation period (Durrieu de Madron, 1997a; 1997b).

In theory, logging companies could log other species that were left during the first harvest but for reasons such as market prices being too low this is often not possible. Also because industries have developed industrial processes for certain species they cannot substitute another species even with very similar qualities.

In FMU 10 030, the area logged illegally was estimated to be as much as 25% of the total FMU surface. The company that currently owns the FMU will probably lose more than 8 years in its rotational management plan due to these illegal harvesting activities. This makes it impossible and less profitable to develop a sustainable logging regimen. The area within which to rotate growth is often reduced drastically through illegal logging. This could end with the abandonment of the concession or in a very limited annual cut, which may not allow the industries to survive under given tax pressure.

One measure that can be implemented to ensure regeneration of logged volumes is to increase the minimum diameter at which a given tree population can be logged. The increase in minimum cutting diameter is going to have a direct effect on the availability for the next cut. Data gathered in FMU 10 030 clearly shows that even after the 20 year rotation necessary in the un-logged part of the FMU, an increase in cutting diameter of up to 130 cm would be required for a recovery of logged volumes close to 100% in the second cut. Actual logging allows the extraction of 9 m³ per ha, felling all the trees over 100 cm dbh. With an increase in minimum harvest diameter, the harvest potential would be reduced to 3 m³ per ha for Sapelli (*Entendophragma cylindricum*) as shown in Figure 8. This would also be the case for all the other valuable timber species.

It is likely that the logging company owning the FMU would not be able to afford a loss of 2/3 of the volume of one of the most valuable species, and illegal logging thus poses a serious threat to the implementation of sustainable forest management.

FIGURE 8. Effect of increased minimum DBH on stand recovery and logging volume.

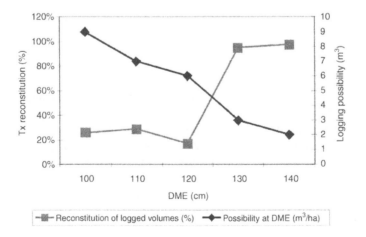

Legal Provisions on Sanctions of Illegal Logging

Illegal logging in a non-allocated concession injures a number of parties who are entitled to claim compensation from the perpetrator of these activities. In the case of FMU 10 303, the State, as owner of the forest, lost potential income in the form of 8.2 years of surface taxes, equaling US$ 3.75 million. The county lost their portion of the taxes allocated to them by the 1994 Forestry Law which is 40% of the surface area tax, equaling US$ 3 million. The neighboring communities lost their share of the forest taxes allocated to them by the 1994 Forestry Law as well, which is 10% of the surface area tax, equaling US$ 0.75 million.

Article 159 of the 1994 Forestry Law states that the financial compensation due to the State as a consequence of illegal logging will be calculated according to the full value of the considered timber (MINEF, 1999 p.81). Every year, the Ministry of Finances publishes a list of timber values for the different commercial species in Cameroon.

For the communities and the municipalities, the civil law principle serving as the legal grounds for liability states that, "the author of a damage should provide compensation" (Civil Code article 1382) (République Française, 1804).

Infractions

The activities conducted by this company operating illegally within the FMU 10 030 resulted in 6 major infractions under the Cameroonian penal code and forestry law. The primary infraction in this context was unauthorized logging in a State forest. A second violation was the fraudulent marking of logs. Although the company was logging with a legal title in a neighboring concession, the marking of the legal title (FMU 10 029) was used for the illegal logs taken from FMU 10 030. The company was also responsible for the fraudulent use of official documents (production, declaration and transportation documents). Consequently, they also committed fraudulent tax declaration. The tax declarations made by the offending company during the considered fiscal years were false, and lead to a loss of income for the State and other beneficiaries. In addition to the former infraction the company also submitted a fraudulent customs declaration. According to the law, the company should indicate in its customs declaration the origin of the timber, and provide a document signed by the local representative of the Ministry of Environment and Forests stating the species, the volumes and the origin of the timber. Given that exports may not be made without such a document, it is likely that the company used a false document. This reveals the final infraction of production and use of false documents.

Legal Sanctions

The idea of imposing sanctions on companies responsible for illegal logging activities is new in many peoples' minds, and therefore in their practices. So far, minor sanctions have been imposed, but they do not adequately reflect the extent and seriousness of fraud in each case.

Considering the forest law and the provisions of the penal code, sanctions should have consisted of fines and/or imprisonment for all illegal activities. These would have included: a withdrawal of the logging agreement used by the company according to articles 65 and 130 of the 1995 decree on forests (meaning the cancellation of the company's right to operate in the forestry sector in Cameroon, a payment of financial damages and interests calculated on the basis of the compromised tax recovery (timber value and unpaid taxes as they should have been in a regular FMU allocation), payment of financial damages and interest to the municipalities and the communities, fiscal adjustment with necessity of payment for unpaid taxes with penalties due to fraud and time loss, and imprisonment for the persons responsible for the production and use of false documents.

In the case of FMU 10 030, only three sanctions were imposed on the company involved in these illegal operations. They were excluded from the June 2000 round of FMU allocation. However, the same company did receive a concession, under a different name. A fine of US$ 150,000 was imposed on the company regarding other charges related to its activities in the neighboring 10 029 FMU. According to a press release issued by the company's lawyer, the company finally paid only US$ 120,000. Under the pressure of bilateral and multilateral donors, the Ministry of the Environment and Forests requested a US$ 3.5 million deposit from the company, before the final evaluation of the damages. It is worth noting that the company publicly denied its involvement in illegal logging in FMU 10 030 despite all evidence. According to the latest information available from MINEF, a final evaluation of damages is pending, and will be carried out by a consultant of international repute at the World Bank's request.

DISCUSSION

Institutionalizing the Gap Between Reality and Declarations?

The fundamental objective of the forest policy reform in Cameroon was to establish a transparent, equitable, and sustainable management system for forest resources. To date, it is more and more obvious that the outcome of the reform process is limited (Essama-Nssah, 2000). A bidding process that has been described as a fine-tuning of Central Africa's first international auction system (Collomb, 2001) has lead to ongoing problems with potentially irreversible consequences. It is now recognized that in the 1997 auction, FMU allocation was decided on a discretionary basis with a succession of problems following the sales (Brunner, 2000). One of the most notable problems is that concessions were not allocated to the most technically and financially capable companies entering the bidding process. Many companies are owned by individuals that have no records and no experience in the forest sector. This important failure in the system is also mentioned in the economic audit that was conducted on the forest sector where it was noted that the financial offer does not always constitute a determinant criteria in a competitive binding process (Karsenty, 2000).

As a result, a few years later, six owners of FMUs allocated in 1997 were threatened with the removal of their concessions. The threat was only carried out for three concessions to preserve the country's stability. The other three FMUs were allocated to key people of the Cameroon Republic. The rescinded concessions presented the threat of serious damages with anarchic logging operations that in some cases led to the redrawing of new FMU boundaries prior to any binding process.

The development of community forests and the split of forest revenues between the State (50%), rural counties (40%) and local communities (10%) lead us to believe that there might be some hope for the ongoing decentralization process wherein the listed parties have an incentive to protect or at least ensure an appropriation of forested areas that would be protected from illegal loggers after they are granted the concessions (Auzel, 2001b). This hope exists even if the state agencies still firmly fight for a centralized monopoly and therefore control of natural resource exploitation. Decentralized revenues coming from area taxes were primarily put back into two counties over the past few years (Lomié and Yokadouma received almost US$ 1 million per year). However, this did not make any difference in the social landscape of these two places (Milol, 2000). This situation has lead to the proposal of a sharing scheme whereby forested counties that disproportionately benefit from most of the revenue will provide resources to other counties that are often more populated and therefore have a greater need for public investment.

Accession of national foresters to FMUs has long been an important source of debate in the reform process (Carret, 2000). It is now obvious that there is access to forest revenue by a few nationals rather than the emergence of national entrepreneurs accessing the forest sector because of their professional skills. It is said that eradicating so-called *"fermage"*–sub-leasing the allocated FMU to secondary parties–would encourage better management while increasing government revenue (Brunner, 2000). Seventy-five percent to 80% of the permits allocated to national individuals used this type of arrangement, often with international logging companies. These permits generated an average of US$ 4.24 to US$ 11.5 per cubic meter logged in their concession. Annually this represents more than US$ 3.5 million split between a few individuals (Fomété, 2000).

From our observations, it is clear that management is not the primary concern for most individuals that were allocated concessions. Moreover, these national individuals often threaten sub-contracted companies, typically foreign companies interested in more careful management, that if their financial demands are not fulfilled they will change partners. In a situation of great scarcity on the raw timber market, these practices will discourage companies pushing for implementation of sustainable forest management.

Sustainable forest management is a topical subject in the debate over the future and conservation of tropical forests. Most of the major international organizations in development and conservation are now involved in SFM programs, primarily at the policy level, and sometimes provide institutional support to a generally centralized administration in charge of forests. It is rare that policy implementation and related management operations in the forest sector are closely followed. The exceptions are a few countries that have important private interests in the forest sector.

The consequences of all these problems might be irreversible, which poses a great threat to the ongoing reform process since the benefit that should be expected from the requested changes are perceived to be small. This is in part because the main actors oppose the reform process at all levels, whether state agents of private companies or not (Carret, 2000). This situation was even more noticeable looking at the Forest and Environment sector program preparations, where MINEF staff responded to discussion of transfer strategies and appropriation by a massive desertion of the numerous meetings.

The FMU 10 030 Case in Context

This empirical study has cleared the way to address new cases of illegal logging. The illegal logging of FMU 10 030 is far from the exception to logging practices in Cameroon. There is a severe threat to community forests from industries seeking to secure a resource base for their activities. This is especially true for all the sawmills built independently of any FMU or other forest title. Social considerations are actually used as a consequence of the construction of these sawmills to put pressure on MINEF officials so that they release cutting rights in community forests that clearly need protection at this stage of their development.

In the case of FMU 10 030, the importance of damage to the forest and the amount of money lost by the State, are not extraordinary in the Cameroonian context. The intensity of illegal logging has been growing since the mid 1990s, and according to some estimates, illegal logging now represents around half of the timber exports from Cameroon (CED, 2001). Large-scale illegal logging is currently occurring in several FMUs located south of the Dja Reserve, on a total area of 127,000 hectares with the involvement of several Cameroonian and French companies.

An attempt to evaluate the extent of illegal logging at the national level, its origins and incidence, reported that at least 36 of the 92 FMUs existing in Cameroon are impacted by illegal logging operations of various magnitudes (Auzel, 2002a).

Detecting and Dealing with Illegal Logging Situations

Previously, some of the opened roads were not detected because they were not large enough and because the natural canopy cover limited detection possibilities, which in this case do not exceed 30% to 50% of the existing roads on average (Auzel, 2002b). In this study the detection was better: up to 80% of roads were detected if we consider the detected area as an index to be compared with what has been verified on the ground. This is because logging is very recent and the roads are largely opened.

The failure to identify incursions detected on Landsat 7 ETM images might be, in part or entirely, due to "mapping problems" by the logging companies operating in the area, and true mapping problems within MINEF departments. It has been observed that for a given title several descriptions of boundaries existed, and sometimes the FMUs shape was unclear due to bearing errors or errors in the legal description that accompany the map of the concession.

Problems can also be related to poor delimitation of a given title in the forest. We have occasionally observed shifts of 500 meters or more in boundaries for several permits with no fraudulent intentions. With the changes in the national legislation, inventories and field materializations were to be done by companies officially accredited for such business. Unfortunately, this accreditation was not synonymous with technical capacity (Durrieu de Madron, 2000), and was often corrupt since many of these firms have connections with actual or former members of the national administration in charge of forests.

The Incidence of Illegal Logging: An Ongoing Debate

There is at present an important debate on illegal logging issues in Cameroon. Approaches differ between those who want to expose prior violations and create a new more informed decision-making process and those more in favor of supporting logging companies in their current management efforts, thereby reducing the general cost of achieving SFM objectives.

In principle, these two approaches could be complementary if there were no long-term results to defend certain parties involved. The fact is that there is no standardized method to address illegal logging issues. This becomes more and more obvious when calculations start showing damages that could reach several million US$, and when logging companies sue the Cameroon government in the national Supreme Court. Further, a national strategy for control of forest activities has not yet been drafted and there is little chance that a standardized method or adequate legal procedure will be available very soon.

Nevertheless, there is a very difficult debate on the dialectic and financial issues related to illegal logging as well as the various sanctions, damages and interests to be applied to the responsible companies. Overall, however, most people agree that SFM is in great danger, and poses the greatest threat to Cameroon's forests. Certification potential for these forests is questionable, and future market opportunities with increasing demand for certified products might be lost.

Sanctions Imposed on the Company

The system of sanctions provided for by the 1994 Forestry Law has at least two major deficiencies. First, the sanctions are less dissuasive and so far it has been more lucrative to log illegally than to obey the law (to date nothing has shown the contrary). Second, MINEF has all the competence to observe the infraction of the law, decide sanctions and ensure they are implemented, but not the human or technical resources to carry these activities.

The fact that MINEF has centralized the management of infractions means that the power to impose sanctions is in the hands of very few people. This may provoke collusion between forestry administration officials and dishonest private sector actors. A more cautious and power-balancing approach is needed. The present system of sanctions has established compromise as the primary method of settling forestry disputes and the refusal or failures to enter into a compromise, observed by MINEF, is a precondition to appeal to a judge for assistance. It is interesting to note that in most of the cases of illegal logging where a compromise was reached, the amount of compensation paid by the offending company appeared to be less than the value of the illegal timber, despite the legal provision in the 1995 forest policy stating that such compromises should be void.

The absence of transparency in the follow-up of forestry disputes leads to extra-legal arrangements between the forestry administration and forestry companies. As a result of both the incompetence of administration officials and the absence of political will, cases of infraction of the law are not always correctly noted and sanctions are not implemented. In FMU 10 030, where infractions of the law were known and documented since December 1999, the World Bank has observed no official report signed by the sworn control team as of May 2002.

Furthermore, even if enforcement mechanisms were fully functional, the modalities for calculating damages are not officially set, and the economic value of timber does not represent the entire value of the forest. At least two important values have been excluded from the calculation of damages: the value of other functions of the forest, and the loss incurred by the population that are beneficiaries of 10% of the area tax.

Legal Prescription: A Major Risk

A major risk of impunity is that the company will most likely not be sanctioned for all the infractions committed in the 10 030 FMU. There is a risk that the company will eventually escape all legal charges in regard to the provisions of the tax law because of a four-year statute of limitations on tax infractions. The company's illegal activities started in 1998, and the four-year time limit is almost over. The only possibility for an extension is to bring legal proceedings against the company, which has not yet been done by the forestry administration. The second risk is that the court case may not be in favor of the forestry administration because of the lack of key evidence that was not prepared by the Ministry at the time of the infraction. Under pressure by the donor community in Cameroon, the Ministry of Environment and Forests seized the financial deposits of the company, as partial payment of the fines. The company challenged the administration's decision at the administrative court, on the grounds of the lack of evidence, and won the case. In its 2003 decision, the court decided that the Ministry of Environment and Forests was not entitled to seize the company's deposit, and had no legal right to suspend the logging titles of all the branches of the charged company. The only solution for the Ministry of Environment and Forests is to provide a court decision condemning SFH for illegal logging. The case is pending in the court in Douala, and a decision is not expected in the coming months due to the slowness of the justice system. The company will have enough time to organize its bankruptcy, and escape all compensation charges.

CONCLUSION

This case study shows the extent of financial and environmental damages by illegal logging in a single FMU, and can be used to imagine the level of losses incurred by the country and perhaps even across Congo Basin. Looking at these figures, a number of questions arise about the misuse of old growth forests based on forest development perspectives. These losses must be compared with international aid processes, ongoing debt relief efforts, and donor-driven reforms in the forest sector.

The magnitude of fiscal losses due to illegal logging was for a number of years far higher than the amount actually recovered by the State. Though this trend is decreasing, Cameroon still has to address the issue of forest law enforcement and governance. With regard to communities, our case study shows that logging taxes funding local development have suffered from illegal activities. The revenues expected by communities have not been paid, and the base for future revenue is being depleted. Furthermore, the high rate of deforestation due to illegal extraction of forest resources is reducing the communities' access to resources for their subsistence and future development. Finally, illegal logging is ruining all the commitments of the ongoing reform of the forest sector in Cameroon. With regard to the reforms aimed at promoting sustainable forest management progress has been made on paper, but the situation in the field still shows very poor capacity of the administration to implement the forest law and policy, and to monitor activities of the various actors.

AUTHORS' NOTE

This study was supported by DFID Cameroon Office, Forest Monitor and IUCN Netherlands Committee, and other NGOs. The field work conducted with agreement of the forest division of the Ministry of Environment and Forests and Direction of Taxes of the Finance Ministry (MINEFI) of Cameroon Government. We also thank MINEPAT (Cellule Environment et Forêt), the UE STABEX fund manager for support of a pilot study, the R. Pallisco company for its collaboration and assistance, and Leanne from the VSO staff at CED.

REFERENCES

Auzel, P., Fétéké, F., Fomété, T., Nguiffo, S., Djeukam, R. 2001a. Incidence de l'exploitation forestière illégale sur la fiscalité, l'aménagement et le développement local: cas de l'UFA 10 030 dans l'arrondissement de Messok, Province de L'Est, Cameroun. Page 39. Nature+/FUSAGx, PAPEL, Université de Dschang, CED, Youndé.
Auzel, P., Fomété, T., Owada, J. C., Odi, J. 2002a. Evolution de l'exploitation des forêts du Cameroun: production nationale, exploitation illégale, perspectives. Page 45. DFID Cameroon Office, Yaoundé, Cameroun.
Auzel, P., Halford, T. 2002b. Contribution à un état des lieux du secteur forestier au Cameroun. Cas des forêts de production permanente. Page 158. MINEF-MINPAT/ Cellule Environment et Forêt, Yaoundé, Cameroun.
Auzel, P., Nguenang, G. M., Fetetke, R. Delvingt, W. 2001b. L'exploitation forestière artisanale des forêts communautaires: vers des compromis écologiquement plus durables et socialement plus acceptables. Rural Development Forestry Network, ODI:1-11.

Brunner, J., Boscolo, M. 2000. Trip report. Page 4. World Resource Institute, Harvard Institute for International Development, Yaounde, Cameroon.

Carret, J. C. 1999. L'industrialisation de la filière bois au Cameroun entre 1994 et 1998: Observations, interprétations et conjectures. Page 67. Ecole des Mines de Paris, Paris, France.

Carret, J. C. 2000. La réforme de la fiscalité forestière au Cameroun. Débat politique et analyse économique. Bois et Forêts de Tropiques 264:37-51.

Carret, J. C., Eba'a Atyi, R., Giraud, P. N., Lazarus, S., Nanko, G., Plan, J. 2002. Etude en vue de la définition d'une politique sectorielle de transformation et de valorisation du bois. Troisième partie: ressource/transformation et les performances des usines. Page 20. CERNA/ONF-Internationale/ERE-Development, Paris, France.

CED. 2001. La valeur de l'illégal. Page 3. Centre pour l'Environnement et le Développement, Paris, France.

CIRAD-Institutions & Développement. 2000. Audit économique et financier du secteur forestier au Cameroun. Ministère de l'Economie et des Finances, Yaoundé.

Collomb, J. G., Bikié, H. 2001. 1999-2000 Allocation of logging permits in Cameroon: Fine-tuning Central Africa's first auction system. Page 39. Global Forest Watch Cameroon, Washington, DC.

Durrieu de Madron, L. 1997a. L'aménagement intégré en forêt dense: les enseignements du projet A.P.I. au Cameroun. CIRAD-Forêt, Montpellier, France.

Durrieu de Madron, L., Forni, E. 1997b. Aménagement forestier dans l'Est du Cameroun: structure du peuplement et périodicité d'exploitation. Bois et Forêts des Tropiques 254:39-49.

Durrieu de Madron, L., Ngaha, J. 2000. Revue technique des concessions forestières. Page 106 p. Comité Technique de Suivi des Programmes Economiques, Yaoundé.

Essama-Nssah, B., Gockowski, J. J. 2000. Cameroon forest sector development in a difficult political economy. Page 123. Evaluation country case study series. The World Bank Operations Evaluation Department, Washington, DC, USA.

Faure, J. J., Njampiep, J. 2000. Etude sur le secteur forestier informel. Ministère de l'Economie et des Finances du Cameroun, Yaoundé.

Fomété, T., Pinta, F. 2000. La structure du secteur. Page 53 in CIRAD, Institution et Développement, editor. Audit économique et financier du secteur forestier au Cameroun. Ministère de l'Economie et des Finances, Yaoundé, Cameroun.

Karsenty, A. 2000. Impact de la fiscalité. Page 22 in CIRAD-Forêt, Institutions et Développement, editor. Audit économique et financier du secteur forestier au Cameroun. Ministère de l'économie de des finances, Yaoundé, Cameroun.

Milol, A. C., Pierre, J. M. 2000. Impact de la fiscalité décentralisée sur le développement local et les pratiques d'utilisation des resources forestières au Cameroun. Page 53 in CIRAD, Institution et Développement, editor. Audit économique et financier du secteur forestier au Cameroun. Ministère de l'Economie et des Finances, Yaoundé, Cameroun.

MINEF. 1993. Document de politique générale: la politique forestière. Page 45, Yaoundé, Cameroun.

République Française. 1804. Article 1382. Code Civil.

Recent Trends in Illegal Logging and a Brief Discussion of Their Causes: A Case Study from Gunung Palung National Park, Indonesia

Marc A. Hiller
Benjamin C. Jarvis
Hikma Lisa
Laura J. Paulson
Edward H. B. Pollard
Scott A. Stanley

Marc A. Hiller is affiliated with the Tropical Forest Trust in Hanoi, Vietnam.

Benjamin C. Jarvis and Hikma Lisa are affiliated with the Tropical Forest Trust in Samarinda, East Kalimantan, Indonesia.

Laura J. Paulson is affiliated with the Nature Conservancy Mexico program based in Tucson, Arizona.

Edward H. B. Pollard is Consultant in protected areas management, The Old Vicarage, DL2 2PF, England.

Scott A. Stanley is Program Director for The Nature Conservancy East Kalimantan Program, Jl Kuranji 1, Samarinda, Indonesia.

Address correspondence to: Edward H. B. Pollard at the above address (E-mail: e_h_b_pollard@yahoo.co.uk).

This work was carried out under the name of Harvard University Laboratory of Tropical Forest Ecology. The authors would like to thank the Department of Forestry of the Republic of Indonesia for their support, Mark Leighton for his guidance and Ronnie Cherry for supervising early parts of the work. Ramsay Ravenel, Charlotte Kaiser, Eko Darmawan and Dessy Ratnasari contributed many hours helping to collect the data and to them the authors are extremely grateful. The work was funded by the Biodiversity Conservation Network, Centre for International Forestry Research and the Gibbon Foundation.

[Haworth co-indexing entry note]: "Recent Trends in Illegal Logging and a Brief Discussion of Their Causes: A Case Study from Gunung Palung National Park, Indonesia." Hiller, Marc A. et al. Co-published simultaneously in *Journal of Sustainable Forestry* (The Haworth Press, Inc.) Vol. 19, No. 1/2/3, 2004, pp. 181-212; and: *Illegal Logging in the Tropics: Strategies for Cutting Crime* (ed: Ramsay M. Ravenel, Ilmi M. E. Granoff, and Carrie A. Magee) The Haworth Press, Inc., 2004, pp. 181-212. Single or multiple copies of this article are available for a fee from The Haworth Document Delivery Service [1-800-HAWORTH, 9:00 a.m. - 5:00 p.m. (EST). E-mail address: docdelivery@haworthpress.com].

SUMMARY. Over the course of three years the ecological and socio-economic effects of illegal logging in a peat swamp forest contiguous with Gunung Palung National Park, West Kalimantan were examined. This paper discusses the key findings as they relate to trends in illegal logging and illustrates some of the most critical factors driving this illicit activity. In 11 villages bordering the Park, census data was collected on population structure and sources of income. A subset of villagers was interviewed to gain more detailed information on the prosperity of loggers and non-loggers. From detailed interviews with 40 loggers mean annual income was estimated and an understanding of the methods used gained. In the forests adjacent to the Park 30 km of extraction trails and 1,200 stumps were measured in one logging network. Data from a more widespread survey of 20% of rivers around the park and a previous study of illegal logging were also evaluated. Study results indicate that 47% of households rely on logging for their primary source of cash income, a 71.7% increase in seven years. Families participating were more likely to be poor than the average villager. Furthermore, loggers are felling more trees and working further from rivers than they did three years ago. From these results and over three years of experience in the region two main factors influencing the increase in illegal logging are identified: (1) easy access to forests, labor, markets and equipment; and (2) local and national economic factors. *[Article copies available for a fee from The Haworth Document Delivery Service: 1-800-HAWORTH. E-mail address: <docdelivery@haworthpress.com> Website: <http://www.HaworthPress.com> © 2004 by The Haworth Press, Inc. All rights reserved.]*

KEYWORDS. Illegal logging, Borneo, Indonesia, peat swamp forests, community-based natural resource management, land tenure

INTRODUCTION

Aim

The aim of this paper is to investigate illegal logging in the Gunung Palung area of West Kalimantan province, Indonesia. The work was carried out as part of the development of a community based forest management (CBFM) project and was designed to help the project understand the local level socio-economic factors that may be behind illegal logging as well as to assess the impact of the logging on the forests. In the course of the work we attempted answer the follow questions:

In the course of the work we attempted answer the follow questions:

1. Has the level of illegal logging in the Gunung Palung area increased over the last ten years?
2. Who is participating in illegal logging? Local communities or recent migrants? Is logging being carried out by people from local communities or from in-migrants?
3. Are the people that participate in logging disproportionately poorer than non-loggers, with few economic alternatives?
4. What is the impact of illegal logging on the local economy?
5. What is the impact of illegal logging on forest resources and the local environment? How does any increase in illegal logging impact the forest and local environment?

Illegal Logging in Indonesia

Forest Resource Development in Indonesia–In contrast to other colonial powers, such as the British in neighbouring Malaya, the Dutch colonial government in the East Indies did not see the exploitation of forest resources as a priority. It was only in 1870 that the Dutch colonial government made state claims over the forest land (Pofenberger, 1997) and even then it was more interested in Java's teak plantations than natural forests on the outer islands (islands other than Java, Madura and Bali). This policy continued into the post-war Independence of Indonesia. (Peluso and Harwell, 2001). Mechanized logging in Indonesia began to boom following the 1967 Forestry Law No. 5 (Peluso and Harwell, 2001).

In the 1970s, Indonesia began to dominate world tropical timber markets and, by 1980, was the world's largest exporter of tropical hardwoods (Pofenberger, 1997). The number of concessions grew dramatically from 1970 (71) through the 1980s (approximately 580). It was during this time that the pattern of widespread corruption and cronyism in the forestry sector took hold. It is estimated that 20 companies held 30% of the leases during this period and that five or six corporate groups dominated the industry. Short lease agreements, the complicated silvicultural practices mandated by the Indonesian selective felling system and labyrinthine bureaucracy made it more attractive for concessionaires to bend the rules than to comply with them. Connections between company directors and the highest levels of government meant that companies were rarely prosecuted for illegal practices (EIA, 2001; Whitten et al., 2001).

During the last fifteen years, Indonesia has undergone dramatic and at times tumultuous changes under a strategy of rapid industrialization. The effects of this strategy can be seen in the forestry sector where plywood and paper processing capacity has risen by 13% and 592% respectively in the last ten years (WB/WWF, 2000; Barr, 2001). Indonesia now ranks as one of the world's leading producers of both commodities. This achievement has come at a high environmental price, including the over-exploitation and deforestation of vast stands of the country's forests. Furthermore, the capacity to process timber is now greatly out of proportion to the forest area designated for management. This over-capacity has been linked as the primary culprit driving much of the illegal logging sweeping through Indonesia (Brown, 1999).

The Emergence of Illegal Logging in Indonesia–The extent of illegal logging is now reaching alarming proportions. At an estimated 30 million cubic meters per year, it accounts for up to half of the national timber production (Brown, 1999). Illegal logging takes many forms: a concession company cutting more than their annual allowable cut or logging in areas outside those designated for logging in annual management plans; unlicensed logging by large industrialised groups or small low-tech groups in closed concessions, water-shed protection areas and protected areas such as national parks. Firms with permits to establish industrial timber or oil palm plantations have frequently profited by clearfelling and selling timber from large areas of natural forest and then disappearing before establishing the plantations (Telapak et al., 2000; WB/WWF, 2000; Dudley, 2001). Indonesia's protected areas have been especially hard hit by this trend because they contain some of the last remaining stands of valuable timber in the country (Jepson et al., 2001; Jepson et al., 2002; EIA, 1999; EIA, 2001; McCarthy, 2001). At the international policy level, donor counties are increasingly concerned about the issue, and some have threatened to cut off forestry-related aid unless the Government of Indonesia (GOI) takes decisive action to curtail illegal logging. At the local level, however, the widespread involvement of practically entire villages in this activity complicates the issue. Furthermore, there is a paucity of data on the trends, ecological impacts, and economics driving the increase in illegal logging at the local level.

While it is important to illustrate the problem of illegal logging in the national context the focus in this paper is at the local level, with village loggers, where we attempt to quantify local participation levels and trends and discuss some of the consequences from illegal logging practiced at this scale. Industrial over-capacity and poverty only partially explain why so many villagers are involved in illegal logging. There are a host of other reasons, which we explore in this paper.

Study Area

Research was conducted in communities and peat swamp forests adjacent to Gunung Palung National Park (GPNP), in the Indonesian province of West Kalimantan (see Map 1). The 90000-hectare park contains a wider range of habitats than any other protected area in Kalimantan and includes coastal mangrove, peat and freshwater swamp, lowland, hill, montane, and cloud forests

MAP 1. Borneo and the Indo-Malayan Archipelago.

(MacKinnon, 1982). The park also contains the largest remaining area of undisturbed lowland forest in West Kalimantan. Due to low hunting pressure (MacKinnon, 1982; Salafsky, 1993), Gunung Palung hosts high densities of endangered mammals such as clouded leopards, sun bears, proboscis monkeys, and the largest population of wild orangutans in Kalimantan (Yeager, 1999). As with many of Indonesia's protected areas, there are numerous threats to GPNP and its associated buffer zone, including overexploitation of forest resources, illegal timber extraction, fire damage, and habitat destruction due to agricultural conversion (MacKinnon et al., 1996).

Until the 20th century, the Gunung Palung region was predominately populated by Dayak indigenous groups, with ethnic Malays living on the coast and in the hills around the town of Sukadana. Human populations have increased substantially since the 1960s due to high birth rates and decreased infant mortality, trans-migration programs, and from people moving to the area to work for timber concession companies (Proyek Pengembangan Kawasan Konservasi di Kalimantan Barat, 1997). In 1994, the total population of the six districts bordering the park was over 170,000 with an average density of 19 people/km^2 (Kab. Ketapang Dalam Angka, 1994).

Our research was conducted in 11 villages bordering the northwest corner of GPNP in the Ketapang regency (see Map 2). A community-based forest management and conservation project to be implemented by the Indonesian Ministry of Forestry and Harvard University planned to work with these communities. The total population size of participating villages in January 1998 was 6,483. Most communities contained around 100 to 150 families with populations of approximately 600 people, although two (Sedahan and Matan Raya) contained over 1000. The majority of villagers consider themselves Malay (81%), with some identifying themselves as Chinese, Balinese, and Javanese. Seventy percent of villagers are under the age of 30 and 39% are younger than 16 (LTFE, 1998).

Historical and Current Land Use–Only in the last three generations has increased migration of Malays lead to the establishment of some of the villages currently found further inland in the study area. Formerly, villagers farmed dry rice and established mixed fruit gardens on the hills surrounding their coastal and riverside communities. In the last 30 years, more interior swamp forests

MAP 2. Gunung Palung National Park and Surrounding Villages.

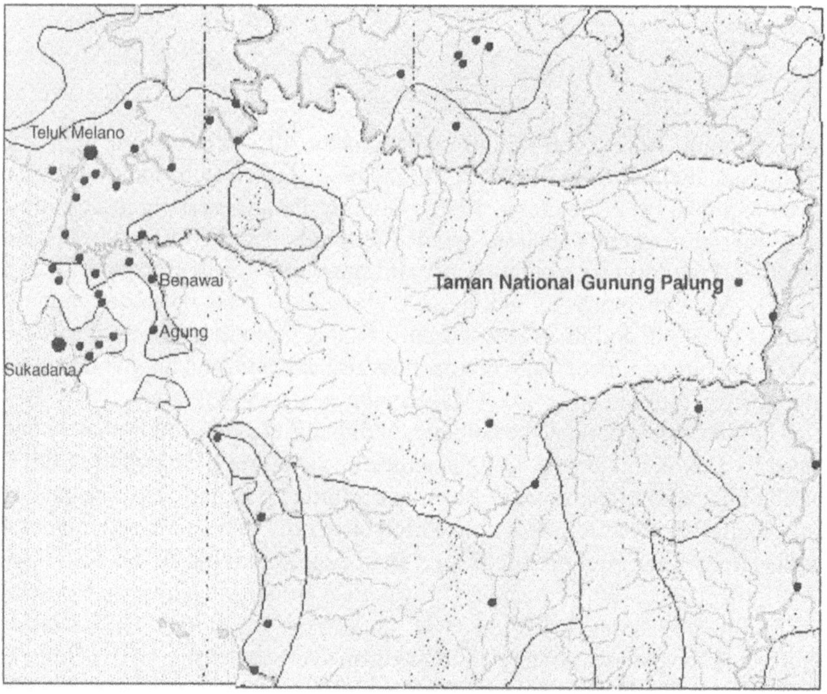

have been opened up for wet rice cultivation, but until very recently this expansion had been confined to the edges of existing villages.

Today, farming continues to be the prevalent economic activity in these communities. More than 85% of households censused grow food for household consumption and income. Common crops include rice, cassava, legumes, and vegetables. Forty-eight percent of households tend forest gardens on the nearby hillsides, which are almost all inside GPNP, harvesting primarily durian, rubber, and coffee. For the people that reported owning rice paddy and forest garden, mean size for either was approximately 1 ha.

For as long as the Malay have lived in the area, they have used nearby forests as a source of timber for building houses and boats. Local community members have also traditionally collected rattan, firewood, resin from agarwood (*Aquilaria* spp.), medang (*Lauraceae* spp.) bark (used for rug manufacturing and insect repellent), wild rubber latex from *Jelutung* trees (*Dyera costulata*), along with various medicinal plants in local forests (Salafsky, Dugelby, and Terborgh, 1993). However, these activities rarely provided a major source of cash income for local villagers.

General Overview of Illegal Logging in the GPNP Region

Beginning in the 1960s, logging concession companies began hiring local villagers to work as laborers in concessions spread throughout the GPNP region. By the late 1980s, timber concessions had removed most of the valuable timber and closed (only one remained in the GPNP region in 1991). Logging frequently continued, albeit illegally, often with former concession employees guiding and providing capital to village loggers. In certain respects, this was advantageous for investors and sawmills still operating in the area, alleviating themselves from onerous forest management activities and planning, substantially reducing their overhead by paying only the log extraction costs. Wealthy local merchants and members of the local police, military, and national park staff also financed logging groups. It is rumoured that these operations were financed and condoned at a high level in the province and nationally, as has been reported in other areas (c.f., EIA, 1999; McCarthy, 2000).

In the Gunung Palung region, illegal logging has been most commonly carried out in small groups of local villagers, using traditional, non-intensive methods and selling timber to wood buyers or nearby sawmills. Community members also fell trees for home use, although most people now buy timber from local traders.

This type of illegal logging is a flexible and adaptable system, providing opportunities for local people to participate when economic needs arise. Roles change frequently and people schedule trips around farming, gardening, cultural activities and family obligations. Logging activities are not fixed to a specific location or time as logging occurs throughout the year. During the dry season, loggers stockpile wood in the forest until it can be later floated out during wet periods.

As many of the Malay villages in the region have only recently been established, few communities have extensive traditional rights/claims to local forest areas. As such, logging for merchantable timber rarely occurs in the small forest tracts managed by communities–i.e., forest gardens. Instead, loggers target areas perceived as ownerless, primarily ex-concession and protected forests. Forests close to communities were logged and degraded in the late 1980s and early 1990s and many have burnt up to three times in the 1990s. As these nearby resources become depleted, loggers move further upriver into areas still rich with merchantable timber. Today, loggers regularly travel more than 20 km from villages to find forest with timber. With low tech extraction techniques (see later in text), loggers can profitably extract selected timber species from areas 15 km from rivers or roads, although until recently most logging in the region occurred within 1 km of riverside.

METHODS

Socio-Economic Study Design

This paper draws on three separate socio-economic studies; (1) a population census undertaken from December 1997 to February 1998, (2) a household survey carried out in March 1998, and (3) comprehensive interviews with village loggers completed in mid-1999.

The population census was conducted in eleven villages bordering GPNP. A total of 1,488 households were interviewed with respect to demographic characteristics and household composition, prosperity and livelihood (household was defined as any homeowner, married couple, or single parent with children). Households were assigned a wealth rank score from different indicators of prosperity, including: land holdings, livestock, transportation, and housing materials.

Following the census, three hundred households were surveyed in greater depth to obtain information on differences in income level, environmental awareness, and perceptions of prosperity between loggers and non-loggers. Fifty households each from six villages were selected by a stratified random sample based on the wealth score generated from the earlier population census.

Interviews with village loggers were conducted from July 1999 through October 1999. Forty-three loggers were surveyed about profit-levels, differences in logging systems, and perceptions regarding trends in illegal logging. The sample of loggers to interview was drawn from four villages: Makmur,

Matan Raya, Sedahan and Pelerang. We chose these villages because of the high number of loggers in the villages, as identified during the census and survey conducted in 1998. The survey was pre-tested with five loggers, questions adapted and then run fully with another 38 loggers.

Ecological Study Design

A 150 ha logging site, active from mid-1997 to mid-1999, was studied in October 1999 through February 2000. The number of loggers working this site varied, at times reaching up to thirty-five local and non-local villagers. We measured the entire log extraction system and all stumps and residual wood left by this group. Age, diameter, and species of each stump were noted. These data were then entered and analyzed in Arcview 3.1™ Geographic Information System.

We installed 52 plots (each 0.2 ha) to assess damage from illegal logging. Plots were stratified by harvesting intensity and distance from river access. For all trees 10 cm diameter at breast height (DBH), we identified tree species, measured DBH, and gave a score for tree health. Additionally, for trees that had died, we subjectively determined if logging damage was responsible for the mortality.

From July to September 2001, we conducted a survey of 12 rivers in GPNP. This represents approximately 20% of the rivers that flow from the park and the survey was stratified to include rivers from the northern, southern and eastern borders of the Park. The western border was not surveyed due to existing data and fears over the safety of the survey team. On each river, the locations of all logging camps were recorded with a global positioning system (GPS) unit. Data on the age and size of the logging operation were gathered from direct observations and interviews with loggers. Information was also collected on the home village of the loggers and the style of wood extraction.

RESULTS

Community Participation and Logging Rates

Illegal logging is now the major industry in the region and for a large proportion of villagers, it is a staple income source. In total, 40% of households had members who stated that logging was their primary source of cash income. This figure should be considered conservative, as some villagers may have been hesitant to admit their involvement in this illicit activity to project researchers. Logging teams reported working on a trip-by-trip basis totally 6 months a year. Each trip lasted an average of 30 days and resulted in an average of 40.1 m^3 of timber per trip. These figures suggest that villagers are more dependent upon logging as an income source than studies from other areas of Kalimantan, where locals were reported working only one to two months a year (Obidzinski and Suramenggala, 2000).

From the above figures and data from the population census, it is estimated that at least 38,000 m^3 of timber was extracted from the national park and surrounding forests by the 11 surveyed communities in 1999. These communities represent 2.2% of the population in districts (*kecematan*) surrounding GPNP, and thus, the total amount of illegal extraction in the area is likely many times higher than the above figure.

The preceding figures also do not consider the percentage of people employed in wood processing industries, approximated at an additional 8-9% of households. Nor, have the merchants that sell chainsaws, spare parts and gas and oil been accounted in the above percentages. Based in on this information, the majority of the village population is directly gaining economic benefits from logging. More importantly, illegal logging has become an integral part of the local economy. These findings corroborate those from a study in East Kalimantan, where illegal logging at the district level had a huge impact on the local economy (Obidzinski and Suramenggala, 2000). Many of these people have been logging all their lives. Some were brought to the area by the logging companies in the 1970s and 1980s and settled there; others were hired and trained locally, either by contractors or by family members. The only trade they have is logging and there are few others available. The only comparable cash generating activity involving large numbers of young men and heads of household is illegal gold mining. Unfortunately, this activity also has extreme environmental and public health impacts that warrant further investigation and remediation.

At the time of the census, timber extraction permits were only granted to companies with active forest concessions, and none were operating in the area after 1997. Thus, all logging in the western region around GPNP is illegal. Despite this inflow of cash to the local economy, it should be remembered that the community as a whole does not receive benefits from this activity; unlike legal logging under the concession system, illegal loggers pay no taxes and do not fund development projects. Eventually, most of the timber will have been extracted in and around the park, and these village loggers will be especially disadvantaged, without sufficient means to earn cash from alternative sources.

Profit Levels and Wealth

From the logger survey in 1999, respondents reported that average earnings for laborers (excluding chainsaw operators) were approximately Rp510,000 per trip or Rp17,300 per day (US$2.30 at time of survey). The minimum daily wage for this area in 1999 was Rp7,000. The mean yearly logging income for laborers was estimated at 3 million Rupiah (US$408). Figures reported are less than those for logging in East Kalimantan, where loggers were reported to earn almost double that (Obidzinski and Suramenggala, 2000). However, that study did not differentiate between wages for laborers and chainsaw operators or *kepala rombongan* (head of the logging group in the field). In our study area, chainsaw operators earned approximately 3-5 times the amount that laborers earned. According to respondents, logging bosses sold illegal timber to sawmills for twice the price paid to villagers.

This system tends to promote indebtedness to moneylenders. Loggers often have to return to the forest again to repay debts and support their families. Data collected by the community based forest management project (LTFE, 1998) indicate that households dependent on logging for their primary source of cash income tend to be the poorer families. In an area with few other options, people log to earn cash. There is a widely held belief that logging is a good way to get rich (possibly originating from seeing the large houses and obvious wealth of logging bosses). These data, however, show that loggers tend to be poorer, and earn less than the average villager. Classifying household wealth by income, land and material goods, the data show that 39% of households not dependent on logging were classified as "poor," whereas 52% of households dependent on logging were "poor." In a more detailed survey of 5 hamlets, the average income of households who indicated that their primary source of income was logging, was considerably less than that of the average household income for the village.

It appears that loggers frequently overestimate income earned from logging. The reasons for this were not examined. However, it is possible that loggers fail to properly account for the debt they have with the "boss," or the interest they are paying on loans (Over 92% of loggers reported being in debt prior to logging trips). Examining the monthly expenditure and spending patterns of a household may be a more useful indicator of wealth as families with a higher income or that simply feel wealthier will spend more on non-essential items. Average household expenditure per month for households dependent on logging is significantly less than other households ($n = 170$, $p < 0.026$). Various authors have stated that poverty is another factor forcing villagers to illegally log (Brown, 1999; WB/WWF, 2000; Obidzinski and Suramenggala, 2000; EIA/Telapak, 2001). We investigated this assumption by examining population census data to determine whether loggers are disproportionately poorer than other villagers.

We chose to use a wealth score generated from a combination of different prosperity indicators, such as amount of such as amount of cultivated and fallow land, livestock, transportation, and type of house construction materials. Even though in the later village survey, income data was recorded, we believe a ranked wealth score more accurately and objectively indicates prosperity.

Furthermore, since many of the prosperity indicators take years to accumulate, the combined wealth score should provide a better picture of villager economic status before the massive rise in illegal logging. Statistical analyses show highly significant differences in wealth scores between loggers and non-loggers (Kruskal-Wallis Tests, ranked sum, $Z = -8.09$, $p < 0.001$).

Thus, loggers do appear to be poorer than other villagers who chose not to participate in this activity. Examining individual prosperity indictors provides some clues in determining why loggers are poorer than other villagers. First, we tested to see if loggers are recent migrants to these communities and thus, would not have had time to accumulate wealth. This did not prove true, and in fact, 73% of loggers censused have resided longer than 20 years in these communities versus 63% for other villagers (Pearson Chi Sq Test, $\chi^2 = 18.9$, $p = 0.0003$). Secondly, we tested differences in mean agricultural land owned by loggers and non-loggers and found that loggers own significantly less land (1.5 ha vs. 2.0 ha for non-loggers, $F = 22.99$, $p = 0.0001$). Since income is earned in selling fruits and vegetables rather than rice, we separately tested the mean size of forest gardens with similar results; non-loggers have larger forest gardens, on average than loggers (0.36 vs. 0.59 ha, $F = 21.2$, $p = 0.0001$). Lastly, we analyzed the livestock scores between the two groups and found that the trend also remained the same, but the difference is much greater (logger 46.1 vs. 140.9 mean score for non-logger, $F = 43.09$, $p = 0.0001$). Five different types of livestock were recorded during the census, from chickens, representing the least valued to cows, the highest valued. Scores were assigned based on a ratio of market price for a chicken versus the four other livestock. The above scores can be interpreted as loggers own the "equivalent" of 46 chickens while non-loggers possess the mean equivalent of 141 chickens. Livestock is a flexible and easily traded commodity, so that if a farmer needs extra money he can decide to sell some of his livestock, or if market prices happen to fall for a period, he can decide to forego selling until prices rise.

In summary, loggers are not newly arrived residents; the majority of them have lived in these communities for over twenty years. But, they do appear to own fewer tangible assets, and besides income from logging, have fewer options for earning cash.

Whether they are poor because they log, or log because they are poor is not clear. Loggers who spend long periods of time in the forest may not be able to manage large areas of farmland or build up numbers of livestock. In debt to bosses and with little or no land they may have no alternative than to continue logging. Respondents reported that with scarcity of timber close to home villages loggers must travel further and stay away from home longer while working in the forest. Prolonged absence of heads of households reduces the ability of families to engage in other activities such as farming. Additionally, many respondents say that logging is what they do and will continue to work as a logger even if they were much more wealthy, indicating that simply providing alternatives or increasing agricultural productivity alone may not stop people from logging. Some people may simply prefer to log for a source of income.

Even if illegal logging is not brought under control and timber supplies dwindle to uneconomical levels, village loggers will be especially disadvantaged and will have few livelihood alternatives.

Socio-Economic Organization of Logging

The origin, size, and structure of illegal logging teams is largely dependent upon different methods of financing. There are currently two basic forms of logging carried out by villagers in and around GPNP, independent and *tauke*, although a third is increasingly frequent, which we term an incentive-based system (Figure 1).

The two prevalent forms, independent and *tauke* (head boss and financier) systems are similar in most practical aspects. In both, small groups of local men construct *jalan kuda-kuda*, a wooden rail system (Whitmore, 1984; Cannon, 1994), to manually push and pull logs on wooden sleds from felling sites to river access. A team is responsible for all of the associated work-building *jalan kuda-kuda*, felling trees, skidding logs to the river, squaring logs into heavy planks, and floating wood to a predetermined point along the river.

The distinction between systems lies in the financing of the trip and the marketing of the wood. Independent loggers have either enough capital saved or access to funds to self-finance logging operations. They often own their own chainsaws and can choose the location, tree species, and form of wood to sell (logs, flitch wood (cut into squared blocks), or planks). They are free to sell their timber to sawmills or on the local market and, during periods of fluctuating prices, can save their wood until more favorable market conditions appear. Independent groups often purchase supplies (petrol and food supplies) on credit.

Loggers working under the *tauke* system normally do not have enough money or access to credit to pay for the necessary equipment and supplies-roughly estimated at Rp1,125,000 (US$150) per group over a 30 day trip in 1999. These men are forced to borrow money from lenders (with interest rates as high as 40% per month) or work for local, and now more commonly, non-local investors. Wealthy businessmen pay for supplies and equipment, give large advances on future wages, and often rent teams a chainsaw in return for having the right to buy their wood production at a set price. Like other debt peonage systems, *tauke* groups receive a price well below that of independent groups and often end up working continually to pay back their debts (Salafsky, 1993) (see Table 1).

FIGURE 1. Flow chart indicating the flow of money and wood from forest to market in a *tauke* system.

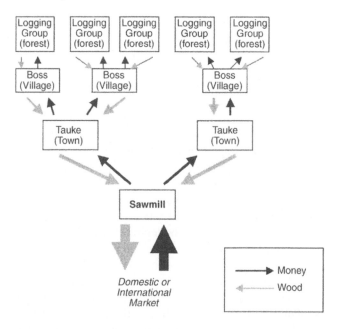

TABLE 1. Comparison of two logging finance systems commonly employed at GPNP in 1999.

DESCRIPTION	INDEPENDENT SYSTEM			TAUKE DEPENDENT SYSTEM		
	MEAN	S DEV	n	MEAN	S DEV	n
Team Size (People)	3.6	0.69	10	5.1	1.35	23
No. Trips (Trips/Year)	9.6	3.7	12	4.9	3.28	31
Trip Length (Days)	20.2	3.44	12	32.9	12.45	31
Volume Sold (m³/Team/Trip)	29.5	13.84	10	45.1	36.9	23
Price Range[a] (1000 Rp/m³)	200-350			80		
Revenue[b] (1000 Rp/Team/Trip)	8,266	3,304	8	4,264	3,196	22

[a] Price obtained represents the 25th and 75th percentiles of reported prices across species.
[b] The authors calculated mean revenue by logging system by first calculating real revenue per respondent (# cubics sold*price for each species sold on the loggers last trip). Individual revenues were then grouped by financing system to calculate mean revenue per system (Rp. 7,500 = US$1).

We find that independent loggers' revenue is almost twice as high as that of *tauke* loggers, despite the smaller volume of wood sold per trip and the close range in prices. There are two explanations for this apparent discrepancy. First, independent groups receive a significantly higher price per unit volume than *tauke* groups. Secondly, there is a price difference between lower-value peat swamp species versus lowland timber species; and the majority of independent groups (67%) choose to work in lowland forests to the north of GPNP for this reason. Interview data with loggers working within both financing systems confirm our findings that independent groups are making more money. More than 75% of surveyed loggers preferred to work independently. Of these, 88% stated that independent loggers earned more money or benefited from the freedom to sell wood anywhere-enabling them to receive a higher price. No loggers interviewed believed villagers were able to obtain larger profits under the *tauke* system.

A third system of financing, which we term "incentive-based logging," is also organized by wealthy financiers and is commonly used in areas far from roads or rivers. In this system, a variant of the *tauke* organization and finance structure is being adopted to obtain higher profits from their investment.

This system of financing is more structured and rigid than the independent and *tauke* systems. Loggers are hired to perform particular tasks and are paid piecemeal-usually per kilometer for trail construction or per m^3/km for extracting or floating out timber. Workers must purchase food from the *tauke* (usually at inflated prices) and are forced to commit to work for months at a time. *Taukes* often bring in outside loggers from other regions of West Kalimantan because they are reputed to work harder and because they present a lower risk of returning home before the end of their commitment.

Incentive-based groups often use more sophisticated extraction methods to remove timber from areas greater than 2 to 3 km from river or roadside. Both methods, called *jalan sepeda* and *jalan gerobak*, are used in areas where forests are heavily depleted, as in other areas of West Kalimantan. However, these methods have only recently been introduced in the GPNP region. One local logger stated that, in fifteen years of working in this area, he had never seen these methods used before. These methods are still labor intensive, but are more efficient over long distances than the *jalan kuda-kuda* system. This system enables wood to be extracted from far deeper in the forest than *jalan kuda-kuda* thus greatly increasing the area accessible to illegal logging. In some cases, financiers pay for the construction of extensive extraction systems and charge local teams a fee per cubic for use.

In one of these new systems, loggers walk out flitch wood loaded on top of modified bicycles on the *jalan sepeda* or "bicycle trail." A single man can carry two to three pieces of small-sized flitch wood several kilometers by this method. The trail is constructed using the residual wood (lapping) from logs squared into large planks or flitch by chainsaws. Few additional trees are felled for skid trail construction. Only about a one-meter wide section of forest is cleared to enable the bicycle and operator to pass through. Thus, this method is considered less damaging than the *jalan kuda-kuda* (personal observation). On a *jalan gerobak*, loggers wheel out three to four pieces of flitch wood using a small cart. This system is built on top of a *jalan kuda-kuda* and requires additional timber for skid trail construction (Table 2).

Markets

Independent loggers can sell illegally harvested timber to numerous sawmills in the regency (*kabupaten*) as well as to wood buyers and villagers in the community. Like other goods and services in the region, information regarding changes in timber prices from local traders and area sawmills is widely known. Although there is an open, free market and an informed population, rarely do local villagers receive a fair market value for the timber they sell. Most significantly, most locals lack the capital to finance their logging trips and are forced to accept low prices from *taukes. Taukes* are their only source of seemingly affordable credit, but to get it they must sell the wood back to the *tauke* at lower prices. While the credit may seem affordable, in effect it is not since the lower prices make their debts more difficult to repay.

In 2000 nine known sawmills were located in the GPNP region and many more located in the Ketapang regency. In 1999, sawmills primarily bought meranti (*Shorea* spp. and *Diptocarpus* spp.), ramin (*Gonystylus bencanus*), keladan (species unknown), jelutung (*Dyera costulata*) and bengkirai (*Shorea*

TABLE 2. Comparison of different extraction methods observed in the GPNP region.

	Kuda-Kuda (Sled)	Jalan Sepeda (Bicycle)	Jalan Gerobak (Small Cart)
History in GPNP	Traditional system, used for decades	Recently Introduced (late 1990s)	Recently Introduced (late 1990s)
# people required to pull timber	5-6	1	1
# m³ per trip	0.75-2.0	0.25-0.4	0.3-0.5
Maximum observed harvesting distance from river/roadside	2.5-3.0 km	15 km	15 km
Volume moved one kilometer per day per person	1.0-3.0	3.75-8	4.5-10

leavis). The wood bought by sawmills is rarely sold in local markets; rather it is sold as flitch or planks in Pontianak (the provincial capital) or beyond, usually to plywood companies.

Villagers can sell a wider variety of wood in local communities, including meranti spp., mengkurai, punak (*Tetramista galabra*), nyato spp. (*Palaquium* spp. and *Ganua* spp.), bedaru (*Xylopia* sp.), pekakal (*Calophyllum* spp.), and mayam (species unknown). Most of these timbers are sold as plank or beamwood and used for home and boat construction. Roundwood is rarely sold in local villages.

Trends in Community Participation

In the 1990s, throughout the forested regions of Indonesia many concession licenses expired and large numbers of other timber companies ceased operations (WB/WWF, 2000). They left behind high-graded forests very low in merchantable timber volume. Logging has continued, albeit illegally, and there has been an explosion of sawmills processing this illicit wood (Obidzinski and Suramenggala, 2000). Many of these new sawmills have permits from regional Department of Industry and Trade offices (World Bank, 2001), but few have legal sources of timber to process.

These changes have occurred in the GPNP region. By 1991, only one concession was still operating on the northwest border of GPNP. However, practically all of the sawmills continue to operate and at least four new sawmills were recently constructed around GPNP in the last two years. These mills are totally dependent on illegal sources of timber for their supply.

Despite the reduction in legal forestry jobs following the pull-out of local concessions and the loss of thousands of hectares of forests due to fire and conversion to agriculture and plantation crops local communities are now more dependent upon logging to meet household economic needs. In the villages of Benawai Agung, both the total number and percentage of households participating in logging substantially increased from 1991-1998 (see Table 3 and Figure 2). The 1991 data include households involved in legal forestry operations as well as illegal logging. The increase in localized illegal logging therefore is higher than the 72.5% increase in number of logging households noted here as all logging activities in 1999 are illegal.

Small local logging teams, those characteristic of the *independent* and *tauke* system, sell more timber and make more money than similar organized groups ten years ago. Groups are now regularly working throughout the year; while in years past logging only occurred during breaks in the agricultural cycle. Logging trips are longer, and, with the currently widespread availability of chainsaws in local communities, are able to produce twice as many cubic meters of wood per trip (Table 4).

TABLE 3. Temporal changes in small-scale logging in Benawai Agung Desa, GPNP region, between 1991 and 1998. (Source data originates from Benawai Agung collected by Salafsky (1993) in 1991, and from the LTFE population census conducted in 1998.)

Description	1991	1998	Percent increase since 1991	Significance (Chi sq.)
Total households	522	614		
Households w/ substantial income from logging	138[a]	237	71.7%	$\chi^2 = 5.17$, $p = 0.022$

[a] This value represents the 'Predicted Total' number of households receiving income from logging, calculated by multiplying the percent of surveyed households involved in logging (26.4%) times the number of households in Benawai Agung at that time (522).

FIGURE 2. Percentage of logging houses in Benawai Agung.

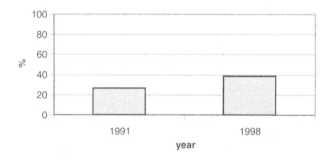

TABLE 4. Temporal changes in logging production and revenue for peat swamp forests in GPNP region between 1991 and 1999. All prices have been corrected for inflation and are reported in 1999 Indonesian Rupiah. 1999 data contains information from both independent and tauke logging teams. (Source data from Salafsky (1993) in 1991, and from a 1999 LTFE logging survey.)

Description	1991	1999
Tools for felling	Chainsaws/Axes	Chainsaws
Team size	6	5.4
Average trip length (days)	14-21	38.3
Average volume sold per team per trip (m³)	10-20	50.3
Price received (Rp per m³)[a]	62,700-104,500	80,000-200,000
Average revenue per team/trip	836,000-1,672,000	5,249,714
Time of year logging occurs	September to January	Year-round

[a] Price obtained represents the 25th and 75th percentiles of reported prices across species.

Increase in Occurrence of Incentive-Based Logging Groups–In the past, illegal logging was organized around small groups of local loggers extracting small amounts of timber from nearby forests. However, following the recent fires and decades of concession and illegal logging, most remaining accessible forests close to river and roadsides have already been high-graded (Salafsky, 1993). Loggers must now go farther to get to forests with harvestable timber and invest heavily in building extensive extraction trails. Large, well-organized groups made up of non-local loggers, and in some cases with Malaysian capital, are now filling this "niche" for illegal logging. These groups, rare even five years ago, can now be found in areas throughout the GPNP region.

Large incentive-based groups have the ability to work in remote and relatively undisturbed forests still rich in timber. Once the investment in time and money has been made constructing the extraction trails, loggers can work in locations for months to years. These groups are much larger–as many as 90 loggers have been seen in one site along the northwest corner of GPNP and sites in the south have had as many as ten teams working at once. Groups focus on a limited range of species and work until few to no valuable trees remain.

Ecological Impacts of Illegal Logging

Park-Wide Situation–The survey of 12 rives in GPNP found 254 active and abandoned logging sites. Ninety-eight percent of these had been logged by people from the local area (within the district). These sites showed an increase in the size of the logging operations and an increase in number in the last few years. Sixty percent of the sites were less than three years old. Additionally there appears to be a trend toward increasingly large logging operations: 70% of the sites that were classified as 'large' were less than 3 years old, and none were more than 6 years old (Figure 3).

FIGURE 3. Proportion of different size networks by age.

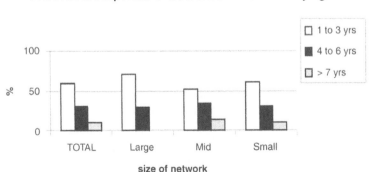

Indonesia has lost millions of hectares of natural forest from numerous large-scale forest fires. The 1983 Borneo forest fires were reported to be the largest in recorded human history, only to be replaced by the 1997 fires in Borneo and Sumatra (WRI, 2000). Three large-scale fires, in 1991, 1994, and 1997, have burned through the GPNP area in the last ten years, destroying thousands of hectares of forest.

Due to drought conditions in 1997, forest fires were particularly destructive, burning several thousand hectares in and around the GPNP. One particularly destructive fire destroyed more than 2,000 ha of peat swamp forest that previously had been selectively logged. Furthermore, 30% of survey respondents stated that their forest gardens were at least partially burned in the 1997 fires. This loss is particularly significant since the majority of farmers' cash income comes from fruit sales from their gardens (LTFE, 1998). Reduced forest coverage from the wildfires caused streams that supplied drinking water to several villages to run dry. In Sumatra, illegal logging was directly linked with water shortage problems in South and North Kluet areas (McCarthy, 2000).

Case Study from One Logging Site–The most notable change in small-scale logging in the past ten years has been the substantial increase in the number of trees per hectare extracted in addition to much larger areas being logged. Ten years ago, direct environmental impacts from this activity were limited to areas close to rivers. Eighty percent of trees felled in a site within GPNP were located within 500 meters from river access (Ravenel, unpublished data). The 1999 survey of one logging operation revealed that one network of extraction trails (*jalan kuda-kuda*) had 30km of trail and nearly 1,200 stumps, 97% of which were located beyond 500 meters.

Damage to the residual stand is commonly used as a rapid measure of the effect logging has on the flora. It has been well documented that as harvest intensity rises, damage to the residual stand increases proportionally (Sist et al., 1998). This is particularly true in lowland Dipterocarp forest where commercial volumes can reach 100 m³/ha and damage can be severe. However, in peat swamp forests, tree heights, crown widths, and wood volumes are lower than in lowland forests. Timber extraction in peat swamp forests is also carried out manually without heavy machinery so skid trail construction and log extraction do not cause as much collateral damage. Logging damage, therefore, is expected to be lower in peat swamp forests than in lowland forests.

An earlier study of illegal logging examined an area of 280 ha immediately across from the study area that was logged by small teams from 1985 until 1997. The mean harvest intensity in this earlier logged area was 2.5 trees per ha (Ravenel, unpublished data). The results from the 1999 location reveal that averaged over the entire area, 8.7 trees/ha were extracted, but harvest intensity varied greatly between plots, ranging from slightly over zero to 35 trees/ha. We chose to install plots in this particular area because the logging reflects a change in pattern whereby larger teams are penetrating further from river access and taking more wood volume.

At a harvest intensity of 2.5 trees per ha, fully two-thirds of the trees remain unaffected by logging (Table 5). Additionally, only 4% of all trees are killed from this operation, while the slightly damaged trees will probably recover. However, as villagers extract an average of 9 trees per ha then only half the stand is left unaffected by logging, and skidding and felling operations killed 13% of the total individuals. As harvest intensity increases to 20 trees per ha, logging damage leaves only 40% of the trees untouched, and more importantly, outright destroys or severely damages 28% of all trees over 10 cm DBH. This intensity covers approximately 10% of the study area and is an indication of what will happen to accessible forest in the near future, if illegal logging is not controlled. Since this research was completed, new groups of loggers have re-entered the study site and are harvesting other species there, thereby increasing the overall harvest intensity, probably close to this 20 trees per ha value (Figure 4).

The illicit nature of this activity means that loggers work as quickly as possible without concern for the residual stand. Because they have no secure ten-

TABLE 5. Harvest intensities and associated damage to residual stands.

Damage Levels	Percent trees from total by harvest intensity (HI)		
	Low HI = 2.5 trees/ha	Medium HI = 9 trees/ha	High HI = 20 trees/ha
Killed or severely damaged	4.0	13.2	27.5
Other[a]	28.7	30.6	32.9
Undamaged	67.3	56.2	39.6
Total	100.0	100.0	100.0

[a] Other includes natural and unidentified mortality, moderate and low damage classes, all trees greater than 10 cm DBH.

FIGURE 4. Distribution of stump distance from river (1996 data from Ravenel).

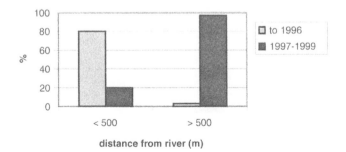

ure over these forests, loggers take little care in building extraction routes and felling trees. Our plot data from a large incentive-based logging group indicates that to extract on average nine merchantable trees/ha, loggers destroyed or severely damaged almost 23% of the future crop trees during skid trail construction and tree felling (Figure 5). If we examine only highly valuable species, mainly *ramin* (*Gonystylus* spp.), then this damage level rises to almost 59% of future crop trees lost due to careless logging practices (Figure 6).

This damage to the residual stand alters the path of stand development such that more non-commercial trees will achieve canopy dominance. This process reduces the economic value of these forests and increases the likelihood of conversion for agricultural use. The 1999 spatial plan for West Kalimantan (RTRWP West Kalimantan, 1999) classified most of the area surrounding GPNP for conversion to oil palm plantation. The land planning agency considered this area as degraded forest with no value and the only viable alternative was large scale plantation crops (BAPPEDA *personal communication*). Additionally, this harvest intensity increases the risk of catastrophic wildfires that could destroy the forest. The increased size and number of openings in the forest allows more sunlight to reach the forest floor. The humidity decreases and litter dries more quickly. Additionally the waste timber from logging greatly increases the quantity of litter in the forest, increasing the amount of potential fuel (Uhl et al., 1988, Whitten et al., 2001).

DISCUSSION OF FACTORS
CONTRIBUTING TO THE INCREASE IN ILLEGAL LOGGING

Based on the analyses described above, discussions with a variety of stakeholders, including loggers and government officials, and three years of experi-

FIGURE 5. Impact of differing harvesting intensities on stems > 10 cm DBH.

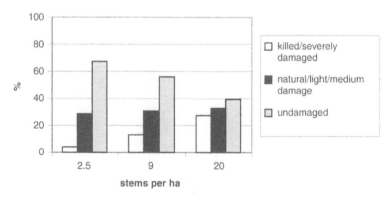

FIGURE 6. Impact on different tree types at harvest intensity of 9 stems/ha.

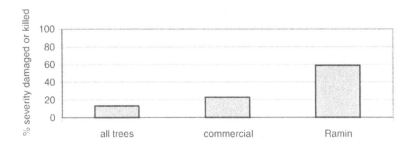

ence in the field searching for solutions to the illegal logging of a National Park, we conclude that there are two key (and several associated), factors driving the illegal logging activity at in GPNP; access and economics. Many other factors also influence illegal logging, especially at other scales (McCarthy, 2000; Smith, this volume). The following discussion focuses on local-level dynamics in the GPNP area, but many of our observations are relevant to other areas of Indonesia as well.

Access

It is clear that logging by villagers requires less organization, capital, and infrastructure than large-scale commercial operations. However, there are still five critical needs for illegal logging to occur. Local loggers need access to all of the following: forest resources, capital, a labor force, equipment, and wood markets. If villagers lack access to any of the above requirements, illegal logging cannot take place.

Most of these requirements have rarely been prohibitive in the past. Village loggers adopted low-technology extraction methods that required little capital investment and cheap labor has been available in months prior to rice harvests. In some respects, access to forest resources has become an increasingly limiting factor. However, in most cases over the past 10 to 15 years, access to each of the five critical access needs have improved or increased. These changes, directly or indirectly, have contributed to increased levels of illegal logging in the GPNP region.

Access to Forest Resources–A combination of physical, legal, and sociopolitical factors determines local communities' access to forest resources. Most importantly, villagers must have forest areas to work in. Large forest concessions granted to timber companies in the GPNP region in the 1970s and 1980s reduced access to local forest areas. Although illegal logging may have occurred in these forests during this period, concessionaires controlled local use to manageable levels. Villagers' access was reduced by fires in the 1980s and 1990s. Tens of thousands of hectares of forest used by locals for timber and as forest gardens were destroyed. By the mid-1980s, local villagers had little or no legal access to timber resources.

By 1997, no forest concessions were operating in the region. As concessionaires pulled out, they left behind three critical ingredients: a group of relatively wealthy people capable of investing in illegal logging, a large contingent of villagers trained as loggers and an extensive network of roads and cleared waterways.

Currently, villagers have no more legal access to their forest resources than they did in the 1980s. What has changed is the perception and effectiveness of law enforcement. The weakening of the central government following the fall of the Suharto regime has led to a climate of near lawlessness in West Kalimantan. Ethnically motivated murder, mob violence, and intimidation occur regularly and with impunity (Peluso and Harwell, 2001). Widespread corruption occurs within every level of government authority. Local military, police, and government officials' active involvement in illegal logging further weakens the legitimacy of national laws and fuels the perception of lawlessness.

Under these circumstances, villagers believe they can "mine" protected and ex-concession forests with impunity. There is little fear of prosecution or punishment. The Department of Forestry and the National Park offices make little effort to manage or protect these areas. Patrols are infrequent; those conducted are often used as opportunities to extort money from loggers. There are no checkpoints along rivers entering TNGP and the few guard posts scattered around the 280-kilometer border are normally unmanned. Logging teams move in and out of the national park freely and regularly use its rivers to transport logs. Forest areas in the national park and in the abandoned timber concessions that surround it have become an open-access resource.

Decentralization initiatives have transferred authority over resource management to regional governments and have led to confusion and have weakened accountability for protecting forest resources. Government departments claim it is unclear which agency is responsible for forest protection. Villagers also use decentralization as an excuse to illegally enter forests, stating that the state does not have the authority to supersede local people's rights to their forests (see McElwee, this volume).

Even among Indonesian conservation organizations and local governments, there is growing acceptance of illegal logging by villagers within protected forests if wood is for "village" use. With the sporadic and unreliable nature of patrolling and monitoring in the region, this exception creates a grey area that is frequently abused and thereby makes efforts to control and curb illegal logging in protected forests even more difficult. In summary, the social, political, and legal climate of the late 1990s and early 2000s has enabled local communities to openly disregard national laws and to seize control of forests with little risk of prosecution.

Access to Labor Force–The construction of a road linking local villages with the regency capital Ketapang led to the transfer from a largely barter system to a cash economy in local villages. This change forced many villagers who did not own fruit, vegetable, or rubber gardens (approximately 50%) to search for sources of cash. Large numbers of loggers employed by timber concessions during the 1970s and 1980s also found themselves without cash incomes in the 1990s and began illegally logging. The rise in prices of most goods caused by the South-East Asian economic crisis in the late 1990s may have also forced locals to try to supplement other income sources with illegal logging.

Access to Capital–Three factors have led to an increase in the capital available to local loggers. Most importantly, the low risk now associated with illegally logging protected forests has encouraged local and outside investors to finance logging teams. Secondly, the cash economy has put more money in the hands of local villagers–giving creditors and suppliers more opportunities to loan money or sell items (e.g., chainsaws) on credit. Finally, the reduction of legal timber producers in the GPNP area has probably led to higher prices for illegal timber from area sawmills, again encouraging investors to lend to local logging teams.

Access to Wood Markets–Marketing illegal wood in local communities has become substantially easier and more profitable in the past 5 to 10 years. The local supply of legal logs has dwindled following the closing of most timber concessions in the region. Mills run by the former logging companies and a number of new illegal mills are now forced to purchase illegal timber and are competing for these smaller quantities.

The improved road connecting local villages to the regency capital has given loggers greater access to forests and a less expensive method of transporting wood. Areas of forests previously un-logged are being harvested from entry points from the road. These forests were previously too far from navigable rivers, so high extraction costs made logging in these forests economically unviable. The improved road serves as an efficient transportation method to move this wood to area sawmills.

Access to Equipment–Finally, the one piece of major equipment used in small-scale logging, the chainsaw, is now widely available in local communities. While ten years ago loggers were using chainsaws, hand saws and hand axes, chainsaws are now available for rent or for purchase on credit. Chainsaws have allowed village loggers to fell and process a far greater amount of wood per trip, increasing both the number of cubic meters sold and revenues earned per trip.

Economic Causes

A range of other causes is associated with the tremendous rise in illegal logging, which we group into macro and micro-economic factors. Many of these economic factors came into effect around the same time, thus multiplying the rising trend in illegal logging.

National Economy–From 1990 to 1997, timber processing rose 13%, while at about the same time Indonesia lost 17% of its forests, or 1,700,000 ha (Brown, 1999). Generous State subsidies, the 1985 log export ban, followed by a 100 percent export tax on sawn timber fostered an expansion in plywood processing capacity far larger than the legal and sustainable supplies of raw logs. This imbalance is even more obvious when examining the pulp and paper industries, where capacity in the last ten years rose by 700% and 590%, respectively (Barr, 2001).

By 1997, logs from forest conversion accounted for over 34% of the total supply. In West Kalimantan, the log deficit is currently 2.9 million cubic meters without including the demand from the vast number of illegal small and mid-sized sawmills. Despite this reliance on imported logs and officially sanctioned forest conversion, demand far outweighs supply, leaving a shortfall of 32 million cubic meters (WB/WWF, 2000). This deficit is made up from illegally obtained logs.

The collapse of the Indonesian currency, the Rupiah (Rp) in 1997 has also had an impact. Initially many sawmills closed due to the prohibitive cost of imported spare parts (McCarthy, 2000). The great increase in the price of wood in Indonesian Rupiah, however, made illegal logging a very profitable enterprise. Furthermore, local inflation did not rise as rapidly as wood prices. In effect, while costs remained relatively constant, the value of the wood increased several fold.

Local Economy–The abovementioned situation of over-capacity exists at the local-level in the Ketapang regency. From 1990-1998, annual plywood processing in this regency rose from 27,700 m³ in to 167,000 m³, a 500% increase (Ketapang Forestry Office *personal communication*). The preceding figure does not account for the large number of unlicensed sawmills operating in Ketapang. Although the total number in the Regency is unknown, at least six were operating on the northwest border of GPNP in 1999. To gain a modicum of legality, many of these sawmills apply for wood-processing permits from the Department of Industry, intended for secondary stage processing, such as furniture or plywood production. However, in reality, illegal mills are only transforming illegal logs into sawn timber. With these permits, mills are not required to provide documentation about the chain of custody of wood sources.

While industrial wood processing shot up during the 1990s, the number of operating timber concessions actually declined sharply. Total production for all types of wood products in 1998 was 368,000 m^3 in Ketapang (Ketapang Forestry Office, *personal communication*) and the maximum production of concession wood was estimated at 100,000 m^3. These figures are for processed wood. To obtain round wood cubic meters, a recovery rate of 50-65% should be added back to the production figures. In 1999, only a few timber concessions continued to operate in the Regency. Therefore, industrial demand for logs is met primarily from forest clear-felling permits. No more than 175,000 m^3 could have been obtained in 1998 (Ketapang Forestry Office, *personal communication*), leaving a deficit in log demand of almost 90,000 m^3. This volume deficit was almost certainly was illegally obtained.

Villagers' increased need for cash income impacted local resource use and altered household economic strategies. A road completed in 1989 connected these communities to the city of Ketapang, enabling villagers to consistently sell fruits and vegetables from their forest gardens. Increased access to markets changed the local economy from one based predominantly on barter to an economy increasingly dependent on cash (Salafsky, 1993). Residents of the area now need cash, for which there are few sources. Almost 80% of censused villagers farm rice. However, few villagers have surpluses to sell. Only 48% own forest gardens, an important cash earning resource. Thus, more than half of the heads of households at the time of the census had few or no means of earning cash.

Illegal logging is an attractive option as it provides a relatively high income. In Ketapang regency, a villager working as an illegal logger will earn far more than the local minimum wage. The collapse of the Indonesian economy in 1997 and the subsequent reforms imposed by the World Bank and IMF have also had an impact at the local level. Inflation of the cost of basic goods such as food and increasing fuel and electricity costs has further exacerbated the need for cash.

CONCLUSIONS

In relation to the questions posed in the introduction of this paper, the following summary conclusions can be made:

1. The level of illegal logging has increased in the Gunung Palung area in the last ten years, with respect to the number of logging operations in the Park, the size of each operation, the volume of wood extracted and the number of villagers involved.
2. The logging is being performed almost exclusively by people from the villages surrounding the Park, but is organised in many cases by people

living far from the Park, and is in some instances financed by people in other provinces or countries.

3. Loggers are disproportionately poorer than non-loggers in terms of annual income and a variety of wealth indicators.

4. It is estimated that more than 50% of local people are financially involved in various aspects of the illegal logging business. After farming, illegal logging is the most important source of cash income in the area.

5. Illegal logging is having a major impact on the forests of Gunung Palung. It is disturbing one of the only remaining megadiverse lowland dipterocarp forests in Kalimantan, perhaps most dangerously by opening them up and leaving them more susceptible to forest fires. Outside the Park, haphazard illegal logging of closed concessions is removing the remaining valuable timber and eliminating the possibility of sustainably managing those natural forests. Government planners perceive that the only land-use option remaining is clearing the natural forest for plantations of oil palm or other monocrops.

6. A complex mix of economic, social and historical factors is driving the increase in illegal logging.

Law enforcement is widely proposed as the solution (Bruner et al., 2001; Jepson et al., 2001) and this is certainly part of the solution. The law is particularly clear in protected areas: no logging is allowed and the apparatus theoretically exists to implement the law. However, in the current economic and political climate these laws are rarely, if ever, enforced. Park officials are underpaid and have little motivation. Even if Park officials do attempt to stop illegal activities they are dependent upon the support of other agencies. In order to prosecute those responsible for illegal logging, support and action is also needed from the police, the judiciary and in many cases local government and the army. Without reform of all these agencies–many of whom have vested interests in maintaining the status quo–improving the efficacy of Park staff will only go so far towards reducing illegal logging. The situation is more complicated outside protected areas. There are many more grey areas in the law. Land tenure is often insecure so it is not even clear exactly what constitutes illegal logging.

Additionally, a large proportion of the rural population in West Kalimantan is now dependant on the income derived from illegal logging, as seen in other areas of Indonesia (Obidzinski and Suramenggala, 2000). Any attempt to control illegal logging by enforcement alone will have severe economic impacts on many communities. This makes enforcement not only a moral problem, but it also makes it more or less impossible. Too many people depend on illegal logging to just give it up with out any alternatives. We believe, therefore, that the proverbial stick of law enforcement needs the carrot of community based forest management. Such a system whereby local communities are given rights to manage small community forest areas for timber harvesting have been proposed by some (Cordes et al., 1997, McCarthy, 2000) as it provides benefits for the local communities as well as maintaining some of the conservation value of the forest.

The nutrient-poor soils characteristic of rainforests result in few viable land-use options after large-scale industrial logging operations have finished. Current Indonesian policy views such "degraded" lands as suitable only for conversion to industrial crops such as *Acacia mangium* or increasingly for oil palm plantations. This widespread clearance of natural forests has far reaching consequences. Clearances deprive local communities of their forest, a source of natural resources and water. On a broader scale, forest loss affects ecosystem functions such as the mast fruiting of dipterocarp forest (Curren et al., 1999) and possibly global climate. Cannon et al. (1998) showed that even logged lowland forests can have significant conservation value as well as large volumes of residual timber. Granting local communities the rights to manage such logged forests could provide long-term incentives to keep the area forested and not convert it to non-forest uses. Small-scale forestry operations would provide local communities legal and sustainable revenues from the logging activities. Local government could receive tax revenue from such operations, whereas now there is no official tax revenue from illegal logging. Community-based forest management could essentially legalize, regulate, and control illegal logging and could reduce the need for local communities to log in protected areas.

This, however, is theory. Unfortunately, in practice this may be very hard to implement in Indonesia. With the absence of political will, the almost total lawlessness in some parts of the country and the corrupt involvement of many people in illegal logging, is it realistic that such whole-scale transformation could take place? This would require the transfer of tenure from large well-connected companies and political elites to village-level groups, as well as the implementation of new laws and regulations in a sector that barely follows the laws and regulations that already exist. In Indonesia today such a proposal is radical at best. But with the rampant pace of illegal logging continuing unabated, radical change is needed soon to stop the ecological, social and economic disaster that would follow the complete exhaustion and conversion of Indonesia's invaluable forest resources.

REFERENCES

Barr, C. 2001. Banking on sustainability: Structural adjustment and forestry reform in post-Suharto Indonesia. Centre for International Forest Research/World Wildlife Fund, Bogor, Indonesia.

Brown, D. 1999. Addicted to rent: Corporate and spatial distribution of forest resources in Indonesia; implications for forest sustainability and government policy. Indonesia-UK tropical forest management program report No. PFM/EC/99/06. Department for International Development, Jakarta, Indonesia.

Bruner, A., R. Gullison, R. Rice and G. de Fonseca. 2001. Effectiveness of parks in protecting tropical biodiversity. *Science* 291:125-128.

Cannon, C., D. Peart and M. Leighton. 1998. Tree species diversity in commercially logged bornean rainforest. *Science* 281: 1366-1368.

Cordes B., R. Cherry, M. Henderson, M. Leighton, N. Salafsky and W. Watt. 1997. Chainsaws as a tool for conservation ? A comparison of community-based timber production enterprises in Papua New Guinea and Indonesia. ODI Rural Development Forestry Network Paper #22b. Overseas Development Institute, London.

Curren, L., I. Caniago, G. Paoli, D. Astianti, M. Kusmeti, M. Leighton, C. Nirarita and H. Haeruman. 1999. Impact of El Nino and logging on canopy tree recruitment in Borneo. *Science* 286: 2033-2220.

Dove, M. 1996. So far from power, so near to the forest: A structural analysis of gain and blame in tropical forest development. In pp. 41-58: C. Padoch and N.L. Peluso (eds.). Borneo in Transition: People, Forests, Conservation and Development. Oxford University Press, Oxford, England.

Dudley, R. 2001. The changing dynamics of illegal logging systems in Indonesia: An initial investigation. In Ch.16: C.J.P. Colfer and I.A.P. Resosudarno (eds.). Which Way Forward ? Forests, Policy and People in Indonesia. RFF, Washington DC.

Environmental Investigation Agency (EIA). 1999. The Final Cut: Illegal Logging in Indonesia's Orangutan Parks. Emmerson Press, London.

Environmental Investigation Agency/Telapak 2001. Timber Trafficking: Illegal Logging in Indonesia, South East Asia and International Consumption of Illegally Sourced Timber. Environmental Investigation Agency, London, England.

Jepson, P., J. Jarvie, K. MacKinnon and K. Monk. 2001. The end for Indonesia's lowland forests? *Science* 292: 859.

Jepson, P., F. Momberg & H. van Noord, 2002. A review of the efficacy of the Protected Area System of East Kalimantan Province, Indonesia. *Natural Areas Journal* 22: 28-42

Johns, A. G. 1997. Timber Production and Biodiversity Conservation in Tropical Rain Forests. Cambridge University Press, Cambridge, England.

Laboratory for Tropical Forest Ecology (LTFE). 1998. Baseline Socioeconomic Status of Communities Participating in the Project 'Community-Based Forest Management of Buffer Zone Forests at Gunung Palung National Park, West Kalimantan.' Unpublished document. Pontianak, Indonesia.

MacKinnon, J. 1982. Gunung Palung National Park Management Plan. Department of Forestry. Jakarta, Indonesia.

MacKinnon, K and E. Sumardja. 1996. Forest for the future: Conservation in Kalimantan. In pp. 59-75: C. Padoch and N.L. Peluso (eds.). Borneo in Transition: People, Forests, Conservation and Development. Oxford University Press, Oxford, England.

MacKinnon, K., G. Hatta, G. Halim, A. Mangalik. 1996. Ecology of Kalimantan. Periplus, Singapore.

McCarthy, J. 2000: 'Wild logging': The rise and fall of logging networks and biodiversity conservation projects on Indonesia's rainforest frontier. CIFOR occasional paper No 31. Centre for International Forest Research, Bogor, Indonesia.

Obidzinski, K. and I. Suramenggala. 2000. Illegal logging in Indonesia–A conceptual approach to the problem. Draft paper to Centre for International Forest Research, Bogor.

Pearce, D., F. Putz and J. Tomclay. 1999. Sustainable forest future. Centre for Social and Economic Research on the Global Environment working paper GEC 99-15. University College London and University of East Anglia, England.

Peluso, N. and E. Harwell. 2001. Territory, custom and the cultural politics of ethnic war in West Kalimantan, Indonesia. In N. Peluso and M. Wells (eds.). Violent Environments. Cornell University Press, Ithaca, New York.

Pofenberger, M. 1997. Rethinking Indonesian forest policy. Beyond the timber barons. *Asian Survey*. Volume XXXVII, number 5.

Potter, L. and J. Lee. 1999. Oil-Palm in Indonesia: Its Role in Forest Conversion and the Fires of 1997/98. WWF Indonesia Forest Fires Project. Jakarta, Indonesia.

Rhee, S. 2000. De facto decentralization and the management of natural resources in East Kalimantan during a period of transition. *Asia-Pacific Community Forestry Newsletter* 13(2): 34-40.

Rhee, S. 2003. De facto decentralization and community conflicts in East Kalimantan, Indonesia: Explanations from local history and implications for community forestry. In K. Abe, W. de Jong, and L. Tuck-Po (eds.). The Political Ecology of Tropical Forests in Southeast Asia: Historical Perspectives. Trans Pacific Press and Kyoto University Press, Melbourne and Kyoto.

Salafsky, N. 1993. The forest garden project: An ecological study of the locally developed land-use system in West Kalimantan, Indonesia. PhD Dissertation submitted to the Department of Environmental studies, Duke University, Durham, North Carolina.

Salafsky, N., B. Dugelby and J. Terborgh. 1993. Can extractive reserves save the rainforest? An ecological and socio-economic of the non-timber forest product extraction systems in Peten, Guatemala and West Kalimantan, Indonesia. *Conservation Biology* 7: 39-51.

Sist P., T. Nolan, J. Bertault amd D. Dykstra. 1998. Harvesting intensity versus sustainability in Indonesia. *Forest Ecology and Management* 108: 251-260.

Telapak, L.B.B., Putrijuji & Madanika. 2000. Planting Disaster: Pemiskinan keragaman hyati, social ekonomi dan pelenggaram HAM dalam perkebunan besar sawit di Indonesia. Telapak, Jakarta, Indonesia.

Uhl, C., D. Nepstad, R. Buschbarcher, K. Clark, B. Kaufman and S. Subler. 1990. Studies of ecosystem response to natural and anthropogenic disturbances provide guidelines for designing sustainable land-use systems in Amazonia. In A.B. Anderson (ed.). Alternatives to Deforestation: Steps Towards Sustainable Use of Amazonian Rainforests. Columbia University Press, New York.

Vayda, A. 1999. Finding Causes of the 1997-98 Indonesian Forest Fires: Problems and Possibilities. World Wildlife Fund Indonesia Forest Fires Project. Jakarta, Indonesia.

Whitten, T., D. Holmes, and K. MacKinnon, 2001. Conservation Biology: A displacement behaviour for academia. *Conservation Biology* 15: 1-3.

World Bank. 2001. Indonesia: Environment and natural resource management in a time of transition. World Bank. Washington, DC.

World Bank/WWF. 2000. Indonesia country paper on illegal logging. Prepared for WB-WWF workshop on the control of illegal logging in East Asia. 28 August 2000. Jakarta, Indonesia.

World Resources Institute 2000. Trial By Fire. World Resources Institute, Washington, DC.

Yeager C. 1999. Orang Utan action plan. World Wildlife Fund, Center for Environmental Research and Conservation and Departemen Kehutanan.

Community-Based Logging
and *De Facto* Decentralization:
Illegal Logging in the Gunung Palung Area
of West Kalimantan, Indonesia

Ramsay M. Ravenel

SUMMARY. This paper considers the emergence of community-based logging in West Kalimantan, Indonesia at the end of the timber boom in the mid-1990s. It begins with an overview of the relevant laws and regulations and describes community-based logging before the *era reformasi* that began with the ouster of President Suharto in May 1998. The paper argues that the Indonesian Selective Felling System (TPTI) both in theory and in practice has severely compromised the quality of Indonesia's natural production forests. The spatial heterogeneity of concession logging has encouraged reentry by teams of small-scale loggers who face lower operational and opportunity costs, and who have appropriated management and financial systems from the departed concessionaires. The paper then contrasts community-based logging with conventional

Ramsay M. Ravenel is an independent consultant in forestland investment.

Address correspondence to: Ramsay M. Ravenel, 101 Mary Street, Mt. Pleasant, SC 29464 (E-mail: rravenel@aya.yale.edu).

The author would like to thank Mark Leighton, Ron Cherry, Charlotte Kaiser, Ed Pollard, and his Indonesian colleagues for their support during this fieldwork. Michael Dove and Andrew Mathews provided valuable comments on an earlier draft of this manuscript.

The fieldwork from which much of this material is drawn was funded by the Biodiversity Support Program.

[Haworth co-indexing entry note]: "Community-Based Logging and *De Facto* Decentralization: Illegal Logging in the Gunung Palung Area of West Kalimantan, Indonesia." Ravenel, Ramsay M. Co-published simultaneously in *Journal of Sustainable Forestry* (The Haworth Press, Inc.) Vol. 19, No. 1/2/3, 2004, pp. 213-237; and: *Illegal Logging in the Tropics: Strategies for Cutting Crime* (ed: Ramsay M. Ravenel, Ilmi M. E. Granoff, and Carrie A. Magee) The Haworth Press, Inc., 2004, pp. 213-237. Single or multiple copies of this article are available for a fee from The Haworth Document Delivery Service [1-800-HAWORTH, 9:00 a.m. - 5:00 p.m. (EST). E-mail address: docdelivery@haworthpress.com].

forest management systems and critiques decentralization and illegal logging discourses. From these critiques, I suggest that the *de facto* decentralization described by Rhee (2000) actually began before *de jure* decentralization. Lastly, I describe how a misreading of the landscape of local-level power and influence led to the failure of a community forest management project undertaken during the period under consideration. *[Article copies available for a fee from The Haworth Document Delivery Service: 1-800-HAWORTH. E-mail address: <docdelivery@haworthpress.com> Website: <http://www.HaworthPress.com> © 2004 by The Haworth Press, Inc. All rights reserved.]*

KEYWORDS. Illegal logging, community forestry, conservation and development, Biodiversity Conservation Network, decentralization, Indonesia, Borneo

INTRODUCTION

Control and access to Indonesia's "political forests" have long been the subject of contestation by state and local interests (Peluso and Vandergeest 2001:762). A complex and shifting mosaic of ethnic groups, government structures, and international interests has wrestled over these lucrative resources since early in the Dutch colonial era (Peluso and Vandergeest 2001: 762). The discourse on this historical struggle has variously invoked legal, social, economic and environmental notions of justice (Peluso 1993; Barber 1998; CIEL 2001; Dove 1995; Salafsky et al. 1995; Wells et al. 1999; Peluso and Vandergeest 2001:762) and in recent years has given increasing attention to the issue of illegal logging (Peluso 1993; McCarthy 2000; EIA 2000; Barr 2000; Barr et al. 2001; Casson 2001a; Casson 2001b; McCarthy 2001a; McCarthy 2001b; Potter and Badcock 2001; Casson and Obidzinski, 2002; Obidzinski, 2003; Hiller et al., this volume; Tacconi et al., this volume).

This paper proposes that the current debate over decentralization and its impacts on forest resources has overlooked two major points. First, the proliferation of illegal logging in Indonesia is attributable in part to the particular manner in which timber concessionaires have exploited the country's forest resources. Most of Indonesia's natural production forests have been selectively logged at least once, leaving commercial timber volumes low in the aggregate, but high in small and disparate patches. For large-scale timber concessionaires, this degraded forest is perceived to be uneconomical to operate by the book. Smaller scale community-based loggers, however, face lower fixed and marginal costs–especially if they are operating illegally or quasi-illegally–and have moved into inactive concessions.

Second, this paper argues that *de facto* decentralization began before 1999, the *de jure* beginning of the process of decentralization that followed the end of the Suharto regime in May 1998. This conclusion is based on my observations of rapid growth in community-based logging and the political economy in which it developed. While decentralization has accelerated community-based forest exploitation, it did not initiate it.

The paper begins by describing the area under consideration (Figure 1) and the community forest management project for which I was worked in 1997-1998. The paper then reviews Indonesian forest history in order to explain how the forests in the area under consideration reached their degraded condition by the mid-1990s. A discussion of the political economy of community-based logging follows and explains how this activity flourished in the political and economic environment of the mid- to late-1990s. The paper considers some of the implications of the growth in community-based logging in the Gunung Palung area before critiquing the discourses of illegal logging and decentralization. This final section concludes with a discussion of some of the challenges to conservation posed by *de facto* decentralization as I saw them.

The failure of the Gunung Palung Community-Based Forest Management project illustrates problems of national level political legibility and local networks of power and influence. The problem currently constructed as illegal logging and recent attempts to address it ultimately illustrate broader failures of governance and civil society occurring in Indonesia today.

SITE AND PROJECT DESCRIPTION

This discussion of community-based logging draws heavily on field work conducted in West Kalimantan, Indonesia from January 1997 to October 1998 and in August 2001. During this first period, I worked for a pilot project entitled "Community-Based Forest Management of Buffer Zone Forests at Gunung Palung National Park, West Kalimantan" (BCN 1999:68-77). The project was being implemented by a small team working under the auspices of the Harvard University Laboratory of Tropical Forest Ecology, an institution originally formed to support rainforest ecology research in the park. Funded by the Biodiversity Conservation Network, the project worked with the Ministry of Forestry (ostensibly) to create a legally recognized and sustainably managed alternative to the illegal and destructive logging that was by then pervasive in the area. This logging was perceived to be a severe threat to the integrity of the park by conservation biologists, international conservation groups, bilateral development agencies, and the state. By organizing communities through a bottom-up approach, the project was designed to achieve "buy-in" from communities by addressing the economic development needs that were perceived

FIGURE 1. Gunung Palung National Park and the island of Borneo.

to drive this illegal logging. The project fit the mold of the Integrated Conservation and Development Project ideology popular in the mid-1990s (Wells et al. 1999).

The Gunung Palung Community-Based Forest Management project worked with villages on the northwestern border of the park in the district (*kecamatan*) of Sukadana and Teluk Melano (Figure 2). These villages were predominantly Melayu, although they also hosted a wide range of Indonesia's ethnic groups including Balinese, Javanese, and Chinese. Unlike many of the small rural villages in the interior of Kalimantan, none of these villages were predominantly Dayak. While there was some intermarriage (with Dayak becoming Melayu), few Dayak lived in these predominantly Muslim villages. According to local history, the Dayak fled to the interior when the Melayu settled the coastal areas many generations ago. The legacy of Dayak inhabitation of the area lives on, however, in the towering durian trees scattered throughout the Sukadana hills.

As descendents from Borneo's original forest dwellers, most Dayak groups hunt more than their Melayu counterparts, who do hunt deer. Although not all Dayak eat orangutans, the Dayak exodus from the Gunung Palung area and the

FIGURE 2. The Gunung Palung Community-Based Forest Management Project (GP-CBFM) Site.

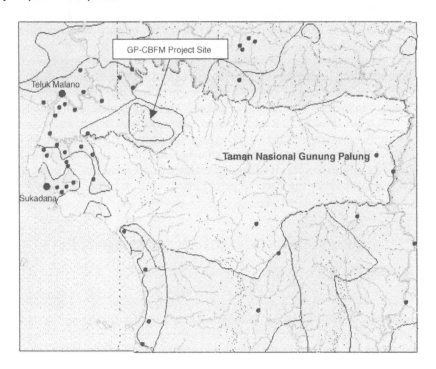

halal diet of their successors may explain the unusually high populations of orangutans and other vertebrates in the area. These vertebrate populations are a major component of Gunung Palung's conservation value and explain the international interest in the area.

Despite the common portrayal of Melayu people as fishing and trading communities, Melayu in the Gunung Palung area engage in many traditional activities in the surrounding forests. While not all Melayu work in the forest– some, in fact, expressed fear of it–local members of project field crews explained that they learned local forest landscape and flora from their parents. When asked about these traditional uses of the forest, these respondents described collection of rattan (*Calamus* spp.), illipe nuts (*Shorea* spp., locally known as *tengkawang*), gutta percha (*Pallaquium* spp.), ironwood (*Eusideroxylon zwageri*, locally known as *belian*), aromatic aloeswood (*Aquilaria malaccensis*, locally known as *gaharu*), as well as poles of many species used to build large fishing traps in the coastal river-mouths. Interestingly, despite this thorough use of the forest, the Gunung Palung Community-Based Forest Management project was criticized by many outsiders (e.g., USAID staff and Ministry of Forestry officials). If this pilot project had any hope of working, they suggested, it should work with Dayak villages since the Dayak are the original forest peoples of Kalimantan. This argument falls into simplistic constructions of indigeneity as static, homogenous, and environmentally benign–arguments that have been thoroughly deconstructed in the literature (Redford 1990; Zerner 1993; Agrawal 1995; Dove 2000; and Li 2000). It also suggests that these actors were more interested in *finding* a successful community forestry project than *developing* one (Seymour 1994:472). Rhetoric aside, the project did fail in the end, but so did a similar project working with Dayak communities in the Sanggau District of West Kalimantan, the German Social Forestry Development Project. Although the contexts were considerably different, both projects experienced violence and arson at the hand of angry community members (Ron Cherry, *personal communication*).

OVERVIEW OF INDONESIAN FOREST HISTORY

Legal Framework

Independence Through Suharto's "New Order" Regime–The illegality of community-based logging in Indonesia has a long history dating back to the Dutch colonial era (Peluso 1991:50). Upon independence in 1945, the new nation's forestry laws were almost entirely translated word-for-word from the Dutch colonial forest laws (Peluso 1991:97). While the Basic Agrarian Law of 1960 was intended to erase remnants of colonial agrarian laws, it failed to initiate any real change as it was interpreted under the New Order regime of President Suharto (CIEL 2001:II,10). Not long after Suharto took power in 1965, the state took control of all forestlands through the Basic Forestry Law of 1967: "all forests within the territory of the Republic of Indonesia, and all the resources they contain, are under the authority of the state" (Article 5, Basic Forestry Law no.5/1967, as cited by CIEL 2001:II,12). In further clarifications of the law, state forests were defined to include all forests except those under private ownership and including those under customary law (Article 2, Official Explanation, General Elucidation, as cited by CIEL 2001:II,12). Under Suharto's New Order regime, forest laws fit the mold of a neocolonial state that perpetuated resource exploitation and rural surplus extraction. As we shall see, the implementation of these laws–or lack thereof–over the subsequent thirty years reflects the underlying political economy of forest exploitation that has proven far more resistant to change than the laws themselves.

After Suharto stepped down from the Presidency in May 1998, the state's iron-fisted control over the nation's resources came under unprecedented attack from an angry public facing considerable economic hardship. One of the first manifestations of decentralization in forest law came shortly after B.J. Habibie took the Presidency, Government Regulation No. 62/1998 Concerning Delegation of Part of Government's Affairs in the Sector of Forestry to the Region. The new ruling gave district governments the authority to oversee activities related to the management of Privately-Owned Forest and Social Forest (Barr et al. 2001:20). The impact of these laws at the local level depended greatly on the initiative of district leaders (*Bupati*) who were the key instruments of indirect rule under the Dutch. In Malinau, East Kalimantan, for example, the *Bupati* quickly issued a number of forestry regulations that cited PP 62/1998 as their legal foundation.

Communities Rights and Indigeneity–Given this paper's focus on illegal logging at the community level, a review of community rights and indigeneity in Indonesian forest law is warranted. As the "era of reform" (e*ra reformasi*) proceeded, the nation's Basic Forestry Law was revised in 1999 (UU 41/ 1999). While the revised law calls for sustainable forest management and equitable distribution of benefits, it also remains firmly rooted in neocolonial state control of forests (CIEL 2001:V, 7-8). Despite numerous provisions for local community involvement, it nevertheless clearly states, "*Adat* forest is *state forest* located in the traditional jurisdiction areas (*Hutan adat adalah hutan negara yang berada dalam wilayah masyarakat hukum adat*)" (UU 41/ 1999, emphasis added). *Adat* forest, then, is a subset of state forest that recognizes traditional community-based forest management institutions but does not give full community-based property rights.

The term *adat*, however, should not be confused with 'community.' Not all communities have the customary legal institutions typically associated with *adat* and *adat* forests. The histories of migration throughout the Indonesian archipelago are complex, so only a subset of Indonesians living in identifiable communities would say that they live under *adat* law. These politics of indigeneity are based on a problematic construction of an indigenous/non-indigenous duality that further marginalizes rural communities that don't fit the "tribal slot" (Li 2000:153; Gupta 1998:289).

> To be an 'ordinary villager' is, therefore, to belong to a homogenized or simplified category of people whose ethnic identity, distinctive forms of social organization, and localized commitments are officially unrecognized and often seen as contrary to national laws, policies, and objectives. (Li 2000:155)

The Gunung Palung Community-Based Forest Management project suffered from this bias because it worked with "non-indigenous" ethnic Malay communities rather than "indigenous" Dayak communities. As discussed in the project description above, this construction of legality and indigeneity led state and development officials to question the rationale of the project and may have constrained their support of it.

Decentralization–The loosening of state control over forest resources, both in practice and later in law, was a key factor in the emergence of community-based logging. Forestry laws passed during the *era reformasi* appear to have advanced the interests and autonomy of provincial and district-level governments at the expense of the central control. More importantly, however, these laws caused confusion over the issue of control and responsibility (McCarthy 2001:46). A new type of community forest concession, a Forest Product Harvest Concession (*Hak Pemungutan Hasil Hutan*, or HPHH), was created by law in 1999 (PP 6/1999). These 100-hectare HPHH can be allocated to community cooperatives (*koperasi*) by the *Bupati* (PP 6/1999, Article 1). By November of that year, however, the new Minister of Forestry and Estate Crops suspended the bill (McCarthy, 2001:46). The process for allocating HPHHs was clarified with Ministerial Decree 310/1999 (Barr et al., 2001:68).

More important than the specifics of which law gave which rights to whom, this discussion of changes in forest law during *era reformasi* highlights the formal uncertainty of forest rights at the time. A Ministry of Forestry study on perceptions of decentralization conducted during December 1999-February 2000 found that regional stakeholders were enthusiastic about decentralization in the forest sector, but had no certainty or guidance on how it might be implemented (MFEC, 2000 as cited by Rhee, 2000:34). Because responses to decentralization by certain stakeholders do not necessarily correspond to actual changes in the legal framework, Rhee (2000:35) argues that *de facto* rather than *de jure* decentralization proves to be the more powerful driver of change.

Decentralization has clearly increased conflicts over forest control and access (Rhee 2000; McCarthy 2000; Barr et al. 2001). In theory, decentralization should have strengthened community claims for local access to forest resources. After decades under the repressive New Order regime, these previously marginalized players were suddenly poised to gain quite a lot from this process. However, entrenched local networks of power and control were not inclined to simply step aside. The process of decentralization, therefore, was not a simple transfer of power to people.

The discrepancy between *de facto* and *de jure* rights is not a new development of the decentralization process. Indeed, this rift and the corruption that enables it has been endemic to Indonesian forest management (Schwartz 1999:140; EIA 1998; EIA 2000). During the course of my fieldwork, most of which was conducted before the beginning of the *era reformasi* in 1998, I witnessed considerable corruption in the forestry sector both directly and indirectly. I directly witnessed many instances of flagrant illegal logging. I also studied the degraded forests that had been left behind, and indirectly saw this corruption as it was "written on the land" (Baviskar, 2001). Going forward, the actual machinations of forest law reform will be less important than how they will be interpreted and implemented on the ground. On both fronts, uncertainty rules the day.

A Short History of Forest Concessions in West Kalimantan

The proliferation of community-based logging under *de facto* decentralization grew out of the degradation of forest resources under Indonesia's forest concession system, so a review of this system follow. In historical terms, the commercial exploitation of timber resources in West-Kalimantan is a relatively new phenomenon (Peluso and Harwell 2001:93). The Sambas region of the province has been a center of gold mining activity for centuries, attracting significant populations of entrepreneurs and traders from Chinese and other ethnic backgrounds (Peluso and Harwell 2001:92). The demarcation and allocation of state forestlands in West Kalimantan did not begin in earnest until the passage of the Basic Forestry Law in 1967. While there may have been adequate markets for West Kalimantan's timber before then, the Indonesian revolution and Sukarno's *Konfrontasi* with Malaysia retarded the development of West Kalimantan's timber industry (Peluso and Harwell 2001:93). Not until the 1970s did Indonesia begin to dominate tropical wood markets (Poffenberger 1997). By 1999, the four provinces of Kalimantan provided 60% of Indonesia's annual $6 billion of annual timber revenue (Curran et al. 1999:2186).

The appropriation of West Kalimantan's forest resources was accomplished by a variety of means. Peluso and Harwell (2001:94) underscore the central role that zoning played in this major reorganization of forest control and access: "certain people–usually *not* the local people–could conduct specific activities." Local land tenure traditions that did not demonstrate continuous use such as fallows in swidden agriculture systems were illegible to the state and were often assigned the category "unclassified" or "conversion" forest (Peluso and Harwell 2001:95). The Indonesian government's transmigration program also radically changed the nature of land tenure in the province. Large tracts of forest land–75% of it from "unclassified" or "conversion" forest (Peluso and Harwell 2001:95)–were given to individuals relocated from other "over-populated" islands such as Java, Bali, and importantly, Madura. Lastly, former soldiers were rewarded for their service with tracts of land through an army resettlement program (Peluso and Harwell 2001:94) that helped spread and ingrain the military's Dual Functions of security and development.

Once mapped and zoned for various land uses, the state began awarding timber concessions (*Hak Pemanfaatan Hutan*, or HPH) in West Kalimantan in the late 1960s. In the ensuing decades the growing timber economy soon penetrated even the most remote of rural villages. Many men and women–both local and extra-local–worked in logging operations and production camps, but often these jobs disappeared once the first cut had been completed. In the course of my fieldwork, local people repeatedly told of logging companies bringing in extra-local laborers, especially from the Sambas region, to carry out logging operations. Many married locally and turned to subsistence agriculture when the logging companies left after completing their first rotation. In some cases, logging contractors returned for second and third cuts when new markets arose for smaller trees or previously ignored species.

During the 1990s, the overall quality of the natural production forest base began to deteriorate. By 2001, many concessions in natural production forests were either abandoned and returned to the Ministry, converted to plantations and/or transmigration sites, of unclear status, or still active (Brown 1999; McCarthy 2001a; McCarthy 2001b; Casson 2001a; Lisa Curran, *personal communication*). The conversion of many of these natural production forests to other land uses and the uncertainty of the tenure of those forests that remained suggests that the resource was degraded and not worthy of commercial exploitation for timber. In an environment where valuable resources are rarely left idle, one can only conclude that Kalimantan's idle natural forests no longer held much value.

A Silvicultural Critique of the Indonesian Selective Felling System (TPTI)

In order to understand how Kalimantan's rich forests lost so much of their commercial value, one must understand the silvicultural implications of the Indonesian Selective Felling System (TPTI). While implementation of forestry regulations was famously lax, the national silvicultural system was technically misguided. Trees in the species-rich dipterocarp family dominate Indonesia's forests (Whitmore 1984; Webb and Peart 2000). Much like mixed moist temperate forests (e.g., oak forests in the Northeastern U.S.), dipterocarp species are highly site-specific (Webb and Peart 2000; Ashton and Peters 1999:16). To be precise, each species can be characterized by a specific mode of regeneration ecology that describes the conditions under which its seedlings will germinate, establish, compete and stratify with its neighbors, and grow to maturity (Ashton and Peters 1999:16; Oliver and Larson 1996). While selective logging-the silvicultural strategy employed by TPTI-may be perceived as having a lower impact on the forest, it actually tends to change the species composition of the forest over time. Dipterocarp species vary widely in terms of their regeneration ecology (Webb and Peart 2000). Selective logging creates understory conditions that favor the regeneration of slow-growing, shade tolerant species (Ashton and Peters 1999:15). Logging that creates larger openings allows more light to reach the forest floor and provides the conditions that favor the regeneration of faster-growing, shade-*in*tolerant species (Ashton and Peters 1999:17). To be successful, however, these larger openings must not be so large as to reduce seed dispersal and soil quality. TPTI, then, favors the regeneration of only a subset of dipterocarp species: shade-tolerant and slow growing species.

The structure of the residual stand after a first cut under TPTI accelerates the growth of those trees that are already in the stand but which are not currently of commercial diameter (Ashton and Peters 1999:15). Unfortunately, these species are often of lower or no commercial value. The shade created by these residual trees furthermore provides for the establishment and succession of a new cohort of slow-growing, shade-tolerant species. If TPTI is designed to ensure the regeneration of a new crop of commercial dipterocarps, it should create instead residual stands that favor the faster-growing, shade intolerant species. TPTI, then, arguably favors the regeneration of the wrong set of species. The second and third cuts made by concessionaires in the 1990s–often before the end of the prescribed 35 year rotation–were the final liquidations of trees that were in the original unlogged stands. TPTI effectively squeezed the commercial value of the existing trees in primary forests at the expense of regenerating commercially viable natural forests. Even if TPTI was not closely followed in practice, the forest management methods employed by concessionaires was similarly problematic in its promotion of these unproductive uneven age-classed forests. The degradation in the value of Indonesian natural production forests witnessed by the turn of the century is therefore not surprising.

It is worth noting that this discussion has avoided the terms "primary" (i.e., unlogged) and "secondary" (i.e., logged-over) forests. The use of these terms is problematic, especially in the Indonesian context. The term "secondary" encompasses such widely ranging forest systems as fallows in swidden systems, forest gardens, and selectively logged forests. Similarly, selectively logged forests can refer to forests logged at a wide range of intensities–e.g., for rare species such as *gaharu* (*Aquilaria malaccensis*) (Paoli et al., in press) or for more common dipterocarps–and over a wide range of return intervals. Many secondary forests have been continuously intact, but may have been logged multiple times by various agents practicing various forms of silviculture. These terms do not provide a useful description of forest ecosystem integrity.

THE POLITICAL ECONOMY OF COMMUNITY-BASED LOGGING IN THE GUNUNG PALUNG AREA

Having described the area and reviewed Indonesian forest history, this section of the paper examines the political economy in which community-based logging proliferated. I will describe the condition of the logged-over forest concessions, the appropriation of concession logging systems, the financing of community logging systems, and the implications of communities reentering logged-over forests. I will conclude this section with discussions of capitalist "articulation" and the community transformation that resulted from the growth of community-based logging.

Spatial Heterogeneity in Logged-Over Forests: A Microeconomic Analysis

Social factors such as legal uncertainty and *de facto* decentralization certainly contributed to the proliferation of small-scale logging in Indonesia. However, biophysical factors-largely ignored in the literature to date-also contributed to this process of change. The driver of change was not simply forest resource depletion and scarcity, as portrayed by Barber (1998), but rather the particular spatial heterogeneity of resource scarcity. During several months of forest inventory work in logged-over forests near Gunung Palung abandoned by PT. SJBU, I found a very patchy pattern of logging similar to that reported by Cannon et al. (1994).

In terms of the forest-level timber volumes legible to concessionaires and state foresters, these selectively logged forests might have appeared commercially unprofitable. Zooming in on a concession-level map to a scale at which a small team of community loggers might operate, however, the same forest might be perceived as commercially profitable. This discrepancy in the legibility of the economic viability of the same logged-over forests deserves further consideration.

These differences in the profitability of logging derive in part from lower marginal costs of community-based logging (i.e., a cubic meter of wood can produced more cheaply by community loggers than concession loggers) and lower opportunity costs for community loggers (i.e., they have no other more profitable opportunities, while timber companies can deploy their capital more profitably elsewhere). Community loggers require little in the way of fixed costs. They only need chainsaws, fuel, food, and transportation to produce timber. Timber companies, however, face very high fixed costs. They must buy heavy machinery, build a base camp and a road network, maintain a large staff, and support community development programs. Both groups must pay bribes in order to operate, but since concessionaires are more visible, they typically pay more.

Similarly, community loggers face lower opportunity costs than concession loggers. Village surveys in the Gunung Palung area found a wide range of sources of cash income for households, but income from logging provided either the primary or secondary (behind agriculture) source (LTFE 1998:5). This evidence suggests a dearth of comparable opportunities in their villages.

Timber companies, however, faced higher opportunity costs. They could simply move their operations to a richer forest area elsewhere–a "hunter-gatherer" strategy followed by timber concessionaires around the world (Mather 1991:30).

Because concessionaires had other opportunities to earn higher profits, many logged-over forests were simply abandoned. Community-based logging took advantage of these open-access resources, but the author's field surveys suggest that before 1995 this activity was limited in scale. As the high-profit forest resource base for concessionaires began to contract–and the opportunity cost of logging in selectively logged forests shrank with it–timber markets turned to lower-cost community logging crews to supply raw material. Informal local and extra-local capital markets sought out community loggers and invested in illegal logging operations.

A new paved highway and electricity also played an important role in the emergence of community-based logging in the area. The road connected the area with the district capital and made it easy and inexpensive to truck timber to sawmills. Also, local people suddenly had much greater exposure and access to motorbikes, satellite television, and chainsaws. These expensive goods required significantly more cash income than these households had previously been able to generate. This exposure increased local people's interest in cash income greatly.

Local Appropriation of Concession Logging Systems

Why did villagers gravitate to community-based logging? In addition to having newfound access to local forest resources they also had leaned how to exploit them commercially. Knowing the forest resources–where the logged-over patches were and, more importantly, where they were not–and having learned the skills of the trade, many local loggers formed small-scale teams to explore the residual stands for valuable logs. Because these small groups required very little capital investment, they were able to profitably log where their predecessors had given up. Informants described how the *Sungai Meliya* (Meliya River) in the Gunung Palung area, for example, was unnavigable by boat before it was cleared of fallen logs and rattan (*duri*) by PT Wanawati in the 1960s. Informants also claimed that in some cases they reused skid trails (*jalan kuda-kuda*) built by logging companies, especially when the companies had logged a narrow range of species.

Under a typical concession logging system, according to former concession loggers, laborers could buy food and supplies for their trip to the forest on credit from the company's shop. They would then repay their debt with the wages they earned in the forest, but often ended up with very little take-home pay. Alternatively, the concessionaire would provide the food and wages would be correspondingly lower. In the forest, company supervisors would allocate blocks (*petak*) of a few hectares along a main skidding line to teams of loggers who would work independently. The company would build the main line–sometimes a small lorry rail–from the access point into the forest. The logging teams would build their own skid trail to bring the logs from their petak to the main line. Under this system, logging crews had relatively little oversight. Once the concessionaire had pulled out, these logging crews were easily capable of conducting this work on their own. Importantly, this system also resulted in a patchy, uneven pattern of logging disturbance since some crews were more diligent than others at getting all of the commercial timber out of their logging blocks.

This logging method was only slightly modified by community groups re-entering residual stands. The operations of community-based logging are divided into stages that often employ different teams for each. These stages include: identifying a suitable stand to log; constructing a skid trail; felling trees; applying chemical preservatives when working with certain species; skidding the logs to the river; collecting rattan from the forest; tying the logs into rafts (*raket kayu*); and floating the *raket*(s) downriver to market. In some cases, logs are cut into 20 cm × 20 cm × 4.2 m pieces (*kayu masak*) or boards either in the forest or at the edge of the river. Such wood can be more accurately examined for defects at the mill and requires fewer cuts, so it commands a higher price. Since chainsaw labor is the most expensive input in the process, this is not always done.

It should be noted that this method of logging is not new. *Jalan kuda-kuda* logging is featured in an exhibit on colonial forest exploitation at the Sarawak Timber Museum in Kuching (Malaysian Borneo). It is also depicted in a 1934 photograph from Indonesia in Dawkins and Philip (1998). The only significant difference in the logging practices themselves is the use of chainsaws instead of hand saws. What community loggers did appropriate from their concessionaire employers were the field operations systems and the systems of borrowing and lending that enabled individuals to organize and earn cash income from working in the forest.

Pinedo-Vasquez et al. (2001) describe the appropriation of logging technology in "post-boom" logging by sawmill owners and smallholders in Brazil. Much of the innovation they describe, however, occurred in sawmilling technology and in the management of small privately owned forests. In the Indonesian context property rights are highly contested and the right to own and operate a sawmill is ostensibly highly regulated. The adaptations in post-boom logging methods considered here focus on access to a new spatial distribution of timber resources and how to get them to the mill. Small sawmills soon began to proliferate as well, but they were not formally studied by the author. For comparison, the appropriation of rubber cultivation by Dayaks in colonial Indonesia described by Dove (2000), will be considered later in the context of logging's transformation of traditional forest management systems.

Financing Community Logging Operations

The financial and political arrangements made in the village that enabled the work to occur in the field were a key component of the growth in community-based logging. The person financing a logging expedition, the *"tauke,"* typically does not join the team in the forest. The *"tauke"* usually finds or is found by an individual who leads the team, the *"kepala rombongan."* In some cases, the *"kepala rombongan"* chooses workers from his village, or from others nearby. In other cases, the *"tauke"* brings in workers from other areas-loggers often cited Sambas as a source for migratory labor. Workers then borrow money for their families to use in their absence and to buy food for their work in the forest. Logging trips are usually 15-20 days long, and often occur during periods when demand for agricultural labor in village is low. One reason that some *"tauke"* prefer extra-local workers is that they believe that if they are too close to home, the workers may leave the forest before the job is done, or more importantly, before they have paid off their debt. Under this system, local workers find that they barely make ends meet (*"hasil kerja pas-pas saja"*).

However, wealthier villagers with enough money to front an operation themselves can capture much more of the value from the timber. Most of the villages in the Gunung Palung area have either a resident *tauke* or loggers with connections to extra-local *tauke*. These *tauke* may be merchants (many of whom are Chinese), or individuals who otherwise have access to cash. One local *tauke* had made his fortune trading aromatic *gaharu* wood in the early 1990s.

A community logging crew, however, needs more than just cash to operate. Depending on the location of the forest to be harvested, bribes must be paid to relevant government officials to win the turning of a blind eye. Loggers responded that these bribes were typically paid to local police, military (*Babinsa*), forestry and sometimes national park officials and village heads (*kepala dusun* and *kepala desa*). The amounts of these bribes were negotiated on the basis of how many cubic meters of wood they extracted. During one visit with the aforementioned logger working in the proposed community forest area, he complained that a team of national park guards had driven up-river to his camp in their speedboat demanding a bribe. Because he was not logging in the National Park, however, they did not have the authority to arrest him themselves, so instead they threatened to report him to the forestry police (*Dinas Kehutanan*) if he did not pay them. He lamented that these bribes were going to make all of his efforts barely profitable and reiterated that he simply wanted to feed his family (*cari makanan*). He may have been downplaying the gains he stood to reap from his work, but this story is only one of a great many reported to me about the extensive networks of power and influence that various players would manipulate to their advantage. These arrangements made this type of community-based logging quasi-legal, blurring the line with "illegal logging," as it is commonly understood.

IMPLICATIONS OF COMMUNITY-BASED LOGGING IN THE GUNUNG PALUNG AREA

Community Logging: Reentry in Natural Production Forests

There is an element of social justice to communities' restored access to forest resources after years of exclusion and exploitation by others. However, reentry by communities into regenerating logged-over forests has also had some unfortunate consequences. The small-scale logging observed in the Gunung Palung area was technically illegal because these selectively logged forests were ostensibly protected so that smaller unlogged trees could grow to commercial size for future harvest. This policy of prohibiting reentry should have been a key feature of the rotation system prescribed by the Indonesian Selective Cutting System (TPTI). As discussed above, the suitability of TPTI as a means for regenerating Indonesia's mixed-dipterocarp forests is dubious. Unregulated reentry by community logging teams into logged-over forests, whether logged according to TPTI or not, further reduced the value of residual stands. Subsequent logging by concessionaires would therefore produce a less bountiful harvest or require a longer rotation.

In the eyes of the state, premature reentry into a residual stand results is a reduction in the viability of that forest for retention and management as natural production forest. It can take decades, however, for a selectively logged forest to return to an attractive level of commercial stocking–especially if the faster growing species have been removed from the system. Most directly, reentry reduces the revenues and taxes from a subsequent cut by a timber concessionaire. In a volatile economic environment such as Indonesia's at the time, however, the future value of any profit opportunity was effectively discounted to nothing. In many cases, timber companies chose to abandon their concessions altogether rather than wait for another cut.

Given the history of state appropriation and exploitation of forests through the 1980s and a lull in activity and attention in the 1990s, it is not surprising that local people would seize this window of opportunity to capture some of the value of those resources. In purely financial terms, reentry by community groups may well have optimized the forest's financial net present value. Local-level uncertainty over the future of newfound access to forest resources had the effect of raising the discount rate–that is, a large discount was applied to any benefits that were far out in the future. Because this implicit discount rate was higher than the rate of appreciation of the forest's commercial timber value, it made financial sense to harvest whatever commercial timber remained. Even with an optimistic assumption that national and provincial land use planners were interested in allowing selectively logged forests to regenerate, reentry between legal rotations would only reduce the economic benefit of doing so. It is little wonder that so much of Kalimantan's natural forest estate was converted to other more profitable land uses during this period (Curran, 1999).

Although conversion to other land uses may have already been a *fait accompli* prior to the onset of local reentry, community-based logging has, on the margin, reduced the alternatives. This is an unfortunate consequence given the variety of ecosystem services that these forests provide. Local community members currently benefit significantly from cash from logging, wood for house building, regulation of water quality and quantity, habitat for fish and game, and the wide range of forest products described above. The commercially focused community logging operations observed by the author were not part of any science-based management system and were far from sustainable, so those forest benefits are sure to decline. Given the increased likelihood of fire after repeated logging, these benefits are at great risk. The Indonesian and international conservation communities have a significant interest in the biodiversity and carbon sequestration provided by these forests. In the eyes of local loggers, however, the decision to reenter selectively logged forests is much simpler: it generates cash income.

Community-Based Logging: Transforming Traditional Forest Management Systems

It should be noted that the community-based logging that supplied informal markets with illegal timber was markedly different from indigenous forest management systems prevalent in Borneo. In indigenous systems, swidden rice cultivation involves the clearing of forests for hill rice and other crops. While these forests are typically fallows that have re-grown into "secondary" forests, "primary" forests are sometimes cleared for these purposes (*ladang*) as well. Traditional forest management systems also include forest gardens that are managed for a highly diverse range of fruits and other non-timber forest products (Peluso and Padoch 1996; Lawrence et al. 1995; de Foresta and Michon 1993). The most notable examples of these forest gardens include those managed for durian, illipe nuts (*tengkawang*), and *damar* resins (Lawrence et al. 1995).

Returning to the notion of local appropriation of new forest management systems, consideration of the colonial rubber boom illustrates a subtle point about cultural transformations. Dove (2000) describes the discrepancy between colonial prescriptions for rubber (*Hevea brasilensis*) plantations and the incorporation of rubber cultivation into indigenous agricultural systems. As industrial rubber plantations grew in Borneo, local people integrated rubber into their complex and highly diversified suite of cultivated crops. The nature of rubber cultivation was such that the work it required could be done when demand for labor for swidden hill rice cultivation was low. Rubber was also typically not aggressively managed. Instead, it could be tapped more frequently if prices were high or the farmer needed cash income. This management system reduced exposure to risk through resource diversification. Local people transformed the industrial systems of rubber cultivation to complement their traditional forest systems.

Unlike the traditional rubber cultivation described by Dove (2000), a survey of 1,500 households in 9 villages the Gunung Palung area between September 1997 and May 1998 found that for 80% of households engaged in logging (*kerja kayu*), logging provided more than half of their cash income (LTFE 1998:1). Unfortunately, this survey does not consider the cash-equivalent value of farming income, so the extent to which logging represents a deviation from the strategy of diversification cannot be precisely stated. Based on the author's observations, however, most village families farmed a number of subsistence and market crops, including rice (paddy), coffee, manioc, bananas, durian, rambutan, breadfruit and jackfruit. Nevertheless, it can be safely argued that the emergence of logging as a way to earn cash income (*cari uang*, LTFE 1998:5) has radically transformed village economies and traditional forest management systems in the Gunung Palung area.

Articulation in Community Logging and the Penetration of Capital Markets

The issue of financing community-based logging mentioned above in the context of appropriation also deserves further study. The expansion of capital markets for community-based logging can be viewed as a process of "articulation," in which a state ideology finds its subject (Li 2000). Indonesia's burgeoning capitalist economy found eager subjects in both community members seeking cash income (*cari uang*, LTFE 1998:5) and wealthy entrepreneurs seeking profitable opportunities to employ and grow their capital.

The author's observations of this growing capital market suggest that this process of articulation began with community members looking for capital to continue generating cash income from working in the forest after logging companies left the area. As the infrastructure and market for illegal timber grew, the capital increasingly sought out the community loggers. A village logger illegally working in the proposed community forest area repeatedly told me of his debt to an extra-local wood trader from Kubu–a town a day away by boat. This investor visited project staff repeatedly to request permission for this logger to continue working until this debt had been repaid. Many villagers were quite happy to have the work.

ILLEGAL LOGGING AND DECENTRALIZATION IN INDONESIA

Indonesia's problems with the proliferation of small-scale illegal logging activities are often attributed to the process of decentralization that began after Suharto stepped down as President in May 1998 (Poffenberger 1999:44). However, this argument directly contradicts my own experience as well as that described in Sumatra by McCarthy (2000). As Rhee (2000) argues, *de jure* decentralization is not necessarily the key driver of the change in logging practices. Such a conclusion would be overly simplistic and ignores the important emergence in the 1990s of a complex, informal, local-level political economy based on illegal logging. This trend preceded the official devolution of control over forest resources. McCarthy (2000) describes these illegal logging networks in Sumatra and below I describe similar networks in West Kalimantan.

Revising De Facto Decentralization

The author's field observations suggest that the proliferation of community-based logging described above began in 1995, well before May 1998–the temporal boundary that formally frames the end of the Suharto era and the beginning of *era reformasi* and the decentralization that followed. The loss of control over state forests witnessed in the mid-1990s suggests that decentralization had in fact begun well before Suharto stepped down and that *de facto* deligitimization of the New Order state prefigured its *de jure* demise. Rhee (2000; 2003) deconstructs the discourse on decentralization by highlighting the fact that *de facto* decentralization can step into the void of emasculated central control. I have further problematized Rhee's *de facto* decentralization by suggesting that the process began approximately four years before *de jure* decentralization began in 1999. The illegal logging that I observed to have accelerated after 1995 suggests a loss of control over state lands and an earlier beginning of the process of *de facto* decentralization.

One might view the *de jure* decentralization as a political concession granted by the state to the increasingly powerful masses. The financial crisis caused by the collapse in the value of the rupiah in 1997 grew into a more widely felt economic crisis by early 1998. This crisis was acutely felt by households and communities throughout the country and called into question the legitimacy of the state. It is in this political context that the ouster of Suharto and the process of *de facto* decentralization took place.

Challenges to Conservation Amidst De Facto Decentralization

The local jockeying for power and control over forest resources described above illustrates the contested nature of these resources and their *de facto* decentralization. In the absence of clear and effective state forest management, an informal network of power and influence grew into the void. While it was not clear exactly who did have legal access to the forest, it was abundantly clear that local people did not. To gain access, local people would have to pay off the multiple competing interests.

It was into this maze of local networks of power and control that the Gunung Palung Community-based Forest Management project attempted to navigate somewhat blindly in the direction of socially equitable and environmentally sustainable community forest management. Although the project was legal in the eyes of the central government–with which the project director had close ties and favorable ears–it was effectively illegal in the eyes of this local power structure. The project grossly underestimated of the power of this illegal but *de facto* legal local network and overestimated the power of *de jure* legal structures. This misreading of the political landscape ultimately buried the project's chances of success.

Eighteen months after meeting the indebted logger described above, he and his crew were still logging in the project site. The project had still not yet acquired the right to implement the sustainable forest management plan, and he refused to wait for it. Because he was extracting large volumes of wood that would ultimately compromise the size of the project's annual sustainable harvest, the project planned a patrol to evict his and a number of other logging crews from the forest. First, however, a group of community members and members of an NGO in the district capital went up-river to explain to the loggers why they shouldn't be there and to warn them that a patrol would soon occur. These patrols had been carried out repeatedly over the previous 2 years and logging crews had been evicted repeatedly. Meanwhile in the village, the envoy was misunderstood to be a proper patrol intending to evict the loggers. Provocateurs (their identities still unknown, despite a police investigation) reportedly rounded up a group of people, got them inebriated and agitated about what was happening in the forest, and sent them up-river to defend the loggers. When they met the pre-patrol in the forest, a fight ensued and two community members were hospitalized and one has since never returned to his home village. According to the victims, the members of this mob were not loggers but were a group of young men run amok.

About a year later, another patrol was conducted and more conflict ensued. In the end, the project's field camp was burned to the ground. Its hopes of supporting the development of a community forestry enterprise dashed, the project was terminated several months thereafter. The unstable environment of conflicting official and *de facto* hierarchies proved to be an insurmountable challenge to the project's conservation efforts.

CONCLUSIONS

This analysis of community-based logging was inspired by the perceived intractability of the illegal logging problem (Barber 1998; EIA 2000). This perception that illegal logging is an obstacle to the real issue of forest conservation can be interpreted as a "signpost" that perhaps the definitions of legality and control over forest resources are in fact the real issues (Thompson et al. 1986). Further investigation has shown that formal and informal networks of power and influence have evolved around uncertainty concerning rights to exploit Indonesia's forest resources. These networks exist within communities and are linked to district, provincial, national and international networks of influence as well (Barr et al. 2000; McCarthy 2000; Hiller et al., this issue). Community-based interventions attempting to change this pattern of resource abuse have largely failed and are destined to fail without corresponding changes in both local and extra-local political economies. Community-based interventions attempt to transform the communities that are currently the losers in the scuffle for these resources into stronger players. Such efforts fail to recognize the intractability of the converse transformation: the current winners–the *taukes*, wood traders, sawmill owners, local police, military, forest service officials, and so on–are not interested in trading down in the hierarchy. Furthermore, they have the power and influence to protect the system in which they enjoy considerable benefits from forest resources. The illegal logging problem reflects the entrenched inequality of access to Indonesia's forest resources and thereby illustrates a much broader failure of governance and civil society occurring in Indonesia.

Given the vacuum left by the forceful presence of the Suharto regime, Indonesia faces the formidable challenge of building effective, powerful, and homegrown civil society institutions. Coupled with increased *de facto* representation in government, civil society institutions could perform the sorely needed role of the government and corporate watchdog. Without fundamental changes in government and civil society institutions, however, community-based conservation interventions–or indeed interventions of any kind–will only spin their wheels and may even promote conflict rather than prevent it.

If the current systems of local power and influence–corrupt or otherwise– are considered to be socially functional, perhaps conservation interventions should focus more squarely on conservation objectives rather than more difficult questions of changing social hierarchies in foreign countries. The trouble is, those systems are *not* terribly functional; and furthermore, conservation problems are inherently social problems, so conservation interventions will invariably rock the boat. The challenge is to affect change on the margins, in ways that are subtle in their social impact (in order to maintain consensus and avoid conflict) but substantial in their conservation impact. Such strategies are less likely to trigger intractable social conflict and are more likely to move society in the direction of improving efforts to safeguard environmental values.

REFERENCES

Agrawal, A. 1995. Dismantling the divide between indigenous and scientific knowledge. Development and Change 26: 413-439.

Ashton, M.S. 1995. Seedling survival and growth of four Shorea species in a Sri Lankan rainforest. Journal of Tropical Ecology 11: 263-279.

Asthon, M.S. and C.M. Peters. 1999. Even-aged silviculture in tropical rainforests of Asia. Journal of Forestry 97: 14-19.

Barber, C.V. 1998. Forest resource scarcity and social conflict in Indonesia. Environment 40(4): 4-9.

Barr, C. 2000. Profits on Paper: The political-economy of fiber, finance, and debt in Indonesia's pulp and paper industries. Forthcoming chapter in Christopher Barr, Banking on Sustainability: A Critical Assessment of Structural Adjustment in Indonesia's Forest and Estate Crop Industries. CIFOR and WWF, Macroeconomics Program Office. Accessed July 2001 from *http://www.cifor.org*.

Barr, C., E. Wollenberg, G. Lmberg, N. Anau, R. Iwan, I.M. Sudana, M. Moeliono, and T. Djogo. The Impacts of Decentralization on Forests and Forest-Dependent Communities in Kabupaten Malinau, East Kalimantan. CIFOR draft paper, September 18, 2001. Accessed December 8, 2001 from *http://www.cifor.org/highlights/ Decentralization.htm*.

Baviskar, A. 2001. Written on the Body, Written on the Land: Violence and Environmental Struggles in Central India, In Violent Environments. Nancy Lee Peluso, Michael Watts, eds. Ithaca: Cornell University Press.

BCN (Biodiversity Conservation Network). 1999. BCN Final Stories from the Field. Washington, DC: Biodiversity Support Program. Available online: *http://www. bsponline. org*.

Brown, D. 1999. Addicted to rent: Corporate and spatial distribution of forest resources in Indonesia; Implications for forest sustainability and government policy. Report No. PFM/EC/99/06, Indonesia-UK Tropical Forest Management Programme, Jakarta.

Cannon, C.H., D.R. Peart, M. Leighton, and K. Karawinata. 1994. The structure of lowland rainforest after selective logging in West Kalimantan, Indonesia. Forest Ecology and Management 67: 49-68.

Cannon, C.H., D.R. Peart, and M. Leighton. 1998. Tree species diversity in commercially logged Bornean rainforest. Science 261:1366-1368.

Casson, A. 2001a. Decentralization of Policy-Making and Administration of Policies Affecting Forests and Estate Crops in Kotawangan Timor. CIFOR draft paper September 18, 2001. Accessed December 8, 2001 from *http://www.cifor.org/highlights/ Decentralization.htm*.

Casson, A. 2001b. Decentralization of Policy-Making and Administration of Policies Affecting Forests and Estate Crops in Kutai Barat. CIFOR draft paper September 18, 2001. Accessed December 8, 2001 from *http://www.cifor.org/highlights/ Decentralization.htm.*

Casson, A. and K. Obidzinski. 2002. From new order to regional autonomy: Shifting dynamics of illegal logging in Kalimantan, Indonesia. World Development 30 (12): 2133-2151.

CIEL. 2001. Whose Resources? Whose Common Good? Towards a New Paradigm of Environmental Justice and the National Interest in Indonesia. Center for International Environmental Law (CIEL) in collaboration with HuMa, ELSAM, ICEL, and ICRAF. Draft, October 2001.

Curran, L.M., I. Caniago, G.D. Paoli, D. Astianti, M. Kusneti, M. Leighton, C.E. Nirarita and H. Haeruman. 1999. Impact of El Nino and logging on canopy tree recruitment in Borneo. Science 286: 2184-2188.

Dawkins, H.C. and M.S. Philip. 1998. Tropical Moist Forest Silviculture and Management. London: CAB International.

de Foresta, H. and G. Michon. 1993. Creation and management of rural agroforests in Indonesia: Potential applications in Africa. In C.M. Hladir et al., eds. Tropical Forests People and Food: Biocultural Interactions and Applications to Development. UNESCO, Man and the Biosphere Series, pp.709-724.

Dove, M.R. 1995. The theory of social forestry intervention: The state of the art in Asia. Agroforestry Systems 30(3): 315-340.

Dove, M.R. 2000. The life-cycle of indigenous knowledge, and the case of natural rubber production. In Roy F. Ellen, Alan Bicker, and Peter Parkes, eds. Indigenous Environmental Knowledge and its Transformations, pp. 213-251. Amsterdam: Harwood.

EIA, Environmental Investigation Agency. 1998. The Politics of Extinction: The Otangutan Crisis, The Destruction of Indonesia's Forests. London: Emmerson Press.

EIA, Environmental Investigation Agency. 2000. Timber Trafficking: Illegal Logging in Indonesia, South East Asia, and International Consumption of Illegally Sourced Timber. London: EIA.

Gupta, A. 1998. 'Indigenous' Knowledges: Ecology. Durham: Duke University Press.

Hall, S.. 1996. "On Postmodernism and Articulation: An Interview with Stuart Hall," edited by Lawrence Grossberg, in David Morley and Kuan-Hsing Chen, eds. Stuart Hall: Critical Dialogues in Cultural Studies. London: Routledge, 131-50. Reprinted from Journal of Communication Inquiry (1986) 10(2): 45-60. As cited by Li (2000).

Lawrence, D.C., M. Leighton, and D.R. Peart. 1995. Availability and extraction of forest producs in managed and primary forest around a Dayak village in West Kalimantan, Indonesia. Conservation Biology 9(1): 76-88.

Li, T. 2000. Articulating indigenous identity in Indonesia: Resource politics and the tribal slot. Comparative Studies in Society and History 42(1): 149-179.

Mather, A.S. 1990. Global Forest Resources. Portland, Oregon: Timber Press.

McCarthy, J.F. 2000. 'Wild logging': The rise and fall of logging networks and biodiversity conservation projects on Sumatra's rainforest frontier. Occasional Paper No. 31. October 2000. Center for International Forestry Research. Bogor, Indonesia.

McCarthy, J.F. 2001a. Decentralization, Local Communities and Forest Management in Barito Selatan. CIFOR draft paper September 18, 2001. Accessed December 8, 2001 from *http://www.cifor.org/highlights/Decentralization.htm.*

McCarthy, J.F. 2001b. Decentralization and Forest Management in Kupuas District. CIFOR draft paper September 18, 2001. Accessed December 8, 2001 from *http://www. cifor.org/highlights/Decentralization.htm.*

MFEC. 2000. Institutional Task Force for Forestry Sector Decentralization. December 1999-February 2000. Ministry of Forestry and Estate Crops. Jakarta, Indonesia.

Obidzinski, K. 2003. Illegal Logging in East Kalimantan, Indonesia: Political History of Expedience and Necessity. PhD Thesis. Amsterdam School for Social Science Research, University of Amsterdam, Amsterdam, Holland.

Oliver, C.D. and B.C. Larson. 1996. Forest Stand Dynamics. New York: John Wiley & Sons.

Paoli, G.D., M. Leighton, D.R. Peart, and I. Samsoedin. 2001. Economic and ecological analysis of gaharu wood (*Aquilaria malaccensis*) in Gunung Palung National Park. Conservation Biology 15: 1721-1732.

Peluso, N.L. 1991. Rich Forests, Poor People: Resource Control and Resistance in Java. Berkeley: University of California Press.

Peluso, N.L. 1993. Coercing conservation? The politics of state resource control. Global Environmental Change, June 1993: 199-217.

Peluso, N.L. and E. Harwell. 2001. Territory, custom, and the cultural politics of ethnic war in West Kalimantan, Indonesia. In: pp. 83-116. N.L. Peluso and M. Watts (eds.). Violent Environments. Ithaca: Cornell University Press.

Peluso, N.L. and C. Padoch. 1996. Changing resource rights in managed forests of West Kalimantan. In: pp. 121-136. C. Padoch and N.L. Peluso (eds.). Borneo in Transition: People, Forests, Conservation, and Development. London: Oxford University Press.

Peluso, N.L. and P. Vandergeest. 2001. Genealogies of the political forest and customary rights in Indonesia, Malaysia, and Thailand. Journal of Asian Studies 60(3): 761-812.

Poffenberger, M. 1997. Rethinking Indonesian forest policy. Asian Survey 37(5): 453-69.

Poffenberger, M. (ed.). 1999. Communities and Forest Management in Southeast Asia. Gland, Switzerland: IUCN.

Potter, L. and S. Badcock. 2001. The Effect of Indonesia's Decentralization on Forests and Estate Crops: Case Study of Riau Province, the Original Districts of Kampar and Indragiri Hulu. CIFOR draft paper September 18, 2001. Accessed December 8, 2001 from *http://www.cifor.org/highlights/Decentralization.htm.*

PP 6/1999. Forest Exploitation and Forest Product Harvest in Production Forests. Peraturan Pemerintah No. 6/1999. Accessed in Indonesian on November 27, 2001 from *http://www.mofrinet.cbn.net.id/informasi/.*

Redford, K.H. 1990. The ecologically noble savage. Cultural Survival Quarterly 15: 46-48.

Rhee, S. 2000. De facto decentralization and the management of natural resources in Eat Kalimantan during a period of transition. Asia-Pacific Community Forestry Newsletter 13(2): 34-40.

Rhee, S. 2003. De facto decentralization and community conflicts in East Kalimantan, Indonesia: Explanations from local history and implications for community forestry. In K. Abe, W. de Jong, and L. Tuck-Po (eds.). The Political Ecology of Tropical Forests in Southeast Asia: Historical Perspectives. Trans Pacific Press and Kyoto University Press, Melbourne and Kyoto.

Rice, R.E. and R.E. Gullison. 1997. Can sustainable management save tropical forests? Scientific American 276(4): 44-49.

Salafsky, N., B.L. Dugelby, and J.W. Terborgh. 1993. Can extractive reserves save the rain forest? An ecological and socioeconomic comparison of non-timber forest product extraction systems in Peten, Guatemala, and West Kalimantan, Indonesia. Conservation Biology 7(1): 39-52.

Schwartz, A. 1999. Nation in Waiting, 2nd edition. St. Leonards, Australia: Allen & Unwin.

Seymour, F.J. 1994. Are successful community-based conservation projects designed or discovered? In David Western and Michael Wright, eds., Natural Connections: Perspectives in Community-Based Conservation. Washington, DC: Island Press.

Thompson, M., M. Wharburton, and T. Hatley. 1986. Uncertainty on a Himalayan Scale: An Institutional Theory of Environmental Perception and a Strategic Framework for the Sustainable Development of the Himalaya. London: Ethnographica.

UU 41/1999. Basic Forestry Law No. 41/1999. Accessed in Indonesian on November 27, 2001 from *http://www.mofrinet.cbn.net.id/informasi/undang2/uu/je_41_99.htm* and in English from *http://www.eu-flb.or.id/htm/english/references.htm*.

Webb, C.O. and D.R. Peart. 2000. Habitat associations of trees and seedlings in a Bornean rain forest. Forest Ecology and Management 88: 464-78.

Wells, M., S. Guggenheim, A. Khan, W. Wardojo, and P. Jepson. 1999. Investing in Biodiversity: A Review of Indonesia's Integrated Conservation and Development Projects. Directions in Development Series. Washington, DC: World Bank (Indonesia and Pacific Islands Country Department).

Whitmore, T.C. 1984. Tropical Forests of the Far East. Oxford: Clarendon.

Zerner, C. 1993. Through a green lens: The construction of customary environmental law and community in Indonesia's Maluku Islands. Law and Society Review 28(5): 1079-1122.

Combating Corruption and Illegal Logging in Bénin, West Africa: Recommendations for Forest Sector Reform

Ute Siebert
Georg Elwert

SUMMARY. The article provides a thorough analysis of the forms of corruption and the actors involved, as well as a critical review of national forest policy and economy in Bénin, West Africa. Empirical data shows that the current state of affairs is not unalterable. Observable differentiations in the local perception of corruption and illegal logging show that a latent majority of people disapproves and that they can be mobilized for sanction activities. The policy-oriented recommendations of this article

Ute Siebert finalized her PhD thesis in February 2003 and submitted it to the Department of Social Anthropology, Drosselweg 1-3, Free University Berlin, 14195 Berlin, Germany. Her research on sacred groves and forest conservation in Bénin was funded by Heinrich Böll Foundation, Germany, and the "Comparative Studies of Societies" Program of the Free University Berlin and Humboldt University Berlin, Germany.

Georg Elwert is Professor of Social Anthropology, Free University Berlin, Germany. He conducted extensive fieldwork in Bénin from the 1970s onwards on forest management, agricultural development and social organization in the Bassila Region and other parts of the country.

The authors thank their colleagues in the Department of Social Anthropology, Free University Berlin, for their helpful comments on the slightly different pre-print version of this article.

[Haworth co-indexing entry note]: "Combating Corruption and Illegal Logging in Bénin, West Africa: Recommendations for Forest Sector Reform." Siebert, Ute, and Georg Elwert. Co-published simultaneously in *Journal of Sustainable Forestry* (The Haworth Press, Inc.) Vol. 19, No. 1/2/3, 2004, pp. 239-261; and: *Illegal Logging in the Tropics: Strategies for Cutting Crime* (ed: Ramsay M. Ravenel, Ilmi M. E. Granoff, and Carrie A. Magee) The Haworth Press, Inc., 2004, pp. 239-261. Single or multiple copies of this article are available for a fee from The Haworth Document Delivery Service [1-800-HAWORTH, 9:00 a.m. - 5:00 p.m. (EST). E-mail address: docdelivery@haworthpress.com].

offer ways to systematically reduce corruption and illegal logging: strengthening civil society (i.e., national NGOs) and promoting media coverage of these issues; introducing local sanctions in the framework of donor-assisted community forestry; adding anti-corruption conditions to foreign-sponsored forestry projects and forest policy reforms; enforcing decentralization in order to decrease the monopoly over forest resources currently enjoyed by state personnel; creating alternative income streams for local actors currently involved in illegal logging; and introducing land tenure reforms in order to increase local incentives for sustainable forest management. *[Article copies available for a fee from The Haworth Document Delivery Service: 1-800-HAWORTH. E-mail address: <docdelivery@ haworthpress.com> Website: <http://www.HaworthPress.com> © 2004 by The Haworth Press, Inc. All rights reserved.]*

KEYWORDS. Corruption, illegal logging, community forest management, land tenure, decentralization, legal pluralism, donor strategies, forest policy, timber, fuel wood, Bénin

INTRODUCTION

The awareness of corruption in forestry and illegal logging has increased within international discussion of environment and development. Today, many international organizations and NGOs (i.e., World Bank, FAO, WWF, Greenpeace) identify corruption as one of the key factors influencing forest decline in all regions of the world.

After looking at current trends in international debates on forest governance, this paper examines the situation of forest policy and forest economy in the West African country of Bénin. At first glance, Bénin does not seem to be a likely candidate for intense corruption in the forestry sector: with a relatively dry climate, it is not a country with vast forest cover and thus rarely exports timber. Unlike countries such as Ghana, Ivory Coast or Cameroon, Bénin does not host foreign logging industries.

However, there is a national timber market for which an estimated 80% to 90% of Bénin's timber resources are illegally logged with the consent of state forest officers (Siebert 2001, unpublished). The result is alarming: If the present deforestation rate continues, the country will be completely depleted of timber resources in seven years (Sulser January 31, 2000 personal communication). The effects are detrimental both for the environment and the climate (degradation and rapid loss of biodiversity), and for the national demand of wood (timber, construction and fuelwood) and non-timber forest products.

FORESTRY CORRUPTION ON THE INTERNATIONAL AGENDA

In May 2000, representatives of international NGOs, academia and international organizations met at Harvard University to discuss the problem of corruption and illegality in the forestry sector worldwide. The anti-corruption NGO Transparency International, together with The World Conservation Union (IUCN), Harvard University's Center for International Development, and the World Bank took the lead in setting up the Forest Integrity Network (FIN), an international network with the exclusive focus on forestry corruption. Over the past 10 years, international organizations, especially those active in development, have gradually become more explicit regarding the problem of corruption. Under the aegis of President James D. Wolfensohn, the World Bank has become more outspoken on the subject in recent years. Programs to reduce corruption within World Bank projects, as well as country-based assistance and learning programs to reduce corruption and improve "good governance" have been launched.

However, openness in addressing corruption issues has not spread quickly to all sectors of the World Bank. The World Bank forestry branch is still hesitant to even use the "c-word" out of fear to antagonize its government members. Countries providing the organization's budget and accepting World Bank loans and projects may be affected by forestry corruption. Circumscriptions are used instead: In September 2001, the World Bank and the government of Indonesia hosted the conference "Forest Law Enforcement and Governance," but the main focus of the meeting was on forestry corruption, illegal logging, and associated illegal trade in the East Asia region.

Another example is the FAO forest sector, which has only recently started to address corruption and illegal logging by using expressions such as "improving law compliance," in light of the sensitivities of its member governments.

The expressions "law compliance" and "law enforcement" show a strong orientation towards legal texts and juridical reform; however, both the World Bank and FAO acknowledged in recent conferences that broader social, economic and political issues are intricately linked with forestry corruption. Therefore, they now ask for "integrated" approaches to combat corruption.

Due to their clientele and membership, the World Bank and FAO presently delegate anti-corruption issues and concrete actions on-the-ground to international NGOs. Both organizations currently stress their interest in "building coalitions with civil society" in combating corruption, but stay vague in assessing their own specific contributions and options of influence.

METHODOLOGY OF THE STUDY

The analysis of forest use and the processes of illegal logging in Bénin are based on empirical fieldwork in the Region of Bassila in Northern Bénin, from April to June 1999 and from January to December 2000.

Research was carried out in an actor-oriented perspective (Long 1993) in order to identify all local actors involved in forest use and forest perceptions. Key informants were "incidentally" asked questions on corruption and illegal logging in 34 semi-structured interviews and many informal conversations with loggers, young male laborers, small scale local and urban entrepreneurs, local forest officers, male and female farmers, and traditional healers and priests. In general, it is not very common to hear comments on corruption or illegal activities from local people. When informants did speak about the situation, the circumstances often were not conducive for taking notes. In addition, statements in the interviews were often biased, depending on the informants' interests and the relation to the researcher. Therefore, there was no attempt to systematically document nor quantify the data.

In addition, valuable information was obtained from three high officers of the State Forest Service (DFRN), from two country experts of the German Technical Cooperation (GTZ), and from five agents of the GTZ forestry project *Projet Restauration des Ressources Forestières dans la Région Bassila* (PRRF).

This oral information was counter-checked by participatory observation of every day logging activities in the villages, which included observation of the number of timber truck loads passing through the villages, recruitment and remuneration of logging labor, as well as observation of procedures of the forest officers and of forest check points on the road.

The fact that this kind of information can be made available through investigation shows that there is room to maneuver in combating corruption in the forest sector.

FOREST RESOURCES AND FOREST USERS

Forest Resources

As one of the coastal West African countries on the Gulf of Guinea, Bénin is situated between Togo and Nigeria and is bordered on the north by Burkina Faso and Niger. Bénin stretches from 6° to 12° northern latitude with a surface of 112,600 km². Due to specific conditions, the climate of Bénin and Togo is dryer than in other coastal West African countries. Therefore, the West African rainforest belt stretching from Guinea to Cameroon is interrupted in Bénin and Togo.

As a result of the dry climate, Bénin's ecosystems and soils are very fragile. The semi-deciduous forests host the richest biodiversity in the country and are the most crucial factor for maintaining soil fertility and regular rainfall. The northern parts of the country have already suffered serious degradation due to forest loss and agricultural over-use. Due to one of the highest amounts of rainfall in the country (1,200-1,300 mm annually), the central part of Northern Bénin is the richest in forest resources. Large areas of dense, semi-dense, semi-deciduous forest and dense gallery forests along the waterways contain valuable timber species such as *Milicia excelsa, Antiaris africana, Diospyros spp., Khaya senegalensis, Khaya grandifolia, Afzelia africana, Isoberlinia doka, Parinari robusta* and *Ceiba pentandra*.

Because exact numbers on forest cover and its changes have not been recorded since the colonial period, the condition and change of Bénin's forests can only be estimated. FAO forest statistics of 1980 estimated a 13% to 20% overall loss of forest in the region of West Africa-the highest deforestation rate worldwide. As for Bénin, FAO statistics (which rely on satellite images and aerial photographs from 1975-76) state that the forest cover in 1975 consisted of 7,602,600 ha (67.5% of Bénin's territory) (FAO et al. 1980). In a 1990 worldwide forest survey by FAO, only 4,947,000 ha remained (44.7% of Bénin's territory) (FAO 1993). Another FAO statistic of 1995 recorded 3,470,000 ha (Fürstenberg and Leinert 1995, p. 5 and General Wood & Veneers Ltd. 1997, p. 40). This would mean a dramatic loss of forest cover totaling 4,132,600 ha between 1975 and 1995. Fairhead and Leach (1998) point out that FAO statistics on deforestation in Benin and West Africa are highly exaggerated. The authors argue that a "deforestation orthodoxy" based on wrong data (on which FAO statistics rely) has primarily served colonial and postcolonial interests for control of local populations' use of forests. They suggest that as a result, local populations have wrongly been accused for unsustainable use patterns. However, Siebert's empirical data shows that for the period from 1960 onwards, the argument of Fairhead and Leach does not hold for Benin and for the Bassila Region in particular. Interviews on vegetation developments in the last 40 years reveal that deforestation has been dramatic and is largely due to commercial logging and agriculture.

For example, the region of Bassila (*Sous-Préfecture de Bassila*) in the central part of Northern Bénin is the richest in forest resources. According to the data of a GTZ expert, 29,500 m³ of wood have been illegally logged between 1996 and 1999 in the project area of the GTZ *Projet de Restauration des Ressources Forestières* (PRRF), with a considerable increase from 6,707 m³ in 1998 to approximately 11,000 m³ in 1999 (Anonymous personal communication 2000). In an international comparison, these numbers seem low, but project consultants estimated in 2000 that if the current exploitation rate continues, timber trees in the region would completely disappear within 5 to 7 years (Sulser January 31, 2000, personal communication).

The maintenance and growth of forest cover in Bénin is therefore a very critical issue today, and also because of the fact that, thus far, there has not been a coordinated nation-wide effort of afforestation and deforestation control.

Forest Users

Many small-scale entrepreneurs and loggers on the local and regional levels engage in mostly illegal timber logging and in the harvesting of construction wood. They recruit other loggers and unskilled labor in the villages close to forested areas. The loggers are mostly middle-aged male farmers and the unskilled laborers are young men between the ages of 16 and 25.

Both male and female farmers close to forest resources engage in commercial charcoal production, which involves the collection of dry firewood, as well as illegal felling. The technological level of charcoal production is rather low, resulting in waste. Small entrepreneurs transport the bags of charcoal from rural areas into the cities. Occasionally, charcoal is sold directly to consumers, who pass by on the big roads.

Male farmers cultivate yams, the most important source of carbohydrates, on forestland because it requires highly fertile soil. The extensive cultivation system of shifting agriculture combined with demographic growth (INSAE 1999, 9), commercial logging and its associated road networks, the introduction of cash crops such as cotton, as well as immigration has led to a considerable expansion of cultivated areas into forests in many regions. In addition, male farmers go into gallery forests in order to harvest oil palms (*Elaeis guineensis*) for palm nuts. In the savanna, they harvest the néré tree (*Parkia biglobosa*) for its fruit, the néré beans, and hunt in forested areas.

In accordance with their domestic duties, female farmers engage in the collection of forest products for household subsistence: firewood, which does not involve the felling of fresh trees, fruit, nuts (most importantly sheanuts from the sheabutter tree *Vitellaria paradoxa*), bark, leaves, mushrooms, and medicinal plants. The latter foodstuffs are crucial complements for the local diet dominated by carbohydrates, providing vitamins and minerals. These forest products are especially important as nutrients at the end of the dry season in April, which is considered the "hungry" period between the harvests in the system of shifting agriculture.

Male and female autochthonous healers harvest forests for herbs, leaves, bark and roots for their various healing qualities. Nomadic pastoralists use forested areas as grazing grounds for their cattle. Since cultivated areas due to growing cotton as a cash crop have increased considerably in the last decades, the pastoralists are being pushed into the forests.

THE ROLE OF TIMBER AND WOOD ON THE NATIONAL AND INTERNATIONAL MARKET

Due to a lack of appropriate forest resources, Bénin is considered a timber exporting country. On the national market, however, timber and wood are an important resource to meet the country's national demand in carpentry, as construction material or as fuel. The official share of 2.8% for timber and wood in Bénin's GNP is very low (General Wood & Veneers 1997, p. 17). However, the figures are low since illegal transactions are not represented in the GNP and the timber and wood market is a part of the country's shadow economy. According to our estimation, the real percentage may attain 16%.

The lack of both forest cover statistics and data on the national wood market is a serious impediment to the control and planning of the national forest economy. It has to be noted that the lack of statistics may be linked with the interest of state forest officials to conceal illegal forest exploitation.

Bénin only has a marginal timber industry with four small sawmills. In addition, there are many small-scale entrepreneurs as well as individual loggers working with chain saws and an estimated 5,500 carpentry shops that operate on a very low technological level (Brüntrup 2000, p. 11).

Most of Bénin's annual demand for wood of 7,779,000 m³ is met by the exploitation of natural forests. Approximately 80% to 90 % of timber is felled illegally with the consent or encouragement of state forest officers.

Timber

The annual national demand for timber consists of an estimated 52,000 m³ of teak (*Tectona grandis*) and 60,000 m³ of indigenous wood species, such as *Khaya senegalensis*, *Ceiba pentandra*, *Khaya grandifolia*, *Antiaris africana*, *Milicia excelsa*, *Isoberlinia doka*, *Pterocarpus erinaceus*, and *Afzelia africana*. The demand for teak can be currently met by state and private plantations. State plantations offer approximately 24,000 ha of trees of which 50% are teak, and non-state plantations offer approximately 11,750 ha of trees which are all teak (General Wood & Veneers 1997, pp. 42-44).

As for red and white wood, 50% of the demand can be sustainably met by national forests and the other 50% are allegedly imported from Nigeria and Togo, according to the study of General Wood & Veneers (1997, 58ff). However, based on empirical data, approximately 90% of "imports" from neighboring countries are, in fact, re-imports of timber illegally felled in Bénin

(Siebert 2000, unpublished). This exceeds the sustainable yield of national timber resources by 27,000 m³.

Construction Wood

Annually, 25,000 m³ of construction wood is consumed at the national level. A shortage in this segment of the forest sector has not yet been identified (General Wood & Veneers 1997, p. 105).

Fuelwood/Charcoal

As one of the poorer developing countries, Bénin is characterized by a very low standard of technology. Due to a lack in new energy technology, the country mainly depends on charcoal and fire wood, both in the cities and in rural areas. Parallel to the rapid growth of Bénin's cities, the demand for charcoal grows annually. The current yearly demand of 7,642,000 m³ can hardly be met and thus a serious fuelwood shortage has to be expected in the near future. At the same time, General Wood & Veneers point to a general lack of strategies for alternative sources of fuel in Bénin (1997, pp. 98-100). The numbers show a need for exploitation and afforestation in state forest reserves, private forests and plantations in the timber segment. In the fuelwood segment, more rational charcoal production and more fuelwood plantations as well as an introduction of alternative energy sources are highly recommended.

NATIONAL FOREST POLITICS AND REGULATION

Bénin's extractive forest economy and policing forest politics of the colonial period continued until the democratization period in 1989 and 1990, when external donors' willingness to provide development aid in Bénin increased notably. German and French donors as well as the World Bank pressed for a reform of the forest policy, aiming at participatory forestry and a more rational use of forest resources. The financing of this reform by the FAO, World Bank, Germany and France started in the early 1990s with the *Projet de Gestion des Ressources Naturelles* (PGRN) and resulted in the new forest law *Loi No. 93-009* in 1993 and its operational guidelines in 1996 (*Décret No. 96-271*).

According to the new forest law of 1993, which replaces the forest code of 1987, the entire forest domain remains state property. There are two forest zones: *forêts classées* and *forêts protégées*. In both types of forests, timber felling is possible through the purchase of permits. The State grants the rights of use to individuals and collectives, which are very restricted in the case of *forêts classées*. In the forests of the *domaine protégé*, the population is granted more extensive rights of utilization regarding agriculture, pastures, as well as the collection and exploitation of forest products.

Even though the forest law of 1993 was supposed to emphasize the participation of local populations and grant more rights of use in accordance with sustainable land management plans (Art. 40), two-thirds or more than 70 Articles of the law regulate timber exploitation and sanctions in a policing manner (Art. 41-112). To summarize regulations defined by this law, loggers have to obtain a professional identity card and felling permits. The profit from felling has to be declared and is taxed by the state. In order to obtain felling permits, a logger has to declare the individual trees he wants to cut and the chief forest officer has to check the respective tree species and diameter. In its annex, the forest law lists all valuable timber species of Bénin as protected species for which there are supposed to be logging limits. Only mature trees are to be felled, and the diameter has to be sufficiently large so that a tree provides a specific number of boards (board standard sizes 30 cm × 8 cm × 4.20 m, or 3.20 m).

After the chief forest officer has again counter-checked logs and the permit after the felling, he has to issue a *laissez-passer* for transport and a declaration of origin of the timber. These documents must be checked at several forest checkpoints on the main roads leading to the wood markets in the cities.

The high number of regulations provides opportunities for state officials to obstruct forest users. The latter, mainly loggers, thus feel encouraged to resort to corruption in order to speed up formalities. Van Klaveren (1957) showed that these types of legal regulations are meant to create illegal income opportunities rather than effectively reduce illegal activities.

According to the law, it is prohibited to cut tree trunks into boards with chain saws after the tree has been felled (Art. 53) thereby curbing the production of timber. It is also illegal to transport timber cut into boards by chain saws unless it is imported. In a country where all forest regions are a maximum distance of 60 km from national borders, this regulation sounds like an invitation to fraud. Timber can easily be cut in Bénin's forests, transported over a border and then be declared as import. This is a current practice in the region of Bassila, where timber is often declared as imported from Togo. In August 2000, this illegal practice was acknowledged by DFRN (*Direction des Forêts et Ressources Naturelles*) high officials when their Togolese colleagues pointed out that their country does not offer the forest potential for all the timber declared as originating from Togo.

In the case of law violations, state forest officers of the DFRN national forest service must enforce various sanctions, ranging from monetary penalties (US$ 9 to US$ 835) to prison sentences (15 days to 6 months) (Art. 88-104). The penalties and sanctions stipulated in the forest law are sufficiently high on their upper limits; however, due to lack of enforcement they are ineffective.

Bénin's current forest politics and policies are largely dependent on external donors and projects, concerning both policies and activities, such as the introduction of participatory forestry and forest management plans. In 1999, the World Bank and other donors agreed to finance the preparatory phase of a new big project PGFTR (*Programme de Gestion des Forêts et Terroirs Riverains*), meant to further enhance participatory management of state forest reserves and coordinate forestry projects on the national level. In this phase, the donors financed an audit of the DFRN in 1999. The audit revealed that the forestry reform of the early 1990s had not been successful due to an insufficient number of personnel and a failed implementation of the participatory approach (MDR and PGFTR 1999, 9). For the first time in DFRN history, problems of corruption and engagement in illegal felling were openly addressed (MDR and PGFTR 1999, 11-13, 15). As a precondition for financing of the new project, PGFTR's main phase, the donors are asking for a restructuring of the DFRN into a "*nouveau service forestier*" (Kakpo December 5, 2000, personal communication). Over all, the audit recommends an annual increase of 80 forest officers from 2000 until 2010, because at the moment, each forest officer controls an area of 8,180 ha while lacking the necessary means of transport. Furthermore, the audit asks for an increased recruitment of sociologists as forest officers in order to build a "*nouvelle génération*" in the line of participatory forestry.

Political and administrative decentralization is another relevant issue in Bénin's forest policies and politics. Since the early 1990s, donors and political opposition parties have been pressing those in power to decentralize in order to increase political participation on the local level. According to the new laws on decentralization such as law 97-029 of 1999 on the organization of the communities, communities will substitute the *Sous Préfectures* and will be vested with financial and administrative autonomy. They can become managers of natural resources similar to the state, following a community development plan in order to organize infrastructure and the management of natural resources. Forests will thus constitute a part of the village and community capital by which local necessities such as road construction, medical units, schools and teachers will be financed.

However, officials within the DFRN are still very much opposed to share responsibilities for forests with the local population. They argue that local populations are too incompetent in management and would quickly deplete the resources. This resistance bears the fear of losing discretionary power over sources of illegal income, since decentralization may lead to radical change of the old power structure by opening the political arena for new players.

Geared towards decentralization, German development aid has recently introduced a compromise structure labeled "participation." The PRRF and *Programme de Gestion des Terroirs et des Ressources Naturelles* (PGTRN), as well as the former *Projet Gestion des Ressources Naturelles* (PGRN), facilitate contracts and management plans (*plans d'aménagement*) between the forest administration and village communities or local individuals. Between 1996 and 2000, five forest management plans (*plans d'aménagement*) for state reserves were implemented by PGRN and PRRF. In 2000, the Banque Africaine du Développement granted finance for the design of two more management plans for state reserves and their implementation.

According to the forest law of 1993, management plans for state forest reserves (*forêts classées*) stipulate restricted logging rights, as well as rights to collect fuelwood, forest fruit and plants for village communities neighboring the forest reserve. According to local land tenure, the so-called "private forests" (*forêts privées*) are usually owned by a lineage. Managers can be lineage members or other individuals upon agreement by the owning lineage. For a private forest, management plans stipulate the same rights as in the case of *forêts classées*, but the beneficiaries are individuals and lineages. For the management of private forests, the PRRF had difficulties in finding local partners because forest owning lineages fear expropriation upon the end of the project phase. This fear is a consequence of former experiences of expropriation in the 1940s and 1950s, when the French colonial administration appropriated state forest reserves from lineage land without local consent.

In order to appease forest managers, the PRRF introduced so-called "*attestations de propriété*" with a signature by the *Sous Préfet* of the Bassila region testifying the ownership of a forest by a given lineage. However, this paper does not have the legal status of a land title, which is officially recognized by state land law. Local forest managers of the villages of Kodowari and Pénélan in the Bassila region know about this fact. Expecting a complete loss of their rights after the PRRF ends, they are not motivated for long-term investment. For village land (*terroir villageois*), management plans (*plans de gestion de terroir*) are made which do not only include forests, but also land used for cultivation, habitation and pasture, and thus stipulate broader use rights. So-called *structure villageoises*, as an ensemble of village-based management committees, are introduced into the villages in order to organize collective management of either *forêts classées* or village land. The benefits stemming from use of the resources such as timber, plantations and other forest products are divided between the DFRN, a fund for forest management, the administration of the *structure villageoise*, and the individual users (PRRF 1998 and PRRF 1999).

However, due to a lack of official recognition of local land tenure arrangements, rights to private forests and village land only have an usufructuary character. As for state forest reserves, the rights conceded appear to be legally better stabilized, but they too merely consist of restricted usufructuary rights.

CORRUPTION AND ILLEGAL ACTIVITIES
IN THE FOREST SECTOR

Within the forest sector, timber production is most prominently affected by corruption. Therefore, this analysis focuses on timber only. Illegal logging is done in both state forest reserves and forests of the *domaine protégé* including *forêts privées*. Although there is corruption in the management of state plantations as well, it is relatively small-scale compared to the corruption in natural forests.

Concepts and Perceptions of Corruption and Illegality

From a social science perspective, we have to distinguish between analytical or etic and endogenous or emic categories of corruption and illegal activities because in international debates on combating corruption, this distinction is rarely made, leading to flawed policies lacking adjustment to local situations.

Corruption as a neutral analytical category in this study is defined as the sale of services by state officials for their personal benefit in spite of formal laws, and the tolerance of illegal acts is such a service. Corruption is accompanied by abuses of power, such as the use of the public officials' authority for committing illegal acts (e.g., forest officers committing illegal logging and engaging in illegal timber trade). Purely illegal acts, such as theft of public or private property, are committed by private individuals and companies. Corruption should also be differentiated from embezzlement because of their different economic dynamics. Embezzlement is limited by the resources accessible to the authority holder. Corruption has, however, a creative potential. In a corrupt system, there are services sold without which one could do very well. Many services for which bribes are paid are artificially created.

Inquiring into endogenous concepts and perceptions of corruption and illegal acts is crucial because in rural Bénin, as in many West African countries, there is a *de facto* legal pluralism which is not officially recognized by the state. What constitutes property, theft, and corruption from the state's point of view may differ from local perspectives. Therefore, we must take into account endogenous concepts of forest-related property rights. We have to ask if forests are conceived as a possible target for theft and what types of duties and expectations are attributed to state officials and local authorities.

According to autochthonous land tenure of the Bassila region, forests and land are collectively owned by patrilinear lineages. The lineage head is the administrator of forests and land for the male lineage members. He does not have exclusive property rights and is not allowed to dispose of the land (i.e., sell it). His duty is to grant all men of his lineage permanent rights of use such as cutting trees, harvesting trees (e.g., oil palm, néré, shea), fishing and hunting rights, and the right to cultivate, build a house and to put out beehives. The men have to concede some use rights to their wives, who married into the lineage. These rights are economically important as they concern the rights to collect tree products and firewood, and to a limited extent cultivation rights.

These local land tenure concepts clash with the legal version of land tenure for the state of Bénin. As shown above, the state defines forests as state property, according to the French *droit domanial*, and gives out restricted usufruct rights, especially in the case of state forest reserves. Thus, for many local people, the state's control over forests, especially regarding forest reserves, constitutes illegal expropriation. Lineage heads often point to the exclusive right of the firstcomers and their kin to own land or give it away to others. They stress that their lineage did not give the land to the colonists or later to the state. Accordingly, locals often disregard the severely restricted use regulations in state forest reserves. Lineage heads are still giving out various use rights for cultivation, collecting and hunting in these areas. Under state law, the local population thus became criminalized as illegal trespassers and thieves.

Due to this legal pluralism, the first officially authorized external loggers coming into the region at the end of the 1950s also had to get permission by lineage heads for logging. When logging was taken over by local actors in the 1980s, they also sought logging agreements with lineage heads if they were to cut on another lineages' land. These lineage heads gradually exceeded their role as administrators by disposing of forests without consent of lineage members. At the same time, local loggers began to offer bribes to the state foresters.

Not surprisingly, our empirical data shows that endogenous concepts of corruption and illegality vary depending on the individual's role and interest. The people who profit from logging positively say that the state foresters "are nice because they let us do our work." They agree that the foresters' behavior is against the law, but say that state law does not mean anything to them apart from the fact that they have to pay bribes because of the law. Regarding lineage headmen, loggers point out that they profit from easy agreements, but acknowledge that the headmen act improperly.

The people excluded from logging profits, which is the majority of the local population, refer more negatively to the activities of foresters. In some cases, entire village populations are excluded from timber profits, but are the targets of extensive logging by actors from other villages. These populations acknowledge an abuse of power by the state foresters and claim that, "They are not doing their jobs." They also criticize the active roles of state foresters in uncontrolled logging. Some people are concerned about vanishing state law, and the privilege given to a few wealthy entrepreneurs to circumvent the law.

Lineage headmen are accused by villagers of the same lineage or other lineages of misusing their position by stealing collective property, and some villagers have recently formed control committees to defend their forest property. Foresters, loggers and lineage headmen are blamed for the current anarchical mode of logging, as well as for the negative effects of rapid deforestation (i.e., degraded soil, less rainfall and lack of non-timber-forest products). Even though most villagers do not recognize the state's control over forests, they consider the misuse of power of state officials worse than that of lineage headmen. In their view, foresters win doubly, first by being paid a government salary, and secondly by accepting or demanding bribes.

Theft of trees and forests constitutes a locally recognized concept. Even in the face of legal pluralism, there are clear-cut endogenous perceptions of proper behavior of both state and local authorities and deviations from it. The description and evaluation of these deviations show many parallels to our analytical categories of corruption and illegal activities.

The Actors

Forest service personnel and loggers are organized into clientelist and corrupt networks encompassing the local, departmental, and national levels of the forest sector. The networks started in the 1980s with a covert market for local individual actors as successors of officially authorized foreign logging entrepreneurs. Corruption of foresters started when they were approached by local loggers, for whom official authorization was costly. They asked the foresters for advice on where to find suitable trees and for support with their illegal transport. Often, corruption also started when foresters caught illegal loggers who offered bribes to them, and developed when foresters took the initiative themselves. By the year 2000, individual contacts in the illegal market became stabilized. Certain individuals on all levels sustained the corrupt network. At the community level, chief forest officers, their aides, and the *Sous Préfecture* administration may be implicated. On the departmental level, the departmental forest service directors and their personnel may be involved. On the national level, agents within the administrative bodies of the DFRN and the overarching ministry for agriculture may play a role.

At the local level, the chief forest officers (*Chef Cantonnement Forestier*) and his aides are theoretically obliged to enforce the forest law by checking timber exploitation. In many cases, however, forest officers bribed by entrepreneurs and loggers ignore their duty of law enforcement and often perpetrate illegal logging themselves. In the latter case, they maintain "business" relations with urban traders.

In current practices, the chief forest officer issues false papers with the consent of the departmental director of the DFRN and the national administration. Benefits of corruption and illegal logging are distributed within the network from the local chief forest officers to the departmental director and from there up to the national administrative level. The forest checkpoints on the main roads to the timber markets are bribed directly by the entrepreneurs.

State officials within the network maintain ambiguous relations with legal norms. They uphold the appearance of order and legality by insisting on authorizations because they can sell their authorization service by way of issuing false papers. In addition, they produce this orderly façade because there are also officials within the administrative DFRN structure who are not implicated in corrupt practices and often maintain good relations with external donors.

The logging of timber is mainly done by locally or regionally based entrepreneurs who hire loggers and laborers, or by independent local loggers. Loggers also often lease chain saws from entrepreneurs. Both types of actors offer the timber to bigger traders from the cities who visit the forested areas. The entrepreneurs organize logging and illegal sawing with chain saws into boards and transport the boards to the timber depots of urban centers. One truckload usually consists of 100 boards. If entrepreneurs and loggers extract wood from forests outside the state reserves, they usually operate in lineage-owned forests (*forêts privées*). According to customary arrangements, they ask the lineage headmen for permission to cut trees.

The traders in the cities own timber depots and sell timber by the board to carpenters or engage in occasional export deals. Often, traders commission timber from corrupt chief forest officers in the forested regions. In areas close to borders, customs agents have to check imports and exports of timber and are often involved in corruption. It is their role to issue false import certificates.

Corrupt Practices and Illegal Activities

Being paid small sums of money by loggers and entrepreneurs for their consent to cut trees on their lineage's land, lineage headmen often appropriate this money for their own profit. Thus, the payment can be considered as embezzlement. After the payment of lineage caretakers, loggers often engage in theft by cutting many more trees than agreed.

In many cases, entrepreneurs lease chain saws to individual loggers for approximately US\$ 1,100 for 12 months and ask for a return commission of another US\$ 1,100 at the end of the year. Many individuals interested in logging depend on such outrageous deals because there is very limited access to chain saws in the country. Consequently, many loggers find themselves under enormous pressure to meet the lease requirements and to earn a profit. Some entrepreneurs enlist violent gangs who threaten non-compliant leasing clients. This threat forces loggers into very unsustainable and wasteful logging. Often young trees that only provide one board are cut.

In addition, the numerous regulations of the forest law are often not sufficiently transparent to loggers and cause them to consent to offers of corruption by forest officers if they want to avoid a blockade or loss of time. Bribing foresters extremely neutralizes the costs of logging authorizations because instead of the usual authorization over approximately five trees, the logger can now cut up to 30 trees.

In their corrupt activities, state forest officers do not differentiate between the extraction from state forest reserves or from lineage forests. The entrepreneur can extract numerous truckloads of timber with one permit, which officially may allow for the felling of approximately five trees. The forest officers also ignore the fact that all loggers use prohibited chain saws in order to produce the boards for trade on the spot.

Bribes for forest officers consist of a considerable share of the entrepreneurs' profit per load. Some chief forest officers have direct business contacts with urban traders and hire local entrepreneurs and loggers for the extraction and delivery of timber. Costs for entrepreneurs for one load of timber (100 boards), without the fee for the permit, and including the salary for workers, fuel and materials amounts to approximately US$ 280. This leaves a profit of US$ 700 to US$ 1,400 per load for the entrepreneur, who will give about 50% to the chief forest officer. Compared to the average domestic income of US$ 15 to US$ 70 per month, this profit is considerable.

The DFRN personnel at the check-points receive approximately US$ 75 to US$ 170 in bribes per load in order to ignore missing or wrong permits, accept false declarations of origin, and allow the transport of boards cut by chain saws. The same applies to custom agents. In the Bassila region, regional entrepreneurs often transport timber over the nearby border to Togo, and re-import it as "*bois togolais.*"

It is easy for forest officers not to do their job because there are no forest inventories or felling statistics that need to be declared to the director of the DFRN at the *Département* level, nor are there any internal or external work controls on the job.

RECOMMENDATIONS FOR COMBATING FOREST CORRUPTION

In Bénin, corruption is an every day reality in many governmental services and domains. This is especially due to a lack of rule of law, information and transparency of decision-making, and deficiencies in the juridical system. Some authors such as Contreras-Hermosilla (2001) argue that if several governmental sectors are infested by corruption, there is little hope for success in combating corruption in one single sector. However, fighting corruption has to begin somewhere, and action in a single sector such as the forest sector can lead to success in other areas as well, if a sufficient number of cooperating actors in these sectors can be identified.

Even though Bénin is in the process of political transformation in the form of democratization and decentralization, it can still be characterized as a Command State, which is a system of politics and administration characterized by the primacy of present authority in daily life interaction and an ambiguous relation with legal norms (Elwert 2001). These norms only serve to define the overall power sharing within the state apparatus. A Command State has a reversed normative hierarchy compared to the state under the rule of law. Whereas the constitution ranks highest in a state under the rule of law, the Command State gives priority to ad-hoc commands often orally voiced and to by-laws. It is important to note that in a Command State, written rules are not only façades. They play an important role in the sectoral and regional division of labor and income, which is to a large extent drawn from corruption.

One of the most salient features of the Command State is clientelism, which is the dominant economic structure for the organization of economic relations. Economic appropriation and distribution is organized by sectoral clientelist and corrupt networks, which can reach from the local to the highest administrative level. Through this clientelist structure, actors who are newcomers or do not have the economic means to participate in corruption, are locked out from access to the controlled resources and from political influence. Therefore, the innovation capacity of a command state is severely reduced, explaining its stability. This also explains why the decentralization process in Bénin, as in many West African countries, has been so slow and has primarily been promoted by external donors, political opposition, and NGOs.

In the perspective of the Command State system, isolated measures such as reforms of legal texts will not be effective since the laws are treated ambiguously or are not applied at all. This aspect should be taken into account by international organizations, which currently focus to a great extent on the reform of forest laws. Thus, more complete political reforms such as decentralization in order to bring about a state under the rule of law, an independent law system, and the strengthening of political transparency and participation of local populations will provide more sustainable solutions.

As Bierschenk and Olivier de Sardan (1998, p. 12) argue, there is a trend in current development discourse to celebrate decentralization as a panacea for all sorts of problems. We agree that it is questionable that formal decentralization in itself will reduce corruption and dismantle the Command State system. Likewise, it has been shown that a strengthening of civil society does not necessarily reduce corruption (Sommer 2000). However, if political decentralization is accompanied by an increasing flow of information by critical mass media and by building new institutions of political participation, it will provide an opening to formerly excluded actors as potential controllers of state performance and a higher level of public information. Both may cause a redefinition of the normative hierarchy of the Command State and strengthen the rule of law as a defense for the weaker political actors against power abuse.

Concentrating on immediate measures of combating corruption in the forest sector, simultaneous action along four lines is recommended:

Publicity and Increase in Norm Awareness

If facts on corruption are made known to a wide public, the systemic lack of information on which the Command State thrives will be reduced. The population will be in a more ready position to hold state officials accountable for their activities. International donors have an important role to play in strengthening a critical and diverse mass media as watchdogs of corruption and provide further training in critical journalism. Given the low literacy rate of 20% in rural areas, investment in more local radio stations will enhance the effective spread of information.

Also, NGOs and civil society organizations are well placed to disseminate legal know-how and information to local populations in the field of local self-governance and forest management. International NGOs and international organizations have the capacity to establish training programs on local mobilization, self-governance and forest control. In addition, a country-wide campaign would be an important asset to inform local populations that forests constitute their capital for village development after decentralization.

Internal Sanctioning

If the local population is engaged in the fight against illegal logging, forests will be more effectively controlled. Often, villagers are divided into actors who profit from illegal logging such as local loggers, young labor, small scale entrepreneurs and a majority for whom forest corruption means a loss like women, excluded male youth, excluded farmers and traditional healers. Once mobilized, this local majority can play an active role in curbing illegal logging and controlling state forest officers. This mobilization can be achieved if donor projects and local NGOs rapidly and systematically inform villagers in forested areas that forests constitute their capital in decentralization. In the case of three villages in the Bassila region, control committees are already defending forests with some success against the interests of a few.

In addition, more pluralistic institutions of natural resource management will decrease the discretionary power of the public sector. Participation of local actors can be broadened by encouraging collaboration and contract management between various stakeholders in the public and private sectors as well as local NGOs and local users (Venema and Van den Breemer 1999; von Stieglitz 2000). In some areas, a first step might be to introduce "*codes locaux*" in the project area (GTZ 2000). These local codes can provide clear norms and sanctions to be applied by the local population and thus reduce the legal monopoly of state officials who ambiguously apply law. Such a code can enforce sustainable forest management plans, sanction forest use on the village level and make local actors on the village level responsible for forest control. After full implementation of decentralization, the codes can be incorporated into the statutes of the communities. Locally accessible independent court systems, new institutions for collaborative management, such as multi-stakeholder management committees and local codes may help local actors to prosecute and penalize illegal activities of loggers, labor, lineage headmen and entrepreneurs by providing the resources needed to bring a case to court. In addition, donors can offer know-how concerning conflict management and sanction procedures.

External Enforcement

The reform of DFRN personnel and structure, as recommended in the PGFTR audit, is not far-reaching enough. Increasing the number of agents and replacing some agents with others will not automatically curb corruption, but rather, it may increase the number of beneficiaries of corruption. Also, it is doubtful whether the measure of sophisticating the DFRN by hiring more sociologists will increase integrity.

It will be more effective to sanction theft and corruption when internationally funded projects are affected. The donors could set an example by adding conditionalities to projects along these lines. Such sanctions will be most powerful if they imply:

1. The publication of facts and persons involved in illegal logging in local and national mass media.
2. The requirement of the state to sanction known cases of corruption.
3. Control of state forest officers as a finance conditionality given the considerable lack of integrity of DFRN personnel. Special committees consisting of international donors' personnel and state agents external to the DFRN can control forest agents and forest check points on the local level and ask for updated forest inventories at least over the next 5 years.
4. The revocation of funds that were used against the spirit of specific agreements. Needless to say, this requires a concerted effort by all donors in the forest sector of Bénin.

The World Bank, Germany, the Netherlands, and others must take a unified stance in order to effectively sanction the state. Contradictory activities and goals of donors generally encourage corruption. Thus, a strict coordination of donor programs in all sectors is necessary. International NGOs are well placed to lobby within the donor community for stricter conditionalities and their enforcement.

Change of Structural Conditions Facilitating Corruption

Current bureaucratic regulations, which are time consuming, allow for arbitrariness and block economic action. If the 1993 forest law is revised in order to increase its transparency for all forest users, room for ambiguity in law enforcement will be decreased.

In addition, if the participation of the population including socially underrepresented groups such as women, pastoralists and immigrants is increased in the new decentralized structures (e.g., community committees); the monopoly of the state personnel will be reduced. A new role of the DFRN should be considered. It could from now on only intervene as an advising body, while communities–as decentralized political units with their forest management committees–could control the forests in their area. Donors and NGOs have an important role to play in local capacity building, in creating local expertise in self-government, and administration.

Furthermore, sustainable management of forests will only be achieved if local actors, especially young men who are the largest group of beneficiaries of illegal logging, are faced with alternative sources of local income. Given the strong local interest in teak and cashew, subsidies and aid for plantations and afforestation might be suitable for initial efforts.

Most importantly, land tenure reform is needed in order to increase security of property and user rights, especially in rural areas. The current insecurity in land tenure, stemming from ambiguous pluralism between state and autochthonous tenure systems, hampers economic investment. For example, poor, credit-seeking farmers and rural women cannot give their land as security for lack of formal land titles. Secure land tenure is especially crucial in forest resource management and afforestation. Donor projects working in resource management often fail to realize that without legal security and predictability, such an investment may not seem economical to local investors. The value of land increases with the planting and protection of trees and becomes an object of interest for those in power. If the project with quasi-legal security ends, local stakeholders will fear expropriation by private and public power holders.

Thus, a combination of resource management and tenure reform is overdue. Such a reform requires the harmonization of state and autochthonous land tenure. Or it could consist of the introduction and controlled keeping of land registers. In this case, multiple use rights of socially vulnerable groups such as women, pastoralists, and immigrants must be secured on registered land, which can be owned by individuals, lineage-based institutions or cooperatives (see also Elwert 1999).

Such reforms will of course not end all problems in this sector. In the case of fuelwood, 90% of Bénin's demand is currently met by charcoal and firewood. This consumption can be considerably reduced by introducing alternative sources of fuel such as solar energy and gas. Less wasteful charcoal production and more fire wood plantations might be additional options.

CONCLUSION

Illegal logging threatens Bénin's last forests. The legitimacy of and the respect for property rights and protective regulations seems flexible. However, theft of trees and forests are locally recognized concepts. Local people differentiate between proper behavior of both state and local authorities and deviations from it.

The corruption of state officials plays a major role in de-legitimizing public property in forests. Protective regulations, which equally affect private forests and wood plantations, seem to be mere annoyances that can be bribed away by corruption money.

Corruption in our case does not seem to be unalterable. Observable differentiations allow us to suggest possible paths for its reduction. Since there is a latent majority that disapproves of corruption, transparency and public information can work as sanctions, and support for material sanctions can be mobilized. If laws with overly complex regulation encourage corruption as "speed money," the laws need to be modified. The state has to monitor where its capacity to observe and sanction illegal acts is insufficient and should proceed to remedy it. If international donors add conditions to their aid requiring effective sanctioning of corruption, this would have a strong effect given the dominant role of the donors in the national budget. However, such conditionalities will only work if the donors are strictly coordinated and leave no loopholes or opportunities for donor-shopping.

REFERENCES

Bierschenk, T. and J.-P. Olivier de Sardan (Ed.). 1998. Les Pouvoirs au Village. Le Bénin rural entre Démocratisation et Décentralisation. Karthala, Paris.

Brüntrup, M. 2000. Synthèse de l'Analyse économique du Projet bénino-allemand de Développement forestier et de l'Économie du bois. GTZ, Cotonou.

Contreras-Hermosilla, A. 2001. Forest Law Enforcement. Retrieved August 15, 2001 from http://wbweb4.worldbank.org/nars/eWorkSpace/eWS004/groupware/WebRes.asp.

Diaby-Pentzlin, F. 2001, May. Rechtsprojekte in der technischen Zusammenarbeit und Rechtspluralismus: Rahmenbedingungen oder Projektaktivität? Paper presented at the German Sociological Society, Section Sociology of Development and Social Anthropology (ESSA). Center for Development Research, Bonn, Germany.

Ehrhardt-Martinez, K. 1998. Social Determinants of Deforestation in Developing Countries: A Cross-National Study. *Social Forces* 77(2):567-586.

Elwert, G. 1987. Ausdehnung der Käuflichkeit und Einbettung der Wirtschaft. Markt und Moralökonomie. In: pp. 300-321. K. Heinemann (Ed.). Soziologie wirtschaftlichen Handelns. Sonderheft Kölner Zeitschrift für Soziologie und Sozialpsychologie 28. Westdeutscher Verlag, Opladen, Germany.

Elwert, G. 1999. Landreform und Rechtssicherheit. In: pp.8-24. E. Alber and J. Eckert (Ed.). Schlichtung von Landkonflikten-ein Workshop. GTZ, Eschborn.

Elwert, G. 2001. The Command State in Africa. State deficiency, Clientelism and power-locked Economies. In: pp. 419-452. I. Cornelssen and S. Wippel (Ed.). Entwicklungsperspektiven im Kontext wachsender Komplexität. Weltforum, Bonn, Germany.

Fairhead, J. and M. Leach. 1998. Reframing Deforestation. Routlegde, London.

Food and Agriculture Organization of the United Nations/United Nations Environmental Programme. 1980. Global Monitoring System Pilot Project on Tropical Forest Cover Monitoring: Benin-Cameroon-Togo Project Implementation. Methodology, Results and Conclusions. FAO Forestry Department, Rome, Italy.

Food and Agriculture Organization of the United Nations. 1993. Forest Resources Assessment 1990. Tropical Countries. FAO Forestry Paper 112. FAO Forestry Department Rome, Italy.

Food and Agriculture Organization of the United Nations. 2002. Retrieved February 2, 2002 from http://www.fao.org/waicent/ois/press_ne/2002/2240-en.html.

Fürstenberg, Peter and S. Leinert. 1995. Möglichkeiten zur Umsetzung der nationalen Forstpolitik in der Republik Benin. Institutionalisierung der forstpolitischen Aufgabenfelder. GTZ, Eschborn, Germany.

General Wood & Veneers. 1997. Etude de la Filière Bois au Bénin. Rapport Final. Ministère du Plan, Cotonou, Bénin.

Gesellschaft für Technische Zusammenarbeit (Ed.). 2000. Codes locaux pour une Gestion durable des Ressources naturelles. GTZ, Eschborn, Germany.

Institut National de la Statistique et de l'Analyse Economique (INSAE). 1999. Projection de la Population de l'ensemble du Bénin de 1997 à 2032. Direction des Etudes Démographiques, Cotonou, Bénin.

Klaveren, J. van. 1957. Die historische Erscheinung der Korruption in ihrem Zusammenhang mit der Staats-und Gesellschaftsstruktur betrachtet. *Vierteljahresschrift für Sozial-und Wirtschaftsgeschichte* 44(4):289-324.

Long, N. 1993. Handlung, Struktur und Schnittstelle: Theoretische Reflektionen. In: pp. 217-248. T. Bierschenk and G. Elwert (Ed.). Entwicklungshilfe und ihre Folgen. Ergebnisse empirischer Untersuchungen in Afrika. Campus, Frankfurt am Main, Germany.

Ministère de Développement Rural/Projet Gestion des forêts et Terroirs Riverains (MDR/PGFTR). 1999. Audit Institutionnel du Secteur Forestier. Diagnostic du Secteur Forestier. MDR, Cotonou, Bénin.

Projet Restauration des Ressources Forestières (PRRF). 1998. Plan d'Aménagement participatif de la Forêt Classée de Pénéssoulou. Bassila, Bénin.

Projet Restauration des Ressources Forestières (PRRF). 1999. Plan de Gestion de Terroir de Pénélan. Bassila, Bénin.

Projet Restauration des Ressources Forestières (PRRF). 2000. Gestion des Ressources Forestières et Décentralisation. Bassila, Bénin.

Sommer, J. 2001. Unterschlagen und Verteilen. Zur Beziehung von Korruption und sozialer Kontrolle im ländlichen Bénin. Campus, Frankfurt am Main, Germany.

Stieglitz, F. von. 2000. Impacts of social forestry and community-based forest management. In: pp. 33-44. Food and Agriculture Organization of the United Nations (Ed.). Proceedings of the International Workshop on Community Forestry in Africa. FAO, Rome, Italy.

Venema, B. and H. van den Breemer (Ed.). 1999. Towards Negotiated Co-Management of Natural Resources in Africa. Lit-Verlag, Münster, Germany.

Laws and Bylaws:

Loi 93-009 du 02 Juillet 1993 portant Régime des Forêts en République du Bénin.

Décret N° 96-271, 1996, portant Modalités d'Application de la Loi N° 93-009.

Loi 97-029, du 15 Janvier 1999, portant Organisation des Communes en République du Bénin.

INTERVENTIONS: THEORY AND PRACTICE

Illegal Actions and the Forest Sector: A Legal Perspective

Kenneth L. Rosenbaum

SUMMARY. This paper offers one lawyer's perspective on illegal acts related to forests. In particular, it considers the role of legal reform in addressing these problems. After cautioning on the danger of looking at illegal acts from too narrow a perspective, this paper lists some of the kinds of failures in law and government institutions associated with ille-

Kenneth L. Rosenbaum is the Principal of Sylvan Environmental Consultants, 1616 P Street NW, Suite 200, Washington, DC 20036 USA (E-mail: kenrosyenco.com).

The United Nations Food and Agriculture Organization (FAO) commissioned the original version of this paper as a background document for a meeting of experts in Rome in January 2002. The opinions expressed, however, are those of the author and not necessarily those of FAO. The author thanks Arnoldo Contreras-Hermosilla, Ali Mekouar, Matti Palo, and the anonymous reviewers for their comments and guidance. Rosalie Parker assisted in research.

[Haworth co-indexing entry note]: "Illegal Actions and the Forest Sector: A Legal Perspective." Rosenbaum, Kenneth L. Co-published simultaneously in *Journal of Sustainable Forestry* (The Haworth Press, Inc.) Vol. 19, No. 1/2/3, 2004, pp. 263-291; and: *Illegal Logging in the Tropics: Strategies for Cutting Crime* (ed: Ramsay M. Ravenel, Ilmi M. E. Granoff, and Carrie A. Magee) The Haworth Press, Inc., 2004, pp. 263-291. Single or multiple copies of this article are available for a fee from The Haworth Document Delivery Service [1-800-HAWORTH, 9:00 a.m. - 5:00 p.m. (EST). E-mail address: docdelivery@haworthpress. com].

gal activity. Next it lists some general approaches to legal reform designed to counter illegal activity. Finally, it offers some observations on the nature of effective change. *[Article copies available for a fee from The Haworth Document Delivery Service: 1-800-HAWORTH. E-mail address: <docdelivery@haworthpress.com> Website: <http://www.HaworthPress.com> © 2004 by The Haworth Press, Inc. All rights reserved.]*

KEYWORDS. Forest law, forest governance, forest policy, illegal acts, corruption, legal reform, institutional reform

INTRODUCTION

This paper offers a legal perspective on illegal activities in the forest sector. It lists some types of illegal activities and problems of governance and gives examples from the literature and the author's experience, from both tropical and temperate forests. It describes problems in the law as written and problems of implementation. It then suggests some possible approaches to improve control over illegal activities. These include respecting local interests, keeping legal structures simple, setting meaningful penalties, increasing transparency, and allowing citizen enforcement. Finally, the paper applies a framework borrowed from systems theory to ask what approaches might really work.

This paper offers *a* legal perspective, but not *the* legal perspective. The author comes to the problem with a background in legislative drafting and process, oversight of government institutions, and advising on institutional reform. That personal experience colors this discussion. A chief inspector of forests, a forest department head, a prosecutor, a judge, a minister of the environment, a minister of finance, or a representative of an international donor would have different views. (For another recent legal perspective, by a group of FAO lawyers who advise on forest law reform in developing countries, see Lindsay et al., 2002.) Further, the analysis frankly reflects the author's politics: A more left-leaning observer would put more emphasis on social conditions, while a more right-leaning observer would give greater weight to individual rights.

This declaration of bias points to a related issue: it is impossible to offer a pure legal analysis of this problem. In fact, there is danger in looking at illegal activities from only a legal point of view. The norms of behavior that law embodies are part of the greater phenomenon of cultural norms, which in turn are affected by social conditions generally, including politics, economy, and social history. Illegal activities are not simply legal problems; they are social problems with legal, political, economic, moral, social, and historical facets. (For some analyses of illegal logging from these broader perspectives, see FAO (2001), Contreras-Hermosilla (2001), and Callister (1999).) Law can play a role in solving these problems. However, the legal reformer must keep in mind the complexity of illegal activities. Appendix I offers some points of orientation that legal reformers should keep in view in trying to solve complex problems like control of illegal activities.

KINDS OF FAILURES OF THE LEGAL SYSTEM

Illegal activities imply a failure of the rule of law. For a legal reformer, this strikes a paradoxical note. How can reforming laws improve the situation when the basic tool–the law–has already failed?

Sometimes the law itself is at fault. In some cases the illegal act ought to be legal. The law should be repealed. In other cases, the law reflects a good standard of behavior, but the law as written is hard to implement. In that case, we need to look for better ways to get at the undesirable behavior. In still other cases, the issue is not the standards but a failure of the institutions that apply the standards. Here, the role of legal reform is to shape the workings of government and to promote public support of the law.

As a first step towards understanding necessary reforms, the discussion below describes kinds of failures of law and legal institutions that lead to illegal activity in the forest sector. It begins with various failures of the law itself, including failure to embody an appropriate norm, failure to offer enforceable standards, and failure to provide sufficient penalties for violations. The paper then explores institutional failings, including poor dispute resolution, unfair application of the law, failure of the agencies to apply the law at all, shortcomings of capacity, lack of intergovernmental coordination, lack of enforcement outside the forest setting, and lack of oversight within the government.

Failures of Law

Clashes of Norms–Clashes of norms occur when the rights set out in law are not the same as the rights that people or communities believe that they are entitled to have. The result is a lack of respect for the law, which can lead to illegal activities and difficulties in enforcement. An historical example comes from the United States (Knowlton, 1972). A great portion of the southwestern United States was once part of Mexico, ceded to the United States in 1848 in the Treaty of Guadalupe Hidalgo, which ended the Mexican American War. Article VIII of the treaty guaranteed that the residents of the ceded lands would retain their property rights. Communities held some of the lands and resources as social properties. The common law that prevailed in most of the United States had no concept equivalent to social property. These social properties, including water and grazing rights, became public properties, owned by the United States. Communities lost lawful access to these resources in favor of settlers coming from the established States who understood how to secure access to public resources under the new legal regime. This loss became a source of enduring resentment among the local communities.

More recent examples are easy to find. In the Philippines prior to the democratic reforms that began in 1986, the law gave the government title to all forests. Rural residents had no guarantee of stable access to forest resources. They thus had little incentive to protect the forest or respect the law. An analysis for FAO concluded that ongoing efforts to give locals a stake in forest resources are key to improving in Philippine forest management (Hammond, 1997, pp. 64 and 66). In Cambodia, local people collect resin from certain forest trees. Resin tapping is an important source of outside income for some communities. The local people recognize individuals' rights to particular resin-producing trees, which can be tapped sustainably over many years. The government does not recognize or record these rights and has regularly sold the resin trees for harvest to loggers (Global Witness, 2001). The sale of trees to loggers has also deprived local residents of access to wood. The law reserves ten to twenty percent of concession harvests for domestic consumption, but the price of concession timber reflects concession fees and taxes and is often too expensive for rural people, who turn to illegal sources (Global Witness, 2001).

Failure to reflect norms goes beyond squaring national laws with rural traditions. On a global level, no one approves of trade in stolen property. Yet no developed country's law allows customs officials to bar or seize imports of illegally harvested timber (Environmental Investigation Agency and Telapak, 2001).

A clash of norms does not always mean that the norm in the law is wrong; the law may represent good policy, and the use of the forest under the social norm may be unsustainable. In the northwestern United States, the northern spotted owl is listed under the federal Endangered Species Act as threatened (The Act is codified at 16 U.S. Code §§1531-1544; the listing regulation is in 7 Code of Federal Regulations §17.11). This puts local forest owners in danger of criminal prosecution for harming habitat occupied by owls. Many locals believe that the government has no right to limit economic uses of private land to protect public values such as biodiversity. They have brought unsuccessful challenges to the law in the courts (for example, Babbitt v. Sweet Home Chapter of Communities for a Greater Oregon, 515 U.S. 687 (1995) (U.S. Supreme Court), and there has allegedly been clandestine illegal hunting of the owl to prevent its occupancy of private lands.

In some areas of the tropics, increasing population densities present situations where the legal norm is sounder than the traditional norm. Shifting agriculture is sustainable when human population densities are low, but the system collapses as densities increase and the time between reuse of plots decreases.

Similar concerns about sustainability arise from increased harvest of firewood or bush meat. Limiting these traditional harvests to sustainable levels is wise, even if it is locally unpopular.

The case of the monarch butterfly in Mexico offers a concrete example of expanding human use and changing law leading to illegal forest use (Environmental Law Institute, 1998). The forested winter habitat of the migratory butterflies lies mostly on high mountain peaks in central Mexico. During the twentieth century, as part of its land reform policy, the government deeded much of this land to landless people, to establish communal farming communities known as *ejidos*. Not until the 1970s did scientists learn that these lands included the butterflies' wintering grounds. The government has reacted by creating reserves to protect the winter habitat. The reserve laws do not change who owns the land; they only change the rights that attach to ownership. The reserve laws reflect good conservation policy, but the laws are difficult to enforce because they conflict with the expectations of the local residents.

In cases like these, the reformer must look beyond the clash of norms to see if social or economic reforms can address the problem or if there are other failures of law that legal reform can address. The discussion of other kinds of legal and institutional failure continues below.

Undetectable Violations–A second type of failure of law is when the law is written in a way that makes it difficult to enforce. Typically, that means it is difficult to spot illegal behavior or to tell the difference between legal and illegal behavior.

Usually this kind of flaw is so obvious that drafters build enforceability into laws from the beginning. The desired policy may be to outlaw transport of illegally harvested wood. A law that simply said, "It is illegal to transport illegally harvested wood on public roads" would be almost impossible to enforce. Legal wood looks much like illegal wood. Instead the law will require legally harvested logs to be marked and be accompanied by permits or other papers. "It is illegal to transport unmarked logs or logs lacking the proper permits" is much easier to enforce.

However, sometimes the flaw is subtle. For example, Russian law prohibits harvest of Korean Pine in the Russian Far East. This kind of ban is easy for a trained forester to enforce in the field, but foresters cannot watch all the pines all the time. Much of the harvested wood is exported. Customs officials can inspect timber passing through the ports, but it takes an expert to distinguish planks of Korean Pine from other conifers. So at the points where the ban might be effectively enforced, it is difficult to enforce (Bureau for Regional Oriental Campaigns et al., 2000).

Even in the forest setting, lawmakers can overlook the possibility of crafting prohibitions that are easy to enforce. For example, criminals often act under cover of darkness. Activities then are harder to detect. The enforcer must not only find the criminal, the enforcer must determine that the particular load of logs is being hauled illegally. The darkness makes it more difficult to notice evidence of fraudulent papers and log markings. The suspect may make claims that cannot be verified with government officers or landowners until morning. Gathering evidence may require seizure of the logs and vehicles and placing them in official custody followed by more investigation the following day. Without a likely conviction the enforcing officer may be reluctant to make this commitment of time and effort. To make law enforcement easier, lawmakers should weigh the option of simply banning transport at night.

Weak Penalties–A fundamental principle of enforcement is that the punishment must be great enough to deter the crime. The punishment should reflect both the potential profit from breaking the law and the probability of being caught. Very profitable crimes require large penalties, and so do crimes that are difficult to detect.

Sometimes the penalties are just too small. In the Russian Far East, timber operators find that potential civil and criminal penalties together are too small to make illegal logging unprofitable. Only the potential of confiscation of timber is an effective deterrent, but this can be sidestepped through bribery (Bureau for Regional Oriental Campaigns et al., 2000). In Cameroon, recent penalties as high as twenty million Communaute Financiere Africaine francs (about 30,000 U.S. dollars) have failed to discourage companies from illegal practices (Greenpeace, 2000c).

Many circumstances can lead to low penalties. One common problem is that law sets penalties as fixed amounts of money, but these become less effective deterrents as inflation lowers their true value. The rates of inflation in some developing countries make this a serious concern. Similarly, rising prices for forest resources can make penalties less effective. This can happen if the supply of the commodity shrinks, as has happened with some forest animals such as tigers or bears whose organs are used in traditional medicines, or if demand grows, as happens when a new market opens up for wood. Sometimes penalties are pegged to the market value of the resources involved, but the mechanism for calculating penalties is a general one, used for thefts or property damage of all kinds. The mechanism may be inappropriate to the forest context where detection is difficult and the damage from illegal activity goes beyond the market values. This paper discusses the design of more effective penalties below, in the section on strategies to combat illegal activities.

Failures of Implementation

Poor Dispute Resolution–One function of the law is to resolve disputes promptly, inexpensively, and fairly. If the legal system fails to do so, it prolongs illegal activity or destroys respect for the law. In Bangladesh, for example, disputes over private rights to occupy or use reserved forests are an ongoing headache for government foresters. The 1927 Forest Act set out an administrative process to resolve those disputes (Forest Act sections 4 to 24). In practice, the process has become bogged down in politics, and many of the disputes have lingered for decades. Until this process is complete, the law cannot separate legitimate property owners from squatters and thieves. In the historical case in the American Southwest that was described above in this paper, the expense of dispute resolution led to unfair results. Small, cash-poor villages had to participate in trials or hearings to assert their rights, which required them to hire attorneys. To pay the attorneys, they often had to sell or hand over some of the resources that they had fought to claim. Regardless of the outcome of the case, they lost rights to the resource.

Unfair Application of the Law–The legal systems of some countries are less than fair. Describing the nature and varieties of injustice could probably fill several volumes, and every legal system is certainly guilty of some. The various arms of the government involved in enforcement-the forest agencies, the police, the customs service, the prosecutors, or the courts-all exercise some discretion in the application of the law and each may play a role in bias. The bias may be overt and intentional, or it may be so subtle and deeply ingrained that the government officials are not aware of it. In the second case, the victims of the bias may be aware of it nevertheless.

Officials may favor their own social or ethnic groups or the powers that can advance the officials' careers. In many countries, the people of the forests are on the margins of society literally and figuratively. They live far from the centers of power and may be socially distinct from them. Officials governing the forest may feel little obligation to the people living there. Forests Monitor (1998) reports that patronage plays a role in award of forest concessions in Papua New Guinea, Solomon Islands, Guyana, Suriname, and the Congo Basin.

Power and discretion can also set the stage for ethical abuses. Global Witness (2001) reports that in Cambodia many low-level government staff are also on the payroll of concession-holders. This amounts to a conflict of interest. Unfortunately, many countries, including Cambodia, do not prohibit or enforce against such conflicts. Bribery is an ethical lapse that civil service laws universally prohibit, but bribery in forest governance remains common. (See, for example, the discussions of bribery of police, foresters, and others in Russia in Bureau for Regional Oriental Campaigns et al., 2000; and in Indonesia in Environmental Investigation Agency, undated, where timber companies were quoted as preferring the "accepted bribing system" to a new law imposing higher logging fees.) Systems open to bribery favor those with wealth and political connections. Bribery is often the catalyst that enables other kinds of illegal activity to proceed unimpeded.

The law cannot eliminate the exercise of discretion. At best, the law can disperse power and set standards for its use. It can also increase transparency and so bring social and legal pressures to bear on unfair exercise of discretion.

The Forest Agencies Do Not Follow the Law–It takes information and a strong rule-of-law tradition to have forest agencies follow the law. In terms of information, the government must have property records and surveys that clearly declare who owns rights to the forests. In rural areas, careful surveys are costly and often unavailable. In many countries, the property records have fallen victim to changes in government or fraudulent record keeping. If the rights are unclear, the forest agencies cannot enforce them fairly and to everyone's satisfaction. Also, legal information must be widely accessible. Forest agencies cannot enforce the law if they do not know what the law is. The author has encountered problems in this area in two recent field projects undertaken for FAO. In Bangladesh, only one forest officer had taken the time to compile all the amendments made to the 1927 Forest Law to produce an authoritative and current version, and rural courts were said to be deciding cases based on versions compiled decades before. In Kosovo, the province had forest laws written in Albanian (from its days as an autonomous region of Yugoslavia), Serbian (from general Yugoslav laws that applied to the autonomous province and from Serbian laws after autonomy was revoked), and English, from the current United Nations administration. English translations of the older laws were often poor, and the newest laws were sometimes written without full understanding of the older laws.

Sometimes the tradition that laws be written, public, and followed uniformly has never developed or has been lost. In Cambodia, Global Witness (2001) reported two examples of failure of the rule of written, public law. The law regarding marking hammers, used to stamp identifying imprints on the ends of legally cut logs, had never been written down. Different officials were following different protocols for hammer marking. A law prohibiting sale or transportation of "old" logs–logs cut in the past without proper permission and now lying in the forest or in storage yards–had been put writing, but officials still ignored it.

Failure to honor written law is typical of the frontier mentality seen in some rural or isolated forest regions, where the local government officials may be more responsive to their own whims or the whims of extra-legal powers than to central authority. The Russian Far East has been called the "Wild East," alluding to the lawlessness of the American "Wild West" in the late nineteen century (Forests Monitor; 2001; Bureau for Regional Oriental Campaigns et al., 2000). The analogy is apt; modern historians have called the race to plunder the resources of the American West in the 1870s "The Great Barbeque" with "fraudulent schemes . . . so varied and colorful that no summary can do them justice" (Wilkinson, 1992, p. 121).

In some countries, several agencies have authority over forestlands and all may not honor the laws equally. In Mexico, for example, the land reform agencies have been known to grant land to land-poor people inside the boundaries of protected natural areas managed by sister agencies (Environmental Law Institute, 1998). Some local politicians in Mexico encourage squatters to settle protected lands and then bargain with the environmental agencies to solve the resulting land-use problems (Jordan, 2001).

Sometimes the law is so out of date or unworkable that it invites people to ignore it. In Bangladesh in the late 1990s, the legal prices for certain non-wood forest products were set by regulations that had not been updated for twenty to forty years. The agency appeared to pay no attention to these regulations and did not feel any strong need to revise them. The agency simply charged what it considered to be reasonable prices.

Poor Enforcement–Law enforcement in forests is difficult. Forests are usually less densely populated than urban and agricultural areas, meaning that there are fewer honest eyes around to spot illegal activity. Often, forestlands have been passed over for urban or agricultural uses because they are too rugged or remote. These characteristics hinder travel and communication. Adding to this, forests are almost always economically less productive than agriculture or urban lands, often meaning that there is less investment in travel and communication infrastructure. Socially, forest may also be separate from the surrounding lands. They may be home to tribal peoples, tradition-bound villages, migrants seeking new lands, or booming communities of transient workers employed to harvest the resources. In a few instances, forests shelter groups in rebellion against central governments. Whatever the case, forest communities may have little interest in helping to enforce laws made in distant cities. The responsibility for enforcement may thus fall heavily on the forest agency. And many forest agencies simply lack the capacity to enforce the law. (For a discussion of how these same difficulties affect the larger field of making forest policy, see chapter one of Environmental Law Institute, 1998.)

Enforcement has three facets. The enforcer hopes to *prevent* illegal acts from happening in the first place. When the acts do happen, the enforcer wants to *detect* the violations. And when a violation is detected, the enforcer wants to prosecute or *suppress* the illegal action. The prevention/detection/suppression nomenclature is from Contreras-Hermosilla, 2001. Other writers have used other terms, such as deterrence instead of prevention or prosecution instead of suppression, but the split into three facets is common. These facets are not entirely independent. A strong and public suppression effort, such as a well-publicized trial, can act to prevent future crimes. So can an obvious presence of officials in the forest pursuing detection tasks. However, considering enforcement as three distinct efforts helps in understanding the enforcer's task.

Prevention starts with making sure people can tell what acts are illegal. People need to know what rights exist before they can be expected to honor them. That may involve educating people on what the law is and why it is in their interests to obey the law. Extension foresters are more effective at this kind of task than traditional forest managers, but extension is an under-funded program in many forest agencies. Practical prevention may also mean locating and marking boundaries and maintaining associated physical objects marking off rights, such as signs, fences, and gates. In remote forest areas, the lack of ability to set boundaries often makes it impossible to determine if a person is acting lawfully.

The next tasks of prevention are to reduce the motives, means, and opportunities to violate the laws. Motives for forest crime are often economic, though they may also have subsistence, social, or traditional roots. (For example, a group of forest dwellers may follow a tradition of setting fire to the forest from time to time to improve hunting.) The most common legal approach to the issue of motive is to increase penalties, making the chance of profit from the crime lower. In the case of bribery, increasing public salaries may decrease the motive to accept bribes. Also, effective prevention of commerce in stolen wood can reduce the motive to cut it in the first place.

The law can directly regulate the means for committing some forest crimes. For example, some Mexican states have experimented with outlawing chain saws. In theory, a state could register or limit ownership of heavy equipment used in logging or milling, though the author is not aware of this approach being used to limit illegal logging.

Because criminals seek the opportunity to act unobserved, the simplest way to reduce opportunity is to increase surveillance. For example, Wardojo et al. (2001) estimate that Indonesia needs 15,000 forest guards, rather than the 9,700 it now employs. Capacity may be also a matter of technology. The Environmental Investigation Agency (2001) declares that Indonesia's Integrated Forest Security Teams are under-funded and under-equipped.

A forest agency can also increase surveillance by recruiting allies among forest residents and NGOs active in the forests. Ironically, the traditional foresters in the government may view these groups as their natural enemies. One surveillance strategy is to hire local people as forest guards or honorary forest officers. Besides reducing opportunity for crime, this strategy provides income to local people and so reduces their need to engage in illegal acts.

Sometimes perverse incentives in the forestry law itself discourage prevention in favor of other aspects of enforcement. The Bureau for Regional Oriental Campaigns et al. (2000) report that Russian Far East forest agencies can keep the proceeds from confiscated timber. Therefore, they have no incentive to prevent crimes. They would far rather detect and seize the timber after the crime occurs. Usually, though, prevention and detection go hand-in-hand. A strong capacity for surveillance helps find evidence of illegal activities.

Finally, note that development affects prevention in complex ways. Development may bring new surveillance tools that aid enforcement. It may attract honest forest businesses that share the government's interest in upholding the rule of law. However, development capital provides motives for theft and the means for bribery. Capital may also bring in heavy equipment and chainsaws, the means for large-scale illegal logging. Construction of forest roads offers new opportunities to employ these means. Construction of sawmills and paper mills increases the means and opportunity to launder stolen goods in the stream of commerce.

Good *detection* requires people trained in detection techniques. Detection officers must have some basic understanding of forests, wood, and the local geography. They must also understand the specific laws applying to the forest. Detection requires tools ranging from basic vehicles and communications equipment to sophisticated geographical information systems. Above all, detection requires managerial talent. The enforcement officers need to allocate limited resources wisely. Akella et al. (unpublished manuscript) describe a case in the Atlantic Forest of Brazil where enforcers were allocating too many resources to detection and too little to suppression to the ultimate detriment of enforcement.

Good *suppression* begins with good detection, but once a crime is detected and a suspect identified, what next? In some countries, the forest agency deals with minor forest offenses administratively, and the detecting officer is effectively also responsible for suppression. (See, for example, the provisions for "compounding" offenses in section 68 of the Forest Act (Act No. XVI of 1927) of British colonial India, which was the original forestry law inherited by India, Pakistan, and Bangladesh.) In most cases, though, the detecting officer is only the first link in a chain that extends through the offices of the police, the prosecutors, and the courts. If the officer's actions are to survive scrutiny down that chain, the officer must understand the laws governing use of force, arrest, detention, interrogation, and prosecution.

Suppression also requires physical resources. Foresters and police need places to hold suspects and confiscated goods. They need means to preserve evidence, with a clear chain of custody, so that it retains probative value to the courts. When the extent of injury to forest resources is an issue in setting penalties or awarding damages, foresters need tools to measure and document that injury.

A final general observation about enforcement: in addressing failures, it is important to look at both capacity and will. In particular, when corruption destroys the will to enforce, increases in personnel or equipment may not make much difference.

Weak Administration–The failure of a forest agency to do ordinary governmental tasks promptly and competently can set the stage for illegal activity. For example, due to inefficiency or understaffing forest agencies can fall behind on their work. A backlog results. Backlogs invite favoritism and create opportunities for bribery. If there is a queue of people waiting for government approvals, there is a possibility of bribing someone to go to the head of the queue, to be considered out of turn. If the backlogged tasks are themselves prerequisites for future actions, the agency may ignore the prerequisites. For example, forest laws typically require long-term management plans before harvesting wood on publicly owned forests. A weak agency may lack the capacity to prepare these plans in a timely fashion. Rather than suspend harvests, the agency may simply manage the forests under short-term operational plans or, worse, no plans at all. These kinds of shortcuts encourage a general disrespect for the requirements of the law. Sometimes basic management tasks like planning are done on time but done badly, and this creates a foundation for illegal acts. In the Russian Far East, the law instructs officials to set levels of cutting based upon existing inventories. The inventories are flawed. In some cases, officials have used clearly bad inventory data to allow overcutting despite legal calls for sustainable harvests (Bureau for Regional Oriental Campaigns et al., 2000).

Lack of Coordination–For effective implementation of the law, government agencies must work together. Forest agencies must coordinate with agencies that do not usually deal with forest matters. For example, to track the financial aspects of illegal forest trade forest officials must work with banking regulators (Kaimowitz, 2001). Russia offers an example of what happens when agencies fail to coordinate; the Ministry of Economy has issued export permits for more timber than can be legally cut, and immigration officials have allowed foreign timber brokers to work without appropriate visas (Bureau for Regional Oriental Campaigns et al., 2000).

Sometimes forest agencies have trouble coordinating with other land management or environmental agencies. Global Witness (2001) reports that in Cambodia, the forest and parks agencies often block each other's attempts to put in place regulations that happen to affect the other agency's lands.

Always, the forest management agency must coordinate with other the on-the-ground enforcers and the prosecutors. Lack of cooperation leads to ineffective enforcement. For example, in Kosovo, the forest agency has had great difficulty in getting police agencies to stop log trucks to verify their permits. The reasons vary in different parts of the province. In one area, there is a dispute between the central forest administration and the local forest managers over which laws apply, and the police want the dispute settled before they become involved. In another region, the police suspect that the foresters are issuing or denying permits to extract bribes, and they do not want to be complicit in that practice. In another area, the police simply have other enforcement tasks that they consider to be higher priorities.

The military, a potential enforcer, can also be a significant illegal actor. In Indonesia, the military has a long-standing role in forest exploitation and has participated in many illegal actions (Barber and Talbott, 2003; Environmental Investigation Agency, undated). The Cambodian military has been accused of harboring wildlife poachers (Global Witness, 2001).

Prosecution of major forest offenses requires cooperation of government lawyers. The prosecutors may be understaffed and the cases backlogged. The prosecutors may be open to bribery. Or, even if they are beyond bribery, they may disfavor forest offenses as relatively unimportant compared to other cases before them. They may seek penalties that are too low or may fail to pursue the cases at all.

Failure of the courts to adjudicate forest-related cases fairly and effectively can make enforcement impossible. Courts fall prey to the same problems that hamper prosecutors. The dockets may be clogged and cases may wait for years to come to trial. Judges may be open to corruption. Or, judges may be reluctant to assess meaningful penalties or damages. They may not understand that forest crimes have environmental consequences that go well beyond the economic value of the wood or wildlife involved. For example, after the devastating forest fires of the 1990s in Indonesia, a group of NGOs brought a civil suit against major forest companies seeking Rp 11 trillion in damages (at summer 2002 exchange rates, roughly 1.2 billion U.S. dollars). The court found the companies liable, but only awarded court costs and ordered the companies to adopt fire prevention measures (Barber and Schweithelm, 2000).

Lack of Enforcement of General Laws Applying to Forests and Forest Products–The focus of this paper has been on the application of forest management laws, but a broader range of laws affects the use of forests and trade in forest products. The previous section mentioned banking, export, and immigration laws. As pressure grows for global "fair trade" in timber, activists will press for application of labor, pollution control, and human rights standards. All of these have already been raised in allegations of illegal forest activity. For example, in the Solomon Islands, Greenpeace (2000b) has reported unlawful labor conditions in logging operations, breaches of pollution control laws, and invasion of sacred sites including burial grounds. Forest officers may be poorly equipped to detect these kinds of offenses. Police and other enforcement officers may be poorly trained to deal with forest matters. As a result, there may be no capacity to handle *enforcement in* these areas.

Lack of Government Oversight–Just as the government must watch the forest to prevent crime by the public, someone must watch the government to prevent inefficient and illegal government acts. In Bangladesh the author found that oversight was non-existent or illusory. Divisional forest officers working outside of headquarters oversaw their own accounts. The Forest Department accounting officer at headquarters had no actual authority to initiate audits. The author was told of regular audits by the financial ministry, but saw no evidence of them. The constitution of the country provided for a national ombudsman to accept and investigate citizen complaints against the government, but that office had never been filled. In Cambodia according to Global Witness (2001), the government has failed to make the regulations governing concessions freely available. It is therefore impossible to know if the government is following its own rules.

Good oversight requires several institutional structures. Government accounts, laws, and records should be open. Government decisions should be made publicly "in the sunshine." Independent offices should exist inside or outside the forest agency with institutional incentives to uncover waste, fraud, and abuse of power. Whistleblowers should have protections under law.

STRATEGIES TO COMBAT ILLEGAL ACTIVITY BY IMPROVING THE LAW

The previous section discussed illegal activities and offered some remedies. This section expands upon the discussion of how legal reform can help combat illegal activities. It argues for respecting local interests, making laws simple, setting effective penalties, increasing transparency, and involving citizens in enforcement. These points all bear either on substantive standards in the law or on processes to apply those standards. In addition, as discussed below in Appendix I, the process of adopting the law is itself critical to effective reform. People are more likely to honor the law if they have had a role in its development.

Respect Local Rights and Interests

Local people are key players in combating illegal acts in the forests. They are the ones with the greatest opportunity to commit illegal acts and the greatest opportunities to deter illegal acts through surveillance. If local people do not understand and support the forest laws, enforcing the laws will be an uphill battle.

Recognize Traditional Rights–The law should recognize traditional rights and expectations of local people if people can exercise those rights sustainably. If people believe that the law has suddenly deprived them of a long-held right, they will respond with anger and contempt. Only fear of detection and the resulting punishment or social stigma will keep them from ignoring the law. If contempt for the law is widespread, social pressures may even encourage people to break the law.

The corollary to this principle is, if the law cannot respect local rights, the government needs to make a case to the local people why that is so. If local support is a concern, the law may need to compensate people in some way for what they see as their lost property.

Resolve Disputes Fairly, Promptly, and Inexpensively–Conflicts over rights to the forest, or indeed to resources generally, are universal. The history of law is a history of disputes, with land at the center of many. A legal framework that aims to end disputes is doomed from the start. A good legal framework must anticipate that disputes will arise and provide for workable ways to settle them.

Often the law has no separate mechanism for forest disputes. They are handled like other legal disputes, within the judicial system. In that case, forest law will say little about how disputes are resolved.

However, the law may prescribe administrative processes for some forest disputes, such as those that arise between the public and the government. Some laws allow people to appeal forest agency decisions to higher authorities within the government. Some laws allow forest officers to set and collect penalties for forest offenses. Administrative dispute resolution offers the possibility of low cost and quick results, but it does not assure justice. Whether the venue is administrative or judicial, the mechanism should be unbiased, quick, and inexpensive to the parties. Rulings should carry the force of law.

The rural poor typically enter governmental dispute resolution processes at a disadvantage. Complex processes, heavy with formalities, will favor the wealthy and educated. Legal systems that are open to bribery also put the poor at a disadvantage. In some cases, the best way to handle local disputes is to use traditional mechanisms rather than to impose complex, costly, or corruptible new processes. In other cases the traditional mechanisms themselves carry biases, and outside mechanisms need to be available.

Give Local People a Stake in Forest Protection–A legal system that protects a resource will have the backing of the people who get to benefit from the resource. If the support of local populations is important, the law must assure those populations benefits from the forests. That benefit may take any number of forms. In some cases, people will want access to forest commodities such as firewood or pasturage. In some cases, they will want a direct share of the income that the forest generates. Or, they may be looking for indirect economic benefits associated with outside investment in the forest, such as jobs and tax revenues. People who wish to preserve the existing social structures may want the right to exclude others from the forest, such as immigrants and developers.

Make Laws Simple and Direct

Make Laws with Clear Standards that Are Easy to Enforce–The discussion above on laws that are difficult to enforce introduced the solution of simplicity and clarity of standards. Here are a few general observations on designing laws to aid enforcement.

Enforcement is easier where the activity is naturally exposed to surveillance. Thus, it is usually easier to enforce laws dealing with transport (on the public highways or rivers) or export (at least through normal commercial channels) than laws dealing with harvest in the woods.

Enforcement is easier if the regulated class is small and obvious. For example, it is easier to control the import, manufacture, or sale of logging equipment than it is to control its use in the woods, because there are fewer importers, manufacturers, and dealers than there are users. Similarly, it may be easier to enforce controls on a few dozen sawmills than on hundreds of loggers. It may be easier to enforce bans on private zoos or public sales of bush meat than to control illegal hunting directly. Remember, though, that regulation of the smaller class does not preclude regulation of the larger. The best approach often is to have laws regulating both but to focus enforcement on the smaller class.

Bans are easier to enforce than regulatory systems. The standard in a ban is inherently clear. For example, if a country regulates the amount of mahogany that can be harvested, every instance that enforcers find of mahogany being cut, transported, or exported is potentially legal. The burden of determining that it is illegal usually falls on the enforcer. Without a reliable system of permits and records, free from forgery or corrupt influences, the burden may be steep. If a country bans the cutting, transport, or export of mahogany, every instance is illegal. The standard could not be clearer, and the facts that the enforcer needs to prove are much simpler.

The disadvantage of bans is that they can be overbroad. Their sweeping nature can have undesirable impacts and side effects. For example, a ban on mahogany harvest may lock up a resource that could sustain a low level of harvest. Local artisans who have used small amounts of the wood for years become outlaws if they continue. Small, sustainable harvesting operations may lose the one item that made them profitable.

Set Clear Standards for the Exercise of Discretion by Officials–Allowing officials to exercise their power without constraints invites two closely related evils. First, officials could use their power to favor one group over another unfairly. Second, officials could accept bribes to influence how they use their authority.

Drafters, who are taught the value of simplicity and flexibility in laws that apply to the public, must learn to write constraints into laws that bind government officials. This is always a challenging task. It is hard to anticipate all of the situations in which a law will be applied and to write a law that will be fair and effective in all cases.

One area where forest law needs to have clear standards is in the sale of forest resources. In a simple world, the law would direct the government to sell the resources to the highest bidder. However, if the law does not set minimum prices for the auction, the government could be cheated through collusion to keep bids low. If the law does not allow officials to throw out bids from unreliable purchasers, the government may end up selling to contractors incapable of careful harvest. If the law gives too much discretion to officials to set the price and award the sale, it opens the door for abuse of that discretion.

Where it is impossible to set standards to cover all contingencies, the drafter can require the exercise of discretion to be transparent. In the case of an auction, all bids should become a matter of public record. A public record of decision should state which bid was selected and why. If the drafter trusts the courts, the law may give them the power to overturn awards that lack a reasonable basis.

Streamline Processes and Eliminate Unnecessary Steps–Especially when applied to private forest owners, complex procedures can make forest management less attractive and can create opportunities for bribery. For example, in Bangladesh in 1996 the author found that private owners needed multiple official approvals for permits to transport logs from private lands over public roads. The owner would begin by submitting "Form A" to the Divisional Forest Officer (DFO). The DFO first had to verify that the land involved was not under management of the nation's Forest Department. The DFO then had to send the application to the Deputy Commissioner (DC) of the district government with jurisdiction over the area. The DC would pass the application on to various officials involved in revenue and land records, down through the sub-district (thana) and sub-sub-district (union) levels to verify ownership. If the private land bordered government land, at this point the Forest Department would have to send someone out to the site to locate the boundary. Once the sub-sub-district officials were satisfied that the trees were on private land, Form A could return back up through channels to the DC, who would pass it back over to the DFO. The DFO would send a ranger to the site to mark the trees. The DFO would also have to make a personal inspection of the site. At this point, if 200 or fewer trees were to be cut, the DFO could issue a transit permit. If more than 200 but less than 500 trees were involved, approval would have to come from the DFO's superior, a Conservator of Forests at the circle level. If 501 to 1000 trees were involved, the application would need to go to the capital for the approval of the head of the Forest Department, the Chief Conservator. If over 1000 trees were involved, final approval was in the hands of the Ministry.

Private land owners found this process daunting and generally turned over the task of getting permits to a wholesale timber buyer familiar with the officials involved. In any given area, only a few buyers would be familiar enough with the local officials, meaning that the market for private timber was in control of a few hands. This depressed prices. Further, the prices had to reflect the costs of shepherding the application. Those costs allegedly included bribes to officials in the chain.

Complex regulatory systems develop in response to perceived needs. Taking them apart is like defusing a bomb; it cannot be done haphazardly. The drafter must understand the role of each step in the process and determine if it is really still necessary. You do not want to cut a link that protects the forest resource, unless you somehow replace it.

Sometimes decentralization is the answer. Why require national or sub-national review for matters that can be competently handled on the local level? If only local matters are at stake, it may make sense to make the decisions locally. However, there must be some oversight of the decision makers to assure uniform application of the law, discourage arbitrary exercise of power, and prevent corruption.

Set Meaningful Penalties

Penalties have many purposes, including deterrence, prevention of recidivism, punishment, rehabilitation (of the offender), and restitution (for the victim). The penalty should reflect its purpose. If penalties are too weak, people will ignore the law no matter how it is enforced. And if penalties are arbitrarily high, the courts may be reluctant to apply them.

Deterrence is the first concern. For illegal acts involving the taking of commercially valuable resources, the penalty must be at least as great as the value of the resource. However, this alone is not enough. In basic economic terms, the penalty must also reflect the risk of being detected. For example, if only half of all timber thieves are caught, and the penalty is a fine of less than twice the value of the stolen goods, it makes economic sense for the thieves to steal. (See Akella et al., 2002, for a mathematical expression of this principle.) The law can impose things besides monetary penalties. These can include imprisonment and public reprimands. They can also include loss of public privileges, such as the right to hold a government position (a potentially important deterrent in bribery cases). If the concern is theft by a forest-sector business, the punishment can include the loss of the right to purchase forest products from public lands or a broader loss of the right to conduct business. This can be a strong tool against fraud by timber purchasers. Corporations may try to get around this kind of penalty by creative use of subsidiary corporations and corporate shells. The drafter must be alert to these possibilities.

The second concern of the drafter should be prevention of recidivism. What can the law do to discourage this person from committing future crimes? Some deterrence tools also work against recidivism. Bribe takers should lose their jobs. Thieves should go to jail, where they cannot steal for a time. Businesses should lose their licenses. All serious offenders should risk the loss of the ability to make future contracts or receive forest-related benefits from the government. Confiscation of equipment used in criminal acts can prevent the equipment from being used in other crimes, but this is a difficult penalty to apply fairly. If a thief borrows a neighbor's tractor to haul stolen firewood, should the neighbor lose the tractor? If a sawmill buys logs knowing that they are stolen, should the mill be forfeit? Giving some officials this kind of discretionary authority would raise a danger of abuse.

Restitution is a basic measure of damages in civil actions and may be available on top of other penalties in criminal actions. Courts used to dealing with injuries to ordinary property will tend to set the damages at the market value of the timber stolen or destroyed. Often, this is inadequate. Unless the theft is from a stand ready for commercial harvest, the theft probably did more harm to the future value of the stand than is represented by the market value of the trees taken. Also, the market value of the lost timber fails to reflect the lost non-commodity values of the land, such as preservation of biodiversity, protection against soil loss, or improvement of water quality. Rather than get courts involved in complex calculations of the present value of future harvests, or of the hypothetical lost market value of the real estate (which may be government or communal lands not subject to sale), or the price of non-market commodities, it may be simpler to direct the courts to award damages equal to double or triple the value of the lost timber. Admittedly, this is an imperfect measure of the injury, but it is usually fairer than simple market value, and the possibility of damages above market value will serve as an extra deterrent to thefts. Another approach is to allow courts to award the cost of restoring the forest rather than the value of the resources damaged or taken. This is especially appropriate where the injury affects non-commodity values of the forest. Restoration costs are also appropriate in timber cases involving young stands. For example, if trespassing cattle destroy a new plantation, it is reasonable to require the cattle's owner to replant the area rather than to pay the value of the lost seedlings.

Increase Transparency

Increasing transparency is an accepted approach to combating corruption. It offers the potential to deter and expose corruption by "letting the sun shine" on otherwise hidden aspects of government decision-making. Alone it does nothing, but it empowers a concerned society to police its government.

Transparency efforts in forestry often run into the problem of piecemeal application of a principle that ought to apply across the entire government. That is, governments ought to have general laws allowing access to government documents, auditing of government accounts and contracts, and so forth. The forest law drafter may be faced with the task of introducing a specialized law for the forest sector, with the expectation that in the future a broader law will apply to the entire government.

Public Access to Agency Documents–Several kinds of forest agency documents ought to be available for public inspection. All documents related to sale of public forest resources should be public. The unsuccessful bidders for the resources will have a natural incentive to take advantage of the open access to police the sale process. Forest plans ought to be open. At the planning stage, the decisions reflected in the plan may suggest favoritism and point the way towards cases of bribery or abuse of power. At the implementation stage, failure to follow the plan may similarly be a symptom of corruption.

The law should require that a written record of decision accompany important government decisions on forest use. A decision in writing is easier to analyze and challenge if it falls outside the law. This requirement fits neatly with environmental impact assessment. Any forest action significant enough to require an impact assessment should have a record of decision justifying the action in light of the assessment.

Agency expenditures ought to be a matter of public record. The public should be able to look into both the amounts spent in ordinary procurement and the expenses paid for government employees. The law should also require the forest agency to prepare an annual report outlining income, expenditures, liabilities, and assets. Such a report provides an orientation to the agency's finances and can serve as a starting point for further inquiries.

The law should exempt some public documents from disclosure to discourage illegal activity. For example, the government may not wish to disclose the location of high-value resources that are particularly prone to theft, such as sandalwood. The government may want to protect the location of rare wildlife habitats that might be damaged by poachers or over-enthusiastic ecotourists. Criminals could use inspection schedules and enforcement strategies to avoid detection, so these should be exempt from disclosure.

Disclosure of Conflicts of Interest–The government forest officer owes loyalty to the government. Any competing claim on the loyalty of the officer poses a conflict of interest. This is a matter of ethics, but the law can reinforce that ethical duty. Ideally, the law can prohibit forest officers from exercising their authority in circumstances where they have a conflict. As a next-best approach, the law can require forest officers to disclose potential conflicts. For example, the law could require the forest officer to file an annual report on ownership of forestlands or forest sector businesses. The requirement could extend to close family members as well. The law can also require forest officers to report on receiving gifts or other things of value from people outside the immediate family. Besides outright gifts, the law can apply to favorable loans, travel, lodging, and sales of property or services at prices below market value.

Note that prosecutors can use disclosure laws as a back-up to bribery laws. Where it may be difficult to prove that money exchanged hands for a particular favor, it may be easy to prove that the recipient failed to report a "gift" as required by law.

Independent Auditing and Oversight–Without an outside eye looking in from time to time, even the most honest person can be tempted to stray from the legal path. A forest agency that does not face audits is a forest agency beyond legal control.

The quality of audits depends on the quality of the auditors as well as the extent of their access to agency records. Audits should go beyond looking at the accuracy of the financial records and should examine whether the agency is following the law. Auditors should have training in the agency's field, and the auditors themselves should face monitoring and evaluation.

Besides traditional auditing, two other oversight mechanisms are worth considering. The first is an oversight or advisory board. This should be a panel of experts in forestry, public administration, or other relevant fields. They should have the tasks of advising the forest agency and making recommendations to the government and the public on matters of forest policy and administration. They should have unrestrained access to forest agency staff and records. The second is a mechanism to respond to citizen and whistleblower complaints. This could be a task of the oversight board. However, it also could be a separate office within the forest bureaucracy, independent of the forest agency it oversees, or it could be an office outside the forest ministry. The office might be known as an inspector general, an ombudsperson, or a forest advocate. This office could investigate allegations of corruption, illegality, and mismanagement.

Allow Citizen Enforcement

If the citizens and courts are up to the challenge, allowing citizens to bring suit against the government or against other citizens is an effective addition to government oversight and enforcement. Suits against the government have had a tremendous influence on forest management in the United States. Sekhsaria (2002) describes a case of successful resort to the courts in India. If a nation has an active civil society and a receptive and impartial court system, the drafter should make provision for citizen enforcement.

Suits against the government often raise technical legal issues, such as–

- Sovereign immunity: can the government be sued in its own courts?
- Standing: can a single citizen bring suit to complain about an issue that affects the public generally?

- Remedies: if the citizen wins, can the court order the government to adopt the action or policy that the citizen seeks?

Legislation can clarify these points. It can further encourage these suits by allowing awards of court costs and attorney fees to the successful citizen or by shielding the citizen from paying the costs of the government if the suit does not succeed.

Suits by citizens against other citizens may run into their own legal hurdles. Usually the law allows citizens to sue to address personal injuries, but not to address injuries to the larger community. That function belongs to the government. However, there are exceptions. Some countries allow injured parties to bring private criminal prosecutions if the government has declined to prosecute. In some, citizens may sue other citizens to recover money owed to the government, with a portion of any recovery going to the citizen as a form of reward. These are known as *qui tam* ("who as well") suits, because the plaintiff sues on behalf of the government as well as himself. In some countries, people may bring "citizen suits" to enforce particular laws. The United States has specific provisions allowing citizen suits in several environmental laws, including the Endangered Species Act. During times of slack enforcement by the government, non-governmental organizations (NGOs) have used these provisions aggressively. Though NGOs lack the funding and legal capacity to replace completely the missing government enforcement, the success of the suits serves to embarrass the government and encourage stronger enforcement.

SETTING PRIORITIES FOR ACTION:
A SYSTEMS FRAMEWORK

The following is an attempt to use a systems theory framework to set priorities for combating illegal forest activities. Like any tool, theory has its limits. Some arise out of the weakness of the theory, and some out of weakness of the analyst. It is offered here to stimulate further thought on how best to approach the problems at hand.

To a systems analyst, the forest sector is a system. Like all systems, it has materials flows, information flows, feedback loops, constraints, and incentives. It is complex, and it functions on many levels–economic, political, social, biological, and geographical. Understanding it and bringing about beneficial change are challenging.

In 1997, the systems analyst Donella Meadows published an informal guide to "Places to Intervene in a System." She identified nine general strategies for bringing about change and ranked them according to their likely strength.

9. The ninth and least likely strategy to be effective is to adjust numbers. That means to change the parameters of the system without affecting its structure. Examples might be to increase the number of forest guards, reduce the volume of timber sales, or increase the frequency of patrols. By themselves, changes in numbers tend to do little unless the system is unstable to begin with or the changes affect structures or feedback mechanisms discussed below. So, for example, if there are no forest guards to patrol a protected area, then the surveillance structure is missing entirely, and changing the number from zero to one or two can make a big difference. But if there are already six guards, adding one more may have limited effect.

8. The eighth strategy is to change material stocks and flows. An example would be to change the way timber is transported, so that logs must move through a single checkpoint or exports must flow through controlled ports. Such changes can be effective, but are often expensive or impractical to achieve. Another approach in this category is to develop buffers (materials storage) or alternative sources of supply. Thus, a buyer with a large stockpile of timber or with many alternative suppliers can more easily afford to turn down an offer of supply from an obviously illegal source.

7. The seventh is to regulate negative feedback loops. Negative feedback loops are features that bring stability to a system. A mechanical example is a thermostat. When the temperature drops, the thermostat detects that and turns the heater on. When the temperature rises past a set point, the thermostat turns the heater off. In economics, rising prices communicate negative feedback to reduce demand, while falling prices tend to decrease the incentives to supply. In forests, negative signals include the physical difficulty of securing the logs (so don't build roads into protected areas); the expense of conducting illegal activities compared against potential income (so decrease the demand for illegal wood or increase the cost of obtaining it); the social stigma of illegal activity (so don't let illegal loggers become folk heroes or civic leaders); and the likelihood, size, and nature of legal sanctions (so don't let the punishment be dwarfed by the profit).

6. The sixth is to drive positive feedback loops. Positive feedback loops provide incentives. Can we reward behavior that is inconsistent with illegal activity? Can we make legal harvest more lucrative? Can we provide lawful employment to the labor base that would otherwise be available for illegal activities? Can we increase the incentives for effective law enforcement? Can we promise honest civil servants a decent standard of living through government wages?

5. The fifth is to alter information flows. Feedback loops, negative and positive, depend on information. Improve information flows, and the feedback gets stronger or entirely new feedback loops arise. Provide more reliable information on property ownership and owners may become bolder in protecting their property. Maintain an accurate forest inventory and over-cutting may become more apparent. Create a system of continuous monitoring of forest cover and the locations of illegal harvests may become painfully obvious. Create a reliable forest certification system and consumers may start to apply pressure to producers to log legally. Increase transparency in government and corruption must run to find the remaining dark corners. Bring in international auditing of forest law enforcement efforts and countries must accept that they will be compared favorably or unfavorably to their peers.

4. The fourth is to change the "rules" of the system. Can we create new incentives, punishments, and constraints? Many potential legal remedies fall in this category: reducing the discretion of government officials; creating new legal avenues such as citizen prosecutions of wrongdoers; creating new punishments such as confiscation of property used to commit crimes; or changing tenure systems to recognize new rights to control forest use, Note that these remedies are potentially powerful, but not guaranteed to be entirely good or effective.

3. The third is to affect the system's power of self-organization. Societies change themselves. How can we encourage changes that fight illegal activity? Some of these actions are quite general but powerful. These include supporting and empowering civil society groups that are pressing for change; empowering the social groups that tend to be harmed by illegal activity; exposing the costs of illegal activity; educating decision makers; and promoting the rule of law.

2. The second is to change the goals of the system. Strong leadership can focus societies on new goals. Sustainable development can overshadow immediate needs. Preserving the nation's forest heritage can outweigh immediate economic gains. Public service can be more important than personal enrichment.

1. The first is to change the mindset out of which the goals grow. What are the shared ideas that underlie the forest system? Does the society believe the forest resource is renewable or worth protecting? Does it see the world as stable and predictable enough to make long-term investments in forests? Does it believe it can afford to protect its forests? Does it believe corruption is inevitable? If we can change underlying assumptions and beliefs, we can change the functioning of the whole society.

CONCLUSION

Meadows's view offers two essential lessons on legal reform as a tool to combat illegal logging. The first lesson is that legal reform cannot do it all. It cannot change people's needs, eliminate poverty, or alter the allure of easy money. Lawmakers need to remember that some tasks remain beyond the capability of law.

But the second and equally important lesson is that legal reform can play a broader role than most people think. Good law encompasses more than good norms. The law shapes processes and institutions. It can alter information flows, change incentives and punishments, or promote avenues of self-organization within society. Over time, it can pave the way for changes in goals and mindsets, which Meadows tells us are the most powerful changes of all.

Social systems are complex, which makes good legal reform difficult. Good reform requires insight into the social and economic roots of the problem. Using legal reform to fight illegal logging requires popular support and works best in countries that already have an effective civil society and respect for the rule of law. But legal reform holds potential rewards that go beyond reducing illegal logging: fairer allocation of forest resources, better forest administration, and an opportunity for people to become more engaged in their own government. Every country fighting illegal logging should take a hard look at legal reform.

REFERENCES

Akella, A.S., J.B. Cannon and H. Orlando. Unpublished manuscript. Enforcement economics and the fight against forest crime: Lessons learned from the Atlantic Forest of Brazil.

Barber, C.V. and J. Schweithelm. 2000. Trial by fire: Forest fires and forestry policy in Indonesia's era of crisis and reform. World Resources Institute, Washington (*http://www.wri.org/pdf/trialbyfire.pdf*).

Barber, C.V. and K. Talbott. 2003. The chainsaw and the gun: The role of the military in deforesting Indonesia. In: ch. 7. S. Price (ed.). War and tropical forests: Conservation in areas of armed conflict. Haworth Press, New York (*Journal of Sustainable Forestry* 16: no. 3-4).

Bureau for Regional Oriental Campaigns, Friends of the Earth-Japan and Pacific Environment & Resources Center. 2000. Plundering Russia's far eastern taiga: Illegal logging, corruption, and trade (*http://www.pacificenvironment.org/PDF/logging_final.pdf*).

Callister, D. 1999. Corrupt and illegal activities in the forest sector: Current understandings, and implications for World Bank forest policy (*http://lnweb18.worldbank.org/eap/eap.nsf/Attachments/FLEG_OB6/$File/OB+6+FPR+Paper+-+Illegal+Actions+-+Debra+Callister.pdf*).

Contreras-Hermosillo, A. 2001. Law compliance in the forestry sector: An overview. (Background paper for Southeast Asia Ministerial Meeting on Forest Law Enforcement and Governance) (*http://lnweb18.worldbank.org/eap/eap.nsf/Attachments/ FLEG_OB3/$File/OB+3+Overview+Paper+-+Arnoldo+Contreras-Hermosilla.pdf*).

Dudley, R.G. 2004. A system dynamics examination of the willingness of villagers to engage in illegal logging. *Journal of Sustainable Forestry* 19(1/2/3): 31-53.

Environmental Investigation Agency. Undated. The politics of extinction web site (*http://www.eia-international.org/Campaigns/Forests/Indonesia/PolExtinction/timber05. html*).

Environmental Investigation Agency and Telapak Indonesia. 2000. Timber trafficking: Illegal logging in Indonesia, South East Asia and international consumption of illegally sourced timber (*http://www.eia-international.org/Campaigns/Forests/ Reports/timber/timber.pdf*).

Environmental Law Institute. 1998. Legal aspects of forest management in Mexico. Environmental Law Institute, Washington. (*http://www.eli.org/pdf/rrmexico.pdf*).

FAO. 2001. Illegal activities and corruption in the forest sector. In pp. 88-101 of FAO, State of the World's Forests 2001. FAO, Rome, Italy.

Forests Monitor. 1998. High stakes, the need to control transnational logging companies: A Malaysian case study (*http://www.forestsmonitor.org/reports/highstakes/ part2b.htm*).

Global Witness. 2001. The credibility gap–and the need to bridge it: Increasing the pace of forestry reform (*http://www.globalwitness.org/campaigns/forests/cambodia/ downloads/credibility.pdf*).

Greenpeace. 2000a. Plundering Cameroon's rainforests: A case study of an illegal logging by the Lebanese logging company Hazim (*http://www.greenpeace.org/~forests/ resources/hazimreport2109.pdf*).

Greenpeace. 2000b. Spotlight on the illegal timber trade: Asia Pacific (*http://www. greenpeace.org/~forests/resources/Asiasupplement.pdf*).

Greenpeace. 2000c. Spotlight on the illegal timber trade: Cameroon (*http://www. greenpeace.org/~forests/resources/Cameroonsupplement.pdf*).

Hammond, D. 1997. Commentary on Forest Policy in the Asia-Pacific Region (A Review for Indonesia, Malaysia, New Zealand, Papua New Guinea, Philippines, Thailand and Western Samoa). FAO Asia-Pacific Forestry Sector Outlook Study Working Paper No: APFSOS/WP/22 (*ftp://ftp.fao.org/docrep/fao/W7730E/W7730E00.pdf*).

Jordan, M. 2001. Unfamiliar turf: Saving the environment. *The Washington Post*, December 28, page A1.

Kaimowitz, D. 2001. Forest law enforcement: Issues and challenges for East Asia. (presented at the Southeast Asia Ministerial Meeting on Forest Law Enforcement and Governance) (*http://wbln0018.worldbank.org/eap/eap.nsf/Attachments/FLEG_S1-2/ $File/1+2+Daivd+Kaimowitz+-+CIFOR.pdf*).

Knowlton, C.S. 1972. Culture conflict and natural resources, pages 109-145 in Social behavior, natural resources, and the environment, W.R. Burch, Jr., N.H. Cheek, Jr., and L. Taylor (eds.). Harper & Row Publishers, New York.

Lindsay, J. et al. 2002. Why law matters: Design principles for strengthening the role of forestry legislation in reducing illegal activities and corrupt practices (FAO Legal Office, Legal Paper Online #27, *http://www.fao.org/Legal/Prs-OL/lpo27.pdf*).

Meadows, D.H. 1997. Places to intervene in a system, *Whole Earth*, Winter 1997 (Issue 91, pp. 78-84) (*http://www.wholeearthmag.com/ArticleBin/109.html*).
Sekhsaria, P. 2004. Illegal logging and deforestation in Andaman and Nicobar Islands, India: The story of little Andaman Island. *Journal of Sustainable Forestry* 19(1/2/3): 319-335.
Wardojo, W., Suhariyanto, and B.M. Purnama. 2001. Law enforcement and forest protection in Indonesia: A retrospect and prospect (presented at the Southeast Asia Ministerial Meeting on Forest Law Enforcement and Governance) (*http://wbln0018. worldbank.org/eap/eap.nsf/Attachments/FLEG_S1-3/$File/1+3+Boen+Purnama+ Speech+-+GOI.pdf*).
Wilkinson, C. 1992. Crossing the next meridian. Island Press, Washington.

APPENDIX I: SOME RULES OF THUMB FOR LEGAL REFORMERS

The following guidelines apply to using law as a tool of social reform:

Root Causes. Problem solvers, whether engineers, physicians, or policymakers, follow the same path: gather information; analyze and diagnose; select a remedy; apply the remedy; and follow-up and evaluate. Ideally, the diagnosis identifies the root causes of the problem as well as the symptoms. The problem solver focuses on the symptoms only if the root causes of the problem remain obscure or cannot be quickly resolved. The root causes of illegal activities are seldom purely matters of law. Economics, politics, and other factors frequently are part of the picture. Law is a far-ranging tool and may influence these factors. However, the solutions to these problems frequently take much more than the passage of a law or two. Law may simply be the first step in a larger program of social reform. And some aspects of these problems are beyond the reach of law. *Look to the root causes of the problem, and understand the power and the limits of law in addressing those causes.*

Interdependence of Actions. Illegal activities include a complex collection of problems, and the solutions to some can make others worse. (See Dudley, 2002, in this collection of papers, for a systems model of illegal logging showing the complexity of causal feedback.) For example, if people are stealing timber from the forests, new requirements for transportation permits or increased powers of arrest for forest guards may help deter thefts. Unfortunately, they may also create new opportunities for bribery of guards and permit issuers. Another example much on the minds of the international community is the effect of harvest bans in large consuming nations. The harvest bans in Thailand in 1989 and China in 1998 have led to increased illegal harvests in neighboring countries (Environmental Investigation Agency and Telapak Indonesia 2001). *Keep in mind the complexity of the system, and try to think through the full consequences of reform.*

Shaping of Institutions. Illegal activities by definition involve the breaking of prohibitions. By instinct lawyers may focus first on the prohibitions themselves and the penalties associated with them. However, more potent of legal reforms may involve the shaping of institutions. If the problem is corruption, for example, the nation may need institutional reforms that increase the transparency of government or that remove control of timber from a corrupt branch of government and transfer it to a cleaner branch. Institutional reform may also help in the case of timber theft. Independent institutions certifying lumber origins or otherwise tracking commerce in timber may make illegal activity more difficult to hide. *Look beyond forest rules, to reform of the institutions that apply those rules.*

Process of Adoption. Lindsay et al. (2002) have made the case that the process of forest law reform is as important as the substance. Law must have public support or it becomes almost impossible to implement. Public participation in the adoption of laws can help educate people about the need for reform, educate them about what is illegal, and create social pressure for the observance of new law. It can also provide important information to the policy makers, improving the substance of the law. In many developing countries, the idea of public participation is still novel. The legal drafter can catalyze public involvement through consultations, workshops, and other forms of outreach. *Pay attention to the process of reforming the law as well as the reforms themselves; look for constructive innovations in both.*

Transplantation of Solutions. When faced with a complex problem, a problem solver's instinct is to look for examples of similar problems and consider how they have been solved. This can be a powerful approach, and for some difficult problems, the most practical approach. Because law has such strong cultural roots, transplanted laws may not function vigorously in their new surroundings. Legal reformers must consider not just the effect of law on social problems, but also the effect of society on the law. *Look to other countries for successful models, but draft with the local conditions in mind.*

Incrementalism. Most bodies of law develop incrementally. The legal system adjusts to the needs of the society in many small steps. Often many changes are happening at the same time with little coordination and no grand design. This process works if the legal system is responding well to feedback. Some changes may be too little, or ineffective, or even counterproductive. The system can eventually react and change again. Faced with a complex problem like illegal forest activities, the legal reformer should take advantage of the power of incremental change to shape legislation. This means that reform may need to occur in several steps. Perhaps this may mean revising the law from time to time. Perhaps it means filling in the gaps of law later with regulations. It also means that the reformer should build feedback into government institutions and processes, to facilitate future reform. *Draft for today and tomorrow; strive for a system of laws that will respond to new problems and be a foundation for future improvement.*

The Role of Monitoring in Cutting Crime

Wynet Smith

SUMMARY. Illegal logging and trade is a global problem that requires a range of solutions, including improved governance and broad-based policy and legal reforms. An immediate step to combat crime is to monitor activities in forest areas. Monitoring offers a short-term solution to detect levels of illegal activities, as well as a long-term preventative means to reduce crime. Currently available methods and tools include trade data analysis, production/consumption analysis, paper audits, remote sensing analysis, and field investigations. This paper provides an overview of these methods and tools, highlighting strengths and weaknesses, and examples of their use. *[Article copies available for a fee from The Haworth Document Delivery Service: 1-800-HAWORTH. E-mail address: <docdelivery@haworthpress.com> Website: <http://www.HaworthPress.com> © 2004 by The Haworth Press, Inc. All rights reserved.]*

KEYWORDS. Illegal logging, sustainable forest management, forest industry

---

Wynet Smith left the World Resources Institute in Washington, DC in September 2003 to start her doctoral studies at the University of Cambridge. She can be contacted c/o Department of Geography, University of Cambridge, Downing Place, Cambridge, UK, CB2 3EN. The original research and writing for this article was undertaken while the author was a Senior Associate at the World Resources Institute.

Many thanks are due to reviewers of an earlier draft report: Janet Abramowitz, Arnoldo Contreras-Hermosilla, John Hudson, David Kaimowitz, Jan McAlpine, Kristof Obidzinski, Charles Palmer, John Spears and various World Resources Institute staff. A special thanks to Steve Johnston of ITTO for his overall input, especially on the use of trade data.

The United States Department of State provided funds to the World Resources Institute (WRI) to prepare a report on illegal logging. This paper is derived from Section 4 of a WRI report on illegal logging.

[Haworth co-indexing entry note]: "The Role of Monitoring in Cutting Crime." Smith, Wynet. Co-published simultaneously in *Journal of Sustainable Forestry* (The Haworth Press, Inc.) Vol. 19, No. 1/2/3, 2004, pp. 293-317; and: *Illegal Logging in the Tropics: Strategies for Cutting Crime* (ed: Ramsay M. Ravenel, Ilmi M. E. Granoff, and Carrie A. Magee) The Haworth Press, Inc., 2004, pp. 293-317. Single or multiple copies of this article are available for a fee from The Haworth Document Delivery Service [1-800-HAWORTH, 9:00 a.m. - 5:00 p.m. (EST). E-mail address: docdelivery@haworthpress.com].

INTRODUCTION

As noted in other papers in this volume, illegal logging is a global problem that can take many forms. Illegal forestry activities can consist of a wide variety of activities, from cutting of trees in unallocated concessions to cutting trees in protected areas, and illegal processing and transport of timber (Contreras-Hermosilla, 2002; Smith, 2003). Solving the problem of illegal logging and trade is difficult given the diversity of illegal activities and the range of underlying causes and contributing factors (Callister, 1999). Combating the problem requires a range of solutions, including improved governance and broad-based policy and legal reforms (Brack, Gray, and Hayman, 2002; Contreras-Hermosilla, 2002).

Monitoring forestry activities provides a short-term means to detect levels of illegal activities, as well as a long-term preventative means to reduce forest crime. As suggested by the notion that that which is measured improves, information can serve as a powerful non-regulatory environmental policy tool. In the United States, the Toxic Release Inventory–required under section 313 of the *Emergency Planning and Community Right to Know Act* of 1986–requires polluters to make their emissions of specific pollutants public. Simply requiring public disclosure has resulted in significant reductions in the emission of those specific pollutants (Hamilton, 1999).

In addition, an effective monitoring system that identifies problems-and perpetrators will lead to reductions in activities. Detection and monitoring activities can highlight not only the scale of the problem, but help to pinpoint where change must occur and help to target measures to reduce illegal logging, including enforcement action (Brack and Hayman, 2001; Contreras-Hermosilla, 2002).

A variety of data collection and monitoring approaches and tools are required to detect and reduce the range of illegal logging activities and a range of methods is currently available. These techniques range from macro-level analysis of trade and production/consumption statistics to intensive site-specific field investigations. Additionally, log-tracking systems offer the ability to monitor forest crime over the long term.

This paper provides an overview of six key methods, including trade data analysis, production/consumption analysis, paper audits, remote sensing analysis, field investigations, and log-tracking systems. It highlights examples of how they are being used to detect and monitor illegal logging activities by governments, non-government organizations (NGOs), research organizations, and communities. The paper points out key challenges of each method as well.

A MONITORING FRAMEWORK

A core set of steps can help to plan an effective program for monitoring illegal logging (see Figure 1). The following questions should be considered:

- What is considered legal and illegal in the national/regional context?
- What are the standards that should be and can be monitored?
- What are the types of illegal logging activities that need to be and can be monitored in the context?
- What tools can be used to help detect and monitor these illegal activities?

Understanding the legal context (what is legal/illegal), as well as the problems in a particular country or area (what needs to be monitored), will help identify the appropriate tools. A systematic program that provides for the ability to detect and monitor a range of activities is best when funding and staff time are sufficient. Where resources are very limited, it is best to focus on priority areas or regions. The specific tools used will also depend on the types of activities that need to be monitored (see Table 1 and Figure 2).

FIGURE 1. Framework for Monitoring Illegal Activities in the Forestry Sector

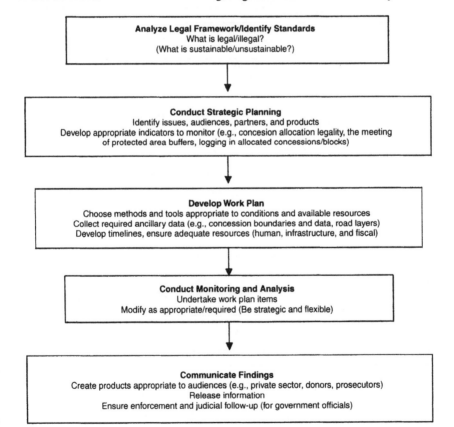

TABLE 1. Types of Illegal Logging and Appropriate Tools for Detection and Monitoring

Type of illegal logging practice	Question	Appropriate methods
Macro level		
Illegal harvesting of logs	Are there major imbalances in supply and demand?	–Analyze production and consumption data
Illegal logging and timber product trade	Are there major discrepancies in trade statistics?	–Analyze trade data between exporter and importer countries
Irregularities or corruption in the allocation of concessions or in issuance of logging permits/ licenses	Are government rules and regulations for allocation being followed? Are permits being allocated correctly?	–Audit tenders and bids against the legal requirements –Audit/review permits and licenses
Corruption in enforcement or judicial system	Are investigations being undertaken? Are enforcement actions being taken and charges being laid? Are charges being followed through on?	–Track judicial system results –Analyze case tracking databases where they exist
Logging in restricted or prohibited areas		
Logging in strictly protected areas	Is there illegal logging in protected areas/reserves where it is banned?	–Conduct remote sensing analysis –Conduct field investigations
Logging in a non-allocated concession or in non-active coupe/block	Where is the illegal harvesting taking place?	–Conduct field investigations –Conduct remote sensing analysis (imagery combined with other land use information, such as where active concessions and road networks exist)
Logging in restricted/ prohibited areas within concessions	Is harvesting taking place in areas within concessions where it is not allowed?	–Conduct field investigations –Conduct remote sensing analysis –Implement log tracking systems
Illegal logging activities within legal concessions		
Cutting above allowable limit, cutting protected species	Are forest regulations regarding cutting practices, species, and volumes being followed?	–Conduct field investigations –Review inspection reports, species trade reports, and case tracking systems –Implement log tracking systems
Illegal reporting activities		
Transfer pricing	Are companies under-reporting value and transferring profits offshore?	–Monitor market information, freight insurance/rates, and import prices
Improper reporting	Is information being reported correctly or incorrectly?	–Monitor official data/statistics to see if there are irregularities
Illegal transport activities		
Illegal transport of wood	Are timber and other wood products being moved illegally?	–Monitor issuance of permits –Conduct field investigations, using check points
Illegal processing activities		
Illegal processing of wood	Are their unlicensed mills? Are licensed mills processing illegally cut timber?	–Conduct remote sensing analysis –Conduct field investigations, using GPS to document sites –Review licenses against sites on ground

FIGURE 2. Corrupt and Illegal Activities in the Forestry Sector Commodity Chain

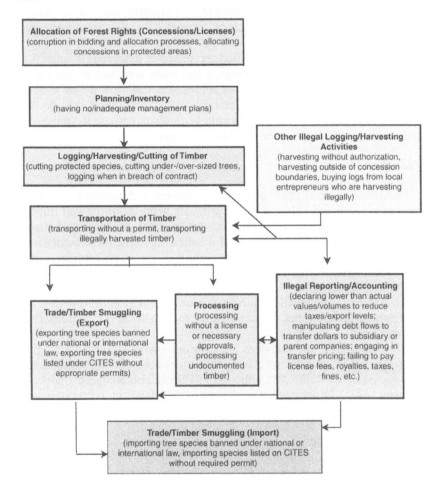

AVAILABLE METHODS

Governments, the private sector, communities and non-governmental organizations are currently undertaking monitoring activities in the forest sector. The formal (or informal) role of the organization and the desired monitoring outcome or product of the monitoring activity both shape the choice of methods and tools and the level of precision required. National and local governments will be responsible for very different monitoring tasks and may require very different tools than nongovernmental organizations. As will be discussed below, some monitoring strategies and tools available are clearly more appropriate for implementation by government. For example, national solutions include the development of good general forestry data and the implementation of wood flow/log tracking systems.

Each of the six monitoring methods reviewed in this paper is suited to detecting specific types of illegal logging and trade activities at specific geographical scales (see Table 2). Other approaches and tools do exist, such as diagnostic surveys (Contreras-Hermosilla, 2001), and others will be developed. It is important that any monitoring system strive to use the most appropriate tools for the context and available resources.

Trade Data

Trade data, collected primarily by government and trade associations, record the volume and value of forest products being exported and imported. Trade flow statistics that document transactions over several years between a variety of trading partners can be useful indicators of illegal or otherwise undocumented trade. Major discrepancies can indicate the occurrence of illegal logging and trade. Forestry trade data are available globally from the Food and Agricultural Organization (FAO) and the Commodity Trade Statistics (Comtrade) database of customs statistics. Information on trade in tropical timber is available from the International Tropical Timber Organization (ITTO) for its member countries.

While trade data analysis does not determine where the illegal logging is occurring, it does help identify areas for further investigation. Analysis typically focuses on volume data although ITTO has also analyzed trade values (Johnson, 2002). Discrepancies tend to be easiest to detect in data on raw logs rather than processed products.

Trade Data–Examples

Several studies have identified major discrepancies in the statistics of exporting and importing countries. The secretariat of the ITTO monitors trade data on an ongoing basis, as part of its mandate under the International Tropical Timber Agreement. For example, the ITTO 2000 annual review notes that the export statistics for Indonesia do not match up with those reported by importing countries. While Indonesia reported a total export of 300,000 cubic meters, Malaysia alone reported almost 600,000 cubic meters and China reported nearly 400,000 cubic meters of industrial roundwood (ITTO, 2001; Johnson, 2002). The data for many other timber producing countries, including Cameroon and Gabon, also indicate major export/import discrepancies (Johnson, 2002).

TABLE 2. Review of Methods for Detecting and Monitoring Illegal Logging

Method	Level of analysis	Questions addressed	Activities that can be detected	Limitations	Relative cost	Requirements and useful tools
Trade Data Analysis	National or regional	How much? To where?	*Flow of illegally cut timber on a macro-level	*Data often of variable quality. *Data do not always exist. *May not detect many types of illegal logging.	$ *Staff time *Data purchase (perhaps)	*Trade data from exporting and importing countries
Production/ Consumption Data Analysis	National	How much?	*Illegal cutting levels	*Data may not always exist. *Data extrapolation may create errors.	$ *Staff time	*Production and consumption data
Paper Audits/ Assessments	National or sub-national	What? Where? How much? Who?	*Corruption *Non-compliance with management plans and operational standards *False reporting *Undervaluing	–Data may be hard to obtain	$ *Staff time *Data purchase (if necessary)	*Laws and regulations *Concession allocation results *Case tracking system *Management plans *Forestry fees database
Remote Sensing Analysis	National or sub-national	Where? When? (if using time series) How much? (to a degree)	*Detection of clearcuts, intensive selective logging, log yards, roads	*High quality images may not be available or difficult to obtain. *Requires specialized skills.	$$ to $$$ *Staff training *Hardware and software *Data layers *Imagery *Flight costs	*Satellite images or air photos *Data layers *Computers and software *Trained staff
Field Investigations	Sub-national	Where? How? Who?	*Logging in unauthorized areas *Illegal logging in legal concessions *Location of illegal log yards, sawmills	*Usually of limited geographical scope. *Accessing remote sites can be difficult and expensive. *Timing of field visits is important. *Work can be dangerous.	$ to $$ *Staff time *Transportation *Recording tools	*Cameras, video cameras, tape recorders *Global positioning systems *Maps and data
Log Tracking	National and regional	Where? How much? Who?	*Unauthorized logs entering legal chain *Specific concessions with problems *Illegal processing of wood	*National systems are expensive, though less expensive per unit than doing on a case-by-case basis.	$ for low-end tools (paint) to $$$ for high-end tools.	*Electronic databases *Tagging system (barcodes, tags, chips, etc.) *Capability to do spot checks (agents/auditors)

Notes: Relative costs are indicated as $ to $$$, with $ being least expensive and $$$ most expensive. The major expenses are indicated below the dollar signs. Many factors influence the cost of these operations in a particular country, from staff wages to ease of transportation. The level of detail and the mode of operation will also influence costs. For example, a nongovernmental organization doing investigative field operations may be able to operate quite inexpensively. Official missions may be more expensive.

Other groups have used trade data to identify potential illegal logging and trade activities. For example, the World Resources Institute used ITTO data to identify major discrepancies in trade reports for Myanmar. They found that for 1995 Myanmar reported a volume of log exports that was 276,000 cubic meters less than that reported by importing countries (Brunner, Talbott, and Elkin, 1998). By 1994, Thailand's reported log imports from Myanmar were four times higher than Myanmar's declared exports (80% illegal), although this was only twice as much in 1995 (50% illegal). Additionally, Myanmar declared no log exports to China in 1995 while China declared imports of 500,000 cubic meters (Brunner, Talbott, and Elkin, 1998).

Comparisons of Georgian and Turkish trade data for 1999 revealed that Georgian customs statistics underestimated timber exports by almost 60%. This number is a conservative calculation as the analysis took account of undocumented wood flowing from the breakaway province of Abkhazia to Turkey. Calculations also showed that official sawn timber production was 120,000 to 150,000 cubic meters below the reported volume of sawn timber exports (Siry, 2000). Many other nongovernmental organizations have used trade data to identify importing countries that may be purchasing illegally cut and traded timber (see FERN, 2001; Matthew, 2001)

Trade Data–Challenges

The results of analyses by ITTO and others illustrate the important role that trade data can play in identifying problem areas. There are some illegal logging and trade problems that will not be easily identified, however, through the analysis of trade statistics. Trade data analysis is not useful in identifying illegal activities in countries with limited external trade. For example, although Brazil has serious illegal logging problems (over 80 percent of logging in the Brazilian Amazon is believed to be illegal), trade data won't reveal much because most of its wood is consumed domestically (FAO, 2001). Fraudulent accounting practices, such as transfer pricing, are also not detected by analysis of trade volumes. For example, trade data from ITTO do not show major discrepancies between the import and export volumes for countries like Papua New Guinea, where transfer pricing appears to be the significant problem. Transfer pricing is the practice of placing artificially low export prices on timber sold to foreign subsidiaries or associates, usually for the purpose of avoiding taxes. Analysis of trade values, however, may help shed light on such problems.

Trade data are also problematic because countries define and measure products differently and make many errors in reporting trade. Some categories of wood products are more difficult to analyze than others. Industrial roundwood–defined by ITTO as all roundwood in the rough–is the most straightforward product to analyze since the product definition is least subject to confusion. Sawn wood, plywood, and other processed wood products are more difficult to analyze because these products are more heterogeneous, which can cause confusion (Johnson, 2002). For example, some countries mistakenly combine trade in mouldings and other further processed sawn wood with rough sawn wood figures.

Of greater significance, however, are problems with statistical reporting and legitimate reasons for discrepancies between trading partner reports; these reduce the usefulness of such analyses for identifying potentially illegal trade flows (Durst, Ingram, and Laarman, 1986; Johnson, 2002). As noted in previous forestry trade data analysis, discrepancies in trade figures are common and can result from the following:

1. Exports that occur at the end of one reporting period (e.g., a calendar year) and do not reach their destination until the next period.
2. Misidentification of the origins and destinations of shipments in trade documents. Sometimes shipments are erroneously identified with shipping registration.
3. Failure to consistently classify, measure and report commodities according to the Standard International Trade Classification System or the Harmonized System of Customs Clarification. Data differences can also result when errors are made in converting between different measurement systems (e.g., board feet to equivalent cubic meters).
4. Simple counting and recording errors caused by carelessness and incompetence of customs officers and reporting officials.
5. Failure to report smuggled goods by one or more trading partners. Serious data inconsistencies follow when traded goods go unreported by only one country (Durst, Ingram, and Laarman, 1986).

The lack of adequate and consistent data is probably the most fundament issue limiting the usefulness of trade data in detecting illegal logging activities. As noted already, comparative analysis of existing data is very difficult due to differences in how the figures are interpreted and compiled. Missing data, however, are just as great a barrier to using trade data for analysis. If data are not available, organizations cannot publish valuable data or conduct useful analyses. In the International Tropical Timber Agreement of 1994, ITTO member countries agreed to report data on an annual basis, but some countries have never submitted information to ITTO or any other organization. For monitoring to be effective, member countries need to supply data and ITTO must be able to follow up on cases where data are not freely offered. The lack of data makes it difficult, if not impossible, for ITTO to fulfill its duties to monitor and report undocumented trade in tropical timber, as required by the

1994 agreement. Missing data make a difficult job harder. The absence of data also leaves member countries and the ITTO vulnerable to criticisms, undercuts attempts to infuse transparency in the tropical timber trade, and makes unbiased assessments across countries impossible. Countries that do not provide numbers cannot be assessed for their own problems or for the role they may play in consuming timber from countries with major illegal logging and trade problems. Until these countries begin to comply, ITTO's data will remain inadequate for obtaining a comprehensive picture of the countries with major problems, and countries trying to create change will be hampered by the lack of information.

There are signs of hope, however, for improved trade data. The Inter-Secretariat Working Group on Forest Statistics, consisting of representatives from Eurostat, FAO, ITTO, and the United Nations Economic Commission for Europe, has developed a joint questionnaire and is working for greater consistency in published statistics. Further work in this area will improve the usefulness of trade data as a tool for detecting and monitoring trade in illegally logged timber and wood products.

Production/Consumption Data Analysis

Production and consumption data are records of how much wood is produced and how much is consumed in a given country. Without this information, it is impossible to track how forests are being used or to predict future trends. Comparison of production and consumption data provides a macro-level tool for identifying imbalances in official supply and demand of industrial roundwood. Deficits in existing supplies of wood for processing and export–that is, where demands for wood are higher than the supply available from domestic and import sources–are problematic. If the required wood demand is not being imported or harvested legally, it is entering the system from undocumented or illegal channels. Similar to trade data, this approach can help identify the overall extent of illegal logging but does not provide any information on where the illegal activities are taking place.

Production/Consumption Data Analysis–Examples

Analysis of consumption and production statistics has been used to great effect in Indonesia. Studies have compared known legal supplies of wood with output from the wood processing industry to quantify the problem. A study by the United Kingdom and Indonesia compared 1997 and 1998 data on known legal supplies of wood (domestic legal production plus imports) with consumption (domestic use plus exports) and revealed that consumption exceeded available supply by 32.6 million cubic meters (Scotland, 2000) (see Table 3). This production deficit amounted to over half of the total domestic

TABLE 3. Estimated Timber Supply and Demand for Indonesia, 1997-98

Sources of Timber Supply and Demand	Volume (Roundwood equivalent) (m³)
Log supply from domestic production	29,500,000
Log supply equivalent from imports	20,427,000
Log supply equivalent from other sources (mainly recycled paper)	1,600,000
Total Supply	51,527,000
Domestic demand	35,267,000
Log equivalent of exports	48,873,000
Total Demand	84,140,000
Net Wood Balance	−32,613,000

Source: Scotland, 2000.

wood production for the years studied. An earlier analysis identified a discrepancy of 56 million cubic meters, which implied that more than 70% of the timber demand was not accounted for by official production (Scotland, Fraser, and Jewell, 1999). An analysis using a modified methodology resulted in estimates of 49 million and 64 million cubic meters in 1997 and 1998 respectively (Palmer, 2001). Yet another study compared the production capacity of plywood and sawmills against the legal supply of wood and estimated a gap of 21 million cubic meters in the log supply for 1998 (Brown, 1999).

Data for the Philippines indicate a major discrepancy between supply and demand. Analysis by the government shows that domestic demand for timber is estimated to be 5 million cubic meters, but only 2.7 million cubic meters are legally available. The legal sources include half a million cubic meters from the Annual Allocable Cut (AAC) from timber license agreement areas, 1 million cubic meters of coconut lumber, 1 million cubic meters imported wood and 200,000 cubic meters from plantations. A deficit of 2.3 million cubic meters annually, or about 45% of domestic requirements, exists (Acosta et al., 2000). This deficit is assumed to come from undocumented sources, most of it illegally extracted from public forests.

The ITTO also uses production statistics to identify potential undocumented or illegal production. It compares industrial roundwood availability (production plus imports minus exports) with production of final products in roundwood equivalent (RWE) volume to determine a log balance for each country (Johnson, 2002). Calculating roundwood equivalent volume involves multiplying each processed product by a factor representing the volume of roundwood required for each unit volume of final product. ITTO's analysis of year 2000 data identified some countries with significant imbalances between industrial roundwood availability and roundwood equivalent volume. Log deficits indicate that there was insufficient log availability to produce the quantity of final products reported. In the cases of Cameroon, Panama, and Peru, either production figures (mostly sawn wood in all three cases) are too high, or the extra logs required came from undocumented sources. Other countries had significant log surpluses in 2000. Since none of these countries has a significant timber industry beyond log and sawn wood production, it is unclear where the excess logs are being utilized. Undocumented sawn wood or other processing mills may be using some of this material, while some may leave the country as undocumented/illegal exports.

Statistical anomalies like these can also arise in importing countries. For example, several tropical timber-importing countries regularly report exports of tropical products in excess of their imports. This can be due to stock accumulation and depletion cycles, but when the quantities involved are substantial and the problem appears regularly, there is cause for concern. This is also the case when the production of tropical sawn wood and plywood, for example, regularly exceeds the availability of imported tropical logs.

Production/Consumption Data Analysis–Challenges

Timber production data are traditionally considered less reliable than trade data, and production and consumption data for countries are often not available. Many countries have no regular industrial survey procedure to establish accurate production figures for forest products and must rely on estimates (Abramowitz, 2002; Johnson, 2002). Figures for consumption often must be extrapolated, which makes drawing conclusions difficult. Johnson (2002) notes that processing efficiencies vary widely between countries, and most countries do not provide roundwood equivalent conversion factors. The use of standard factors (e.g., 1.82 for non-coniferous sawnwood, 1.9 for veneer, 2.3 for plywood) can lead to large apparent log imbalances for more or less efficient processors. The analysis therefore only highlights for further investigation very large imbalances that are not easily explained by differences in processing.

Countries must further develop their capacity to establish accurate production figures for forest products. Reliable production and consumption data would improve the ability to detect major illegal logging problems and improve the ability of governments to sustainably manage their forests.

Paper Audits/Assessments

Paper audits or assessments of compliance with existing legislation and regulations involve the comparison of reported and actual practices against what is required in legislation. Paperwork audits can identify illegal and corrupt activities by both government and private sector entities. The following are some of the paper analyses that can be performed:

1. *Audits of concession and license allocation processes and results (where they exist):* A review of the licenses and concession issuance processes and results can identify where awards were made in violation of bid and allocation processes. This type of review can also identify licenses that have been issued in areas where they should not have been, such as parks and other protected areas.
2. *Audits of forest management plans:* Audits of management plans can determine whether they have been filed, whether they meet legal requirements, and whether and to what extent they are being implemented.
3. *Audits of enforcement files and case tracking systems:* Analyzing enforcement files and databases can identify charges and infractions, as well as provide environmental governance information about the illegal activities tracked and acted upon by officials. This type of assessment can identify where follow-up has not occurred; a lack of follow-up may indicate that high-level officials have interfered in the enforcement of forestry laws and regulations.
4. *Audits of forestry payments databases:* Assessments of the forestry fees and taxes paid can help identify irregular reporting and payment activities. This type of auditing also provides environmental governance information.

The collection of these data can help pinpoint areas that may require further investigation as well as providing a sense of the overall governance situation.

Paper Audits/Assessments–Examples

Global Forest Watch Cameroon partners used a two-tier process to assess compliance with existing forest regulations. The first step was a coarse-scale assessment of the legal status of concessions in which national government data were collected on logging rights allocation and activity. The next step was a fine-scale assessment, which examined the types and numbers of citations issued for non-compliance with regulations. Provincial government data were collected as well as field data on specific logging violations and the ensuing judicial processes (Global Forest Watch Cameroon, 2000).

The results of both assessments indicated significant and widespread irregularities in 1997. The coarse assessment results concluded that 29 (or 56%) of the licenses issued under the old legal system were operating illegally. In addition, fewer than one-third of concessions allocated under the new 1994 law fully complied with new regulations. The fine scale assessment indicated that 96% of violations reported in 1992-93 (the latest year for which data was available) were followed by incomplete judicial procedures. Twenty percent of all violations received no follow-up because an influential person interfered in the process (Global Forest Watch Cameroon, 2000). Global Forest Watch Cameroon partners later reviewed the 1999 and 2000 timber concession allocation process. They found more transparency than in the 1997 allocation process, which indicated that environmental governance had improved. However, the June 2000 allocation process raised several unanswered questions about Cameroonian concession allocation policy, particularly regarding bids offered by companies that had been censured for illegal logging (Cameroon, 2001). The results of these assessments indicate the need for significant improvement in the allocation process and qualification and ranking criteria (Cameroon, 2001).

In the Canadian province of British Columbia, the Sierra Legal Defense Fund (SLDF) audited government forestry payment data to review variations between forecasted government revenues for a given forest management area and the funds that were actually received (and reported to have been received) for that area from timber companies. Companies are required to scale the timber they log, which involves both grading and classing, at the end of their first year of operation. Payments are then recalculated for the next 5 years based on this reporting. SLDF's analysis revealed substantial variations in the reported scaling of species in relation to the original assessments done by the British Columbian government. The organization's findings indicated that the government loses at least Can$224 million (US$146 million) per year through irregularities in how logs are scaled and graded by companies. SLDF also found that the stumpage manipulation technique known as "grade setting" had cost the provincial government more than Can$138 million (US$90 million) between the first quarter of 1998 and the second quarter of 2000. While it is not clear if these irregularities are illegal, there do appear to be major problems in how the system functions (SLDF, 2001).

Paper Audits/Assessments–Challenges

Government records on concession and license allocation, forest management, and forestry fees and taxes are key to tracking overall decision making and management of forests. Often, records and databases on allocation results, forestry payments, and case tracking of infractions are not available. Without these records, it is very difficult to monitor activities and to improve transparency in the decision-making system. While the above examples show ways that paper audits can identify problems, such analyses can be difficult to undertake in regions or countries where corruption and violence are endemic, transparency is lacking, and access to information is severely constrained.

Governments need to keep records of concession allocation and bidding processes and outcomes. These records need to be publicly available. Management plans developed by companies need to be filed and made available to the public. Government officials should assess these plans to ensure that appropriate standards are being met. Their assessments also need to be made available to the public. Payments of forestry fees and taxes need to be carefully recorded. Electronic databases, where they exist, allow easy access and monitoring of fee payments. Files and databases of infractions and judicial follow-up are needed for auditing and will help ensure transparency. All of these files should be made available to the public for review.

Remote Sensing

Remote sensing involves the use of remote observation of on the ground activities. Remote sensing tools, such as aerial over-flights, aerial photographs, and various types of satellite imagery, can detect where and when illegal logging is occurring. Remote sensing is particularly helpful for monitoring forest use activities in remote areas that are difficult to visit. It is also a useful tool for rapid assessments of both land cover and land use (Smith, Meredith, and Johns, 1999).

Aerial flights can be useful for on-the-spot detection of activities and, if undertaken by enforcement agencies, can allow quick legal action. Aerial photographs can be taken at high resolution, enabling researchers to perform detailed analysis. Satellite imagery can detect indirect evidence of illegal logging activities, such as roads or illegal log-yards, or direct evidence, such as clearcuts or intensive selective logging in prohibited or unallocated areas. Analysis of time-sequenced satellite imagery can provide clear indications of the level of activity in specific places (World Bank, 2000). The integration of satellite imagery into a monitoring process normally consists of screening lower-resolution images for features that can serve as proxy indicators of logging. In most cases, these proxy indicators require confirmation through ground or aerial surveys, so the screening is done in order to narrow down the choice of locations to be inspected in detail, and to establish priorities (World Bank, 2000).

A range of satellite imagery is available to the public. Medium and high-resolution satellite imagery offer the best means to map and monitor large areas cost-effectively. High-resolution imagery, such as IKONOS, shows much greater detail but its value in an illegal logging monitoring program has not yet been determined.

Remote Sensing–Examples

Aerial flights have been used by a number of countries to investigate remote forest regions. For example, the Philippines has recently used aerial raids to help find and confiscate logs, worth several thousands of dollars, on the

Agus River in Quezon Province and on the Agusan River in the Caraga region (DENR, 2002a, 2002b, 2002c, 2002d). In Cameroon, several groups are using remote sensing to study illegal logging. Global Forest Watch is combining spatial data on concessions with Landsat 7 ETM+ satellite imagery to detect areas with unauthorized activities. The analysis has detected numerous roads in unauthorized areas, as well as clear indications of forest degradation from intensive selective logging.

A World Bank mission in Laos used satellite imagery to analyze whether recent logging activities were legal (World Bank, 2000). In this mission, a recent spot black and white image and old Landsat images were used to identify areas of recent human disturbance. The logging survey mission discovered four areas of serious logging infractions, each including several individual logging sites, that violated existing agreements. Several less substantial infractions were found as well. Without the use of satellite imagery, the inventory of illegal logging activities would have been far less complete. For example, it can be difficult to detect features that are only a few hundred or even only a few dozen meters perpendicular to the helicopter flight path.

In Brazil, the State Foundation of the Environment in Mato Grosso has used Landsat 7 to help identify fires and changes in land cover as part of its overall environmental control system. The technicians pay particular attention to extensive deforestation and deforested riverbanks (State Department of Mato Grosso, 2001). Environmental officers use this data to help them prioritize areas for field visits. In Cambodia, remote sensing has also been used extensively as part of the new independent forest monitoring system to detect changes in land cover, new roads, stockpiles of logs and recent illegal logging activity (Ouen, Sokhun, and Savet, 2001; Savet, 2000). Unfortunately, the government appears to have rarely used its remote sensing data to proceed with prosecutions, and it rarely shares these data with other interested parties (P. Alley, pers. comm., 2002).

Remote Sensing–Challenges

There are a number of challenges related to the use of remote sensing approaches. First, access to other geo-referenced information is essential (Dehqanzada and Florini, 2000) for combating illegal logging. Flying over a remote region may detect activities, but determining whether those activities are legal or illegal will depend on knowledge of the permitted land uses. This information can be combined with satellite imagery analysis to provide accurate assessments of illegal activity. The following are other challenges presented by remote sensing methods and tools:

1. *Limited availability:* Aerial photographs must be ordered for an area and are not available immediately. Cloud-free satellite imagery and aerial photographs for humid and mountainous areas can be difficult to obtain.
2. *High resource requirements:* Remote sensing requires sophisticated computer resources. Specialized software programs for imagery analyses and geographic information interpretation are required. Personnel must be trained to run the systems, maintain the data, perform the analysis, and produce useful maps and evidence for use by enforcement officers and courts. These items can drain small or already over-taxed departments and organizations.
3. *High costs:* High-resolution imagery, such as IKONOS, and aerial photographs are expensive to accumulate and analyze over time. The costs associated with setting up a computer system and acquiring the necessary staff are also great.
4. *Reduction in detailed knowledge:* Once a system is in place, the use of these tools can be much cheaper than intensive field investigations. However, the information obtained through remote sensing is generally less detailed than that which is obtained through field investigations. For example, remote sensing instruments currently available for such uses cannot distinguish tree species or determine exactly how much is being cut. Therefore, it is very important to carefully assess what level of detail is required from the monitoring program.

The development of basic information layers on logging rights is an essential first step to using remote sensing. Information is required on logging concessions, existing road systems, and protected areas, among other things. Governments should make concerted efforts to create these data sets where they do not exist and to share them publicly where they do exist. Countries (especially the G8 countries who have committed to taking action on illegal logging) should consider providing resources to help producer countries fight illegal logging.

Field Investigations

Field investigations consist of on-the-ground assessments of illegal activities and are a core part of any illegal logging detection and monitoring program. Field investigations are used as a data collection and monitoring tool by enforcement agencies, nongovernmental organizations, and academic researchers. Approaches to field investigations tend to differ by sector; these differences in approaches affect the type of information collected and how it can be used.

On-the-ground field visits can result in comprehensive, valid assessments of illegal activity for legal enforcement and other purposes. They are also important for verifying data collected through other methods, including remote sensing analyses. Field investigations can be used to detect logging in unallocated areas, within protected areas, and within allocated but inactive areas. They can also be helpful in detecting illegal sawmills and log yards and illegal transportation of logs. Investigators require basic baseline information, including information about protected area locations, active concession allocations, pre-existing road networks, and legal sawmills.

Key methods include field observations, interviews, and investigative undercover work. Investigators set up control points to monitor traffic flow on roads or rivers, to survey mill capacity, and to conduct timber audits of port facilities. Nongovernmental organizations, including Global Witness and the Environmental Investigation Agency, specialize in using undercover investigative techniques. Field research can also include extensive surveys and interviews by academic researchers (Obidzinski and Palmer, 2002). Government enforcement agencies undertake official missions or investigations to verify compliance with laws and regulations, but they tend not to use undercover techniques, as they must ensure that evidence can be used in court. This is not to say that nongovernmental organizations do not collect careful evidence that can be used in court. In fact, evidence by nongovernmental organizations has been used in many cases to pursue actions against illegal logging activity. For example, Greenpeace fieldwork has been used in court cases in Brazil. Global Witness has carefully documented evidence in Cambodia and the Cambodian government used such data in mid-2002 to cancel a company's contracts.

Several tools are available to help with field investigations, including low-technology tracking tools, such as ultraviolet paint, to identify and track timber coming from illegal sources (see the next section for further details). Investigators also make frequent use of global positioning systems (GPS) to compare the location of activities with official data on maps and concession permits, to survey logging roads, and to determine the geographical locations of log yards and sawmills. Video cameras, cameras, and tape recorders also are used to help record evidence of illegal activities.

Field Investigations–Examples

A number of key field studies on illegal logging have been undertaken by NGOs. Recent research by Greenpeace in Brazil explores the illegal harvesting and trade of mahogany. Greenpeace has worked closely with IBAMA, the Brazilian environmental agency. Together, they have conducted several inspections that in 5 days netted the largest volume of illegal mahogany logs in

Brazilian history: 7,165 cubic meters, valued at US$7 million (Contreras-Hermosilla, 2002).

Another important example is the work of Global Witness in Cambodia, where its investigative approach to detecting illegal activities was key to raising awareness of the problem (Global Witness, 1999, 2000). Since then, the organization has played the role of an independent monitor as part of a forest crimes monitoring unit to help track illegal forestry activities. EIA and Telepak, an Indonesian non-governmental organization, also have conducted intensive undercover fieldwork to monitor illegal logging activities in national parks in Indonesia (Environmental Investigation Agency and Telepak, 1999, 2000) Their work has helped to raise awareness of the issue of illegal logging and to create political space for change.

In Canada, the British Columbia Forest Watch program, a special project of SLDF, and a member of the Global Forest Watch Canada network, has conducted numerous audits of forest concessions (tenures) to monitor compliance with provincial forestry legislation and regulations. In Ontario, SLDF has partnered with the Wildlands League, an Ontario nongovernmental organization, to conduct similar work. In these field audits, the groups have compared forest operational practices with what is required by law. For example, they examine whether companies are respecting no-logging zones, including streams and their buffers and environmentally sensitive sites. They also inspect whether logging roads meet required standards. The most recent results from an audit in Ontario found that 35% of the "areas of concern" (environmentally sensitive sites) audited were logged and 88% of the stream buffers required by law were not retained (SLDF and The Wildlands League, 2001). Earlier reports cover audits in other tenures in Ontario, all of which documented similar types of infractions (SLDF, 1999; SLDF and The Wildlands League, 2000).

Field Investigations–Challenges

Field investigations tend to be limited in scope, geographic area, and time because areas can be difficult to access, and resources to conduct investigations are often lacking. While field investigations can be relatively inexpensive if conducted by community groups, official missions by government tend to be more costly. In areas with corrupt and undemocratic conditions, baseline information required by investigators is often not available–or is withheld–making research difficult. In addition, investigators, whether from government or the nonprofit sector, may be at risk in these conditions. A journalist investigating corruption and land rights in Indonesia was killed in August 1996 (Pacific Media Watch, 1996). According to the Foundation for the Philippine Environment, more than 60 people have been killed in the Philippines in the last several years trying to fight illegal logging and fishing, including 8 forest rangers killed by a grenade blast at a checkpoint (Human Rights Watch, 1996). A forestry official was killed and another wounded in northern Cambodia in 2000 by armed, suspected illegal loggers (Richardson, 2000). Global Witness staff in Cambodia were assaulted in Phnom Penh in April 2002 (Richardson, 2002). In January 2000, 2 employees of Telepak and the Environmental Investigation Agency were kidnapped and held captive for 3 days in south Kalimantan, Indonesia. They were subject to physical and verbal abuse before being released (Environmental Investigation Agency, 2002).

The data collected during field investigations are not always directly available to the general public and others, as groups may keep it confidential to protect sources, their own investigators, or to build legal cases. By default, much of their information is garnered from sources that may require anonymity and is not readily attributable to government sources or databases. When enforcement officers (e.g., police officers, forestry inspectors, or environmental protection officers) undertake field investigations, they need to carefully record and document their work. Meticulously kept files and databases of infractions and follow-up are needed to ensure that forestry laws and regulations are met, and that there is transparency in the process.

Field investigations, while not producing statistically meaningful results, continue to play a key role in documenting the existence of the problem, building awareness, and helping to create pressure for positive government action. Data collected during field investigations can help create transparency and accountability in both the forestry industry and the governments that oversee forest resources.

Some standardization in methods and recording of data would simplify the use of this data for long-term monitoring of changes in a given geographical location. Political space needs to be maintained, and created where it does not exist, for civil society to undertake independent monitoring. Collaborative partnerships between different stakeholders should be pursued where possible as well.

Log Tracking Systems

To obtain better production, consumption, and trade data, timber or log tracking systems can be used. These chain of custody systems are tools for tracking and monitoring the flow of forest products from their point of origin to their destination. The use of log tracking systems offers an opportunity to tackle the problem of illegal logging and trade, as well as unsustainable harvesting, on a more comprehensive basis. To be truly effective, these systems need to be implemented on a nation-wide level. They are considered an essential component for any comprehensive procurement policy (Toyne, O'Brien, and Nelson, 2002).

A variety of systems and specific tools are available, including low technology alternatives such as paint and ultraviolet paint. While relatively inexpensive and useful for detecting individual cases of illegal logging, these tools are not particularly useful in broader-based or national-level efforts to combat illegal logging and timber product trade. More high-tech tools are required in national tracking systems. These tracking systems usually involve a combination of databases, the physical tagging of logs, and some form of spot checking to ensure the system is being implemented. Numbers are assigned to identify companies and areas of operation. These numbers are entered into the database for tracking. Identification tools, such as bar codes, microchips, or micro-tags, are issued bearing the identification numbers. These identifiers are then attached physically to the logs being cut so that the wood can be tracked from the point of harvesting through transportation, processing, and export.

These tags can be issued so as to identify not only the company but also the source concession and the legal volume of wood. This automated system of timber control combines advanced tracking technology and physical inspections in order to identify, inspect, and monitor the domestic and/or export flows of forest products (SGS, 2001). Such tracking of wood can greatly improve the chain-of-custody record-keeping system. Other research is being undertaken on genetic fingerprinting for tracking purposes as well, although this option is still very expensive and requires baseline data that is not easy to develop. Use of these systems can help to identify where illegally logged wood enters a system. They can help eliminate the problem of fraudulent transport and harvesting licenses that are often used to launder timber into the legal stream of wood.

Log Tracking Systems–Examples

Some nongovernmental organizations have used low technology tools, such as ultraviolet paint, to mark and track the flow of logs. For example, Greenpeace used ultraviolet paint in Brazil to track timber from known illegal sources to trading areas, where the organization was then able to document that particular companies were buying illegally harvested wood (Anonymous, 1999). The United States Forest Service, a land management agency within the US Department of Agriculture uses a low-technology log-tracking system that relies on personnel who are assigned to administer timber sales. They conduct informal monitoring, which involves making periodic visits to timber sale areas by supervisors and law-enforcement personnel. They also do formal monitoring where they make unannounced, in-depth reviews of all timber and log accountability activities undertaken by each administrative unit (such as a national forest district office). The agency has a variety of specific techniques, including log identification using paint and hammers (Dykstra et al., 2002).

Log Tracking Systems–Challenges

While relatively inexpensive and useful for detecting specific cases of illegal logging, low technology systems are not the most appropriate in a broader-based national-level strategy to combat illegal logging and trade. In terms of high tech systems, there are a number of issues that need to be considered, including the need for periodic spot checks to ensure that the system is working. Setting up these systems is time-consuming, expensive and personnel-intensive, and they require ongoing monitoring to ensure that they are implemented properly. At a national level, the cost should not be onerous: the company SGS estimates that a national system would cost on average about US$1 per cubic meter to implement (de la Rochefordiere, pers. comm.). The expense will vary, however, according to such factors as the size of the system and the institutional set-up. These approaches are not likely to work or be cost-effective unless they are undertaken at a national scale. Further details on log tracking systems are available in a report from a WWF meeting on log tracking systems held in Cambodia in March 2002 (Dykstra et al., 2002).

CONCLUSION

Detecting and monitoring illegal logging offers one means of helping to cut forest crime. Two key areas that require improvements and further development if monitoring is to play a truly effective role are:

1. *Data:* Improved data on forests is needed, including forest inventory, forest management data, production and consumption data, trade data, concession allocation data, and compliance data. Without adequate information on the baseline condition of forests, it is impossible to assess whether forestry activities are in compliance with existing laws, let alone whether they are sustainable. The data must be available to the general public.
2. *Detection and Monitoring Systems:* Systems for detecting and monitoring illegal logging must be developed and implemented or improved where they already exist.

Available methods and tools offer a means, both in the short and long term, to help identify the scope and nature of the illegal logging problem. Examples of monitoring underway in Cambodia, Indonesia, Brazil and Cameroon illustrate how methods and tools may help to both detect illegal activities and provide disincentives that will reduce the problem of illegal logging and trade. Until concerted action is taken, illegal logging will continue to rob governments and local communities of significant revenues and resources, cause damage to forest ecosystems, distort timber markets, and act as a disincentive to sustainable forest management (G8, 1998).

REFERENCES

Abramowitz, J. 2002. "From Rio to Johannesburg: Too much talk, too little action on forests." Washington, DC: World Watch Institute.

Acosta, R. T., E. S. Guiang, L. A. Paat, and W. S. Pollisco. 2000. "The control of illegal logging: The Philippine experience." *World Bank/WWF workshop, Controlling Illegal Logging, Jakarta, Indonesia, 2000.*

Anonymous. 1999. "Greenpeace turns UV light on illegal Amazon timber trade." Environment News Service. December 8. Available online at *http://ens-news.com/ens/dec1999/1999-12-08-03.asp.*

Brack, D., K. Gray, and G. Hayman. 2002. *Controlling the International Trade in Illegally Logged Timber and Wood Products.* London: Royal Institute for International Affairs.

Brack, D. and G. Hayman. 2001. *Intergovernmental Actions on Illegal Logging: Options for Intergovernmental Action to Help Combat Illegal Logging and Illegal Trade in Timber and Forest Products.* London: Royal Institute of International Affairs.

Brown, D. W. 1999. *Addicted to Rent: Corporate and Spatial Distribution of Forest Resources in Indonesia; Implications for Forest Sustainability and Government Policy.* Jakarta: Indonesia-UK Tropical Forest Management Programme.

Brunner, J., K. Talbott, and C. Elkin. 1998. *Logging Burma's Frontier Forests: Resources and the Regime.* Washington, DC: World Resources Institute.

Callister, D. 1999. *Corrupt and Illegal Activities in the Forestry Sector: Current Understandings and Implications for World Bank Forest Policy.* World Bank.

Cameroon, G. F. W. 2001. *1999-2000 Allocation of Logging Permits in Cameroon: Fine-Tuning Central Africa's First Auction System.* Washington, DC: World Resources Institute.

Contreras-Hermosilla, A. 2001. *Law Compliance in the Forestry Sector: An Overview.* Washington, DC: World Bank.

Contreras-Hermosilla, A. 2002. *Policy and Legal Options to Improve Law Compliance in the Forest Sector.* Rome: FAO.

Dehqanzada, Y. A. and A. M. Florini. 2000. *Secrets for Sale: How Commercial Satellite Imagery Will Change the World.* Washington, DC: Carnegie Endowment for International Peace.

DENR. 2002a. "Alvarez orders crackdown on illegal logging in Quezon." Department of Environment and Natural Resources, the Philippines. June 27. Available online at *http://www1.denr.gov.ph/article/articleview/241/1/106/.*

DENR. 2002b. "DENR nabs P3.5M hot logs in Quezon." Department of Environment and Natural Resources, the Republic of the Philippines. August 23. Available online at *http://www1.denr.gov.ph/article/articleview/383/1/138/.*

DENR. 2002c. "DENR nabs P7-M 'Hot' logs in Agusan Del Norte as anti-illegal logging operations goes hi-tech." Department of Environment and Natural Resources, the Philippines. June 19. Available online at *http://www1.denr.gov.ph/article/articleview/236/1/106/.*

DENR. 2002d. "DENR seizes hot logs in Negros." Department of Environment and Natural Resources, the Republic of the Philippines. August 7. Available online at *http://www1.denr.gov.ph/article/articleview/333/1/138/.*

Durst, P. B., C. D. Ingram, and J. G. Laarman. 1986. Inaccuracies in forest products trade statistics. *Forest Products Journal* 36: 55-59.

Dykstra, D., G. Kuru, R. Taylor, R. Nussbaum, W. Magrath, and J. Story. 2002. *Technologies for Wood Tracking: Verifying and Monitoring the Chain of Custody and Legal Compliance in the Timber Industry.* Washington, DC: World Bank.

Environmental Investigation Agency. 2002. "Press release: Australian film on Indonesian illegal logging causes outrage." Environmental Investigation Agency. July 31. Available online at *http://www.salvonet.com/eia/cgi/news/news.cgi?a=86&t=template. htm.*

Environmental Investigation Agency and Telepak. 1999. *The Final Cut: Illegal Logging in Indonesia's Orangutan Parks.* Washington, DC: Environmental Investigation Agency.

Environmental Investigation Agency and Telepak. 2000. *Illegal Logging in Tanjung Puting National Park: An Update on the Final Cut Report.* London: Environmental Investigation Agency.

FAO. 2001. *State of the World's Forests 2001.* Rome: Food and Agricultural Organization.

FERN. 2001. *Special Report: EU Illegal Timber Imports.* Moreton-in-Marsh, Gloucestershire: Forests and the European Union Resource Network.

G8. 1998. *Action Programme on Forests.* Tokyo: G8.

Global Forest Watch Cameroon. 2000. *An Overview of Logging in Cameroon.* Washington, DC: World Resources Institute.

Global Witness. 1999. *The Untouchables: Forest Crimes and the Concessionaires–Can Cambodia Afford to Keep Them?* London: Global Witness.

Global Witness. 2000. *Chainsaws Speak Louder than Words.* London: Global Witness.

Hamilton, J. T. 1999. Exercising property rights to pollute: Do cancer risks and politics affect plant emission reductions? *Journal of Risk and Uncertainty* 18: 105-124.

Human Rights Watch. 1996. "The Philippines: Human rights and forest management in the 1990s." Vol. 8, no. 3. Available online at *http://www.hrw.org.*

ITTO. 2001. *Annual Review and Assessment of the World Timber Situation 2000.* Yokohama: International Tropical Timber Organization.

Johnson, S. 2002. Documenting the undocumented. *Tropical Forest Update* 12: 6-9.

Matthew, E. 2001. *Import of Tropical Timber into the UK.* London: FOE-UK.

Obidzinski, K. and C. Palmer. 2002. *"How Much Do You Wanna Buy?"–A Methodology for Estimating the Level of Illegal Logging in East Kalimantan.* Bogor, Indonesia: CIFOR.

Ouen, M., S. T. Sokhun, and E. Savet. 2001. "Cambodia paper on forest and wildlife law enforcement experience." *Forest Law Enforcement and Governance Conference, Bali, Indonesia, 2001.*

Pacific Media Watch. 1996. "Indonesia journalist murdered." Pacific Media Watch. 26 Nov. Available online at *http://acij.uts.edu.au.*

Palmer, C. E. 2001. *The Extent and Causes of Illegal Logging: An Analysis of a Major Cause of Tropical Deforestation in Indonesia.* London: University College London.

Richardson, M. 2000. "Where forests disappear: Illegal loggers defy conservationists in Cambodia." *International Herald Tribune,* March 8.

Richardson, M. 2002. "Illegal logging topples Cambodia's forests." *International Herald Tribune*, June 21.

Savet, E. 2000. "Country paper on forest law enforcement in Cambodia." *World Bank/ WWF Workshop, Controlling Illegal Logging, Jakarta, Indonesia, 2000.*

Scotland, N. 2000. "Indonesian country paper on illegal logging-DRAFT," pp. 45. Jakarta.

Scotland, N., A. Fraser, and N. Jewell. 1999. *Roundwood Supply and Demand in the Forest Sector in Indonesia, Draft Report.* Jakarta: Indonesia-UK Tropical Forest Management Programme. PFM/EC/99/08.

SGS. 2001. "Leaflets: Timber flow control, log tracking systems, and log tracking technology." Available online at *http://www.sgs.com/SGSead.nsf/pages/forestry. html.*

Siry, J. 2000. "Annex A: Georgia timber exports assessment based on Turkey import data," in *Total Economic Valuation of Georgian Forests: Under the Current Resource Management Regime.* Edited by T. Arin and J. Siry, pp. 47-58. Washington, DC: World Bank.

SLDF. 1999. *Cutting Around the Rules.* Toronto: Sierra Legal Defense Fund.

SLDF. 2001. *Stumpage Sellout: How Forest Company Abuses of the Stumpage System Is Costing B.C. Taxpayers Millions.* Vancouver: Sierra Legal Defense Fund.

SLDF and The Wildlands League. 2000. *Grounds for Concern.* Toronto: Sierra Legal Defense Fund and the Wildlands League.

SLDF and The Wildlands League. 2001. *Improving Practices, Reducing Harm.* Toronto: Sierra Legal Defense Fund and the Wildlands League.

Smith, W. 2003. *Combating Illegal Logging: A Review of Initiatives and Monitoring Tools.* Washington, DC: World Resources Institute.

Smith, W., T. Meredith, and T. Johns. 1999. Exploring methods for rapid assessment of woody vegetation in the Batemi Valley, North-central Tanzania. *Biodiversity and Conservation* 8: 447-470.

State Department of Mato Grosso. 2001. *Environmental Control System on Rural Properties in Mato Grosso.* Cuiaba, Brazil: State Foundation of the Environment, Mato Grosso.

Toyne, P., C. O'Brien, and R. Nelson. 2002. *The Timber Footprint of the G8 and China: Making the Case for Green Procurement by Government.* London: WWF.

World Bank. 2000. *World Bank Logging Survey Mission: Technical Report.* Washington, DC: World Bank.

Illegal Logging and Deforestation in Andaman and Nicobar Islands, India: The Story of Little Andaman Island

Pankaj Sekhsaria

SUMMARY. This study looks at deforestation and legal rights to forest resources on the island of Little Andaman, in the Andaman and Nicobar Island group, an Indian territory located in the Bay of Bengal. The island is the home of the Onge–a threatened indigenous community of Negrito origin. The island was clothed in tropical evergreen forests and inhabited only by the Onge through the middle of the 20th Century. In the 1960s, the Government of India introduced major development and coloniza-tion programs for the island. These programs have destroyed vast areas of forest and have severely affected the Onge. Over the last 35 years, roughly 30% of the island has been taken over by outsiders for settle-ments, agriculture, timber extraction, and plantations. These develop-

Pankaj Sekhsaria is affiliated with Kalpavriksh, Apartment 5, Sri Dutta Krupa, 908 Deccan Gymkhana, Pune, 411004, India (E-mail: kvriksh@vsnl.com).

This study began as an environmental investigation carried out in the Andaman Is-lands by the author in 1998 on behalf of the environmental action group, Kalpavriksh.

The study was funded by the Bombay Natural History Society. For all their inputs and support, the authors would like to thank colleagues in the organisation, Kalpavriksh. In addition the author would also like to thank the Society for Andaman and Nicobar Ecology (SANE), the Bombay Natural History Society (BNHS), the Andaman and Nicobar Islands Environmental Team (ANET), Association for India's Development (AID), and the Human Rights Law Network (HRLN).

[Haworth co-indexing entry note]: "Illegal Logging and Deforestation in Andaman and Nicobar Islands, India: The Story of Little Andaman Island." Sekhsaria, Pankaj. Co-published simultaneously in *Journal of Sustainable Forestry* (The Haworth Press, Inc.) Vol. 19, No. 1/2/3, 2004, pp. 319-335; and: *Illegal Logging in the Tropics: Strategies for Cutting Crime* (ed: Ramsay M. Ravenel, Ilmi M. E. Granoff, and Carrie A. Magee) The Haworth Press, Inc., 2004, pp. 319-335. Single or multiple copies of this article are available for a fee from The Haworth Document Delivery Service [1-800-HAWORTH, 9:00 a.m. - 5:00 p.m. (EST). E-mail ad-dress: docdelivery@haworthpress.com].

ments have adversely affected the island's flora and fauna as well as the Onge themselves. This paper reports findings from a series of investigations starting in 1998. Logging operators and government agencies have systematically violated both the laws and the resources of Little Andaman. A coalition of Indian social and environmental non-governmental organizations (NGOs) initiated legal action to stop the deforestation in 1999. These efforts led to landmark injunctions on all timber felling operations in the islands by the Supreme Court of India in October 2001 and May 2002. *[Article copies available for a fee from The Haworth Document Delivery Service: 1-800-HAWORTH. E-mail address: <docdelivery@haworthpress.com> Website: <http://www.HaworthPress.com> © 2004 by The Haworth Press, Inc. All rights reserved.]*

KEYWORDS. India, Andaman and Nicobar Islands, illegal logging, deforestation, forest law, indigenous people, forest-dependent communities, Shompen, Nicobari, Great Andamanese, Onge, Jarawa, Sentinelese

Forestry (in the Andaman islands) is a comparatively new department for utilizing convict labour and is now the chief source of revenue in cash.

Lt. Col. Richard C. Temple
Chief Commissioner of the A&N Islands
in the Census of India 1901

The cutting of naturally grown trees in any (on)going projects (in the Andaman and Nicobar islands) . . . except plantation wood is prohibited.

Supreme Court of India
Order dated October 10, 2001

INTRODUCTION

The history of timber extraction in the Andaman and Nicobar Islands dates back over a century, when the islands were used as a penal colony by the British and forestry provided the primary means of employment for resident convicts. Over the past 100 years, forestry has played a significant role in island income, and has thus been heavily supported by the Indian government through subsidization of various aspects of production. It has expanded hand in hand with efforts by the Indian government to populate the island with continental Indians. The timber industry has grown to be unavoidably linked with the cultural and political rights of the islands' original inhabitants, which include six distinct cultural groups some of which still have minimal contact with outside cultures, and on whose land stands much of the remaining merchantable timber.

This paper will begin with an overview of the cultural and biological heritage of the Andaman and Nicobar Islands. It will then provide a historical account of the role of timber extraction on the islands and the associated illegal activities that have been uncovered in the islands in the recent few years. Finally, the paper will document the efforts and initial success of local NGOs to halt illegal activities on the island through India's legal system.

ENVIRONMENTAL HISTORY
OF THE ANDAMAN AND NICOBAR ISLANDS

Biogeography

The Andaman and Nicobar islands are located in the Bay of Bengal to the West of Thailand. The islands are summits of a submerged mountain range connecting the Arakan Yoma (Manipur-Burma) ranges, through the Coco and Preparis Islands of Burma; to Banda Aceh in Sumatra and the Lesser Sundas. They swing out as an arc into the Bay of Bengal (Andrews and Sankaran, 2002). The main features of the islands include steep hills, and generally poor soil with little water holding capacity. Flat terrain is extremely limited and the larger islands have long meandering creeks along the coast. The islands are clothed in thick, biodiverse, evergreen forests and also host some of the finest mangroves and coral reefs in the world.

There are more than 350 islands in the Andaman group and over 24 islands in the Nicobar. Of all these islands roughly only 10% are inhabited by humans. The islands cover a total area of 8,249 km^2 and have 1,962 km of coastline, equivalent to 25% of the coastline of mainland India.

The Andaman and Nicobar Islands are biogeoraphically unique. The flora is related to mainland Indian flora, but shows much closer similarity to that of Indo-Malayan origin. There is also a high degree of faunal endemism, with at least 32 species and subspecies of mammals, 99 species of birds and 24 species of reptiles endemic to the group of islands. Conservation organizations have identified the islands as one of the hotspots of biodiversity in India (Saldanha, 1989; Andrews and Sankaran, 2002).

Indigenous Tribal Communities

The islands are also home to six indigenous tribal communities, two of which–the Shompen and the Nicobari–are of Mongloid origin and reside in the Nicobar group. The other four communities are of Negrito origin and live in the Andaman group: the Great Andamanese, the Onge, the Jarawa, and the Sentinelese. These latter tribes are were hunter-gatherer communities and have successfully survived in these islands for centuries. Their knowledge and understanding of the forests is extensive and they share a close relationship with it.

While the Nicobarese have adapted to a more "modern" lifestyle, the four tribes living in the Andaman group continue to live as hunter-gatherers that depend on the forest for their survival. Today, they are faced with the grim prospect of extinction as insensitive policies of the state, pressures from modern civilization and a large timber industry adversely impacted their environment.

Colonial History of Forestry Operations

The development of commercial forestry operations in the islands has been limited to the Andaman Islands. Although there has been some deforestation in the Nicobars for the establishment of settlements, the timber industry has thus far avoided the area. As in the rest of India, the British started the practice of forestry in the islands. In fact, the British were the first outsiders to successfully establish their settlements on the islands.

In 1789 Lieutenant Archibald Blair of the Indian (British) Navy was appointed to survey the islands with a view to finding a harbor, " where fleets in the time of war can refit by any means . . . or to which any part or the whole may retire in the event of a disastrous conflict with the enemy" (Whitaker, 1985). A small settlement was established on the island of North Andaman, but it proved disastrous due to the prevalence of various diseases, particularly cerebral malaria. The attempt was abandoned soon thereafter.

Attempts to settle the area were made again by the British in the late 1850s. They created Port Blair as a penal settlement in 1858, primarily for criminals from mainland India but it would later host freedom fighters as well. Large tracts of land were first cleared in 1858 for the penal settlement itself. By 1870 the limited exploitation of hardwoods had already begun. The Forest department was started in 1883 (Whitaker, 1985) and the Chatham sawmill was set up around this time. It was long considered the largest sawmill in the whole of Asia and is operational even today.

The 1901 census report for the islands was the first to list forestry as a source of employment, with about 900 individuals involved in activities such as extraction, sawmills and firewood. In 1929 the Swedish Western Match Company (WIMCO) started a match splint factory in Port Blair (Whitaker, 1985). This remained the only private forest-based industry until the early 1960s, when plywood mills were first set up.

Post-Independence History

With India gaining independence from the British in 1947, a new phase began for the islands too. Independent India began a 'colonisation scheme' through which thousands of people were brought from mainland India to settle on the islands. The Government of India's interest in the islands derives from their strategic proximity to Southeast Asia and one of the region's most important commercial shipping lanes, the Straits of Malacca. To strengthen its claim over the islands, the Government of India has encouraged mainlanders to settle on these islands.

These resettlement schemes explain the bulk of the total population growth experienced since the 1950s (see Table 1).

In fact, in the early years various incentives were offered to citizens in mainland India to come and settle here. Each settler household was given four acres of flat land for paddy, five acres of hilly land for tree crops and one acre to build a homestead. Twelve tons of royalty-free timber was given for house construction and an additional five tons for house repairs every five years (Saldanha, 1989).

The growth in population meant that the pressures on the forest, both direct and indirect, also increased (see Table 2).

It is clear that the growth in the timber extraction operations corresponds directly to the growth in the population of the islands. This destruction of the forests for the extraction of timber was in addition to the clear felling that was done for the settlements themselves. It has been estimated that between 9 and 13% of the total land area of the islands has been clear-felled in little more than

TABLE 1. Population Figures for Indigenous Groups in Andaman and Nicobar Islands.

Year	Total Population	Andamanese	Onge
1901	24,499	625	678
1911	26,459	455	631
1921	27,080	209	346
1931	29,476	90	250
1951	30,971	23	150
1961	63,548	19	129
1971	115,133	19	112
1981	188,745	25	103
1991	280,661	28	101
1998	400,000 (estimate)		

Sources: Census of India, 1995; Saldanha, 1989; Anon., 1998a

TABLE 2. Annual Extraction of Timber in the Andaman and Nicobar Islands.

Date	Average Annual Extraction in m^3
Pre-1950	
1869-1929	15,300
1930-1950	49,700
Post-1950	
1951-1962	88,800
1968-1983	118,800
1990-1995	112,000
1997-1998	75,000
2000-2001	~40,000

Sources: Saldanha, 1989 and A&N Forest Department

a century of operations (Pande et al., 1991). There is no consensus on the actual area that remains forested, and some observers are of the opinion that even a large part of the remaining forest is degraded and under secondary growth (Whitaker, 1985).

The Development of Modern Timber-Based Industry

With the growing population of migrants on the islands the government realized that it needed to create opportunities for the people. The abundant forests and the timber within them became the primary source for income and employment. The British had already started the operations, but their expansion had been limited. As late as the 1960s, the Chatham sawmill and the WIMCO match splint factory were the only major timber units in the islands. The growing influx of people, however, forced the government to expand the industry.

In the initial years following independence, the local administration offered huge incentives to industry in the form of subsidies to attract investment to the islands. The private-sector plywood industry that began in the late 1950s took advantage of these subsidies and continued to do so until recently. The private plywood mills were the largest consumers of timber in the islands; they consumed nearly 70% of all the timber harvested in the islands (Anonymous, 1993-1994). By early 2001, however, two of the three large plywood mills had already ceased operations, citing inadequate timber supplies and other financial difficulties. Following landmark orders passed by the Supreme Court of India, first in October 2001 and then in greater detail in May 2002, all timber extraction operations in the islands have stopped.

It is significant that nearly 98 percent of the plywood manufactured in the Andaman Islands was being exported to mainland India and its ever-increasing markets (Anonymous, 1998b). The Government of India gave these mills low stumpage prices, subsidized as much as 90% of the cost of transporting goods to and from the mainland, and subsidized over 80% of energy costs, to name a few (Anonymous, 1991). These subsidies allowed the timber industry to survive and even to make substantial profits. In effect, the exploitation of the forests in the islands was being subsidized for the use and benefit of a far away population of mainland India that had no real stake in the islands or the conservation of the forests.

Current Threats to Tribal Communities

The people who have suffered the most in these islands are the indigenous communities for whom the forests are home. These communities have suffered from the combined impacts of the destruction and degradation of the forests and the imposition of an alien and insensitive culture that brought along with it various diseases and vices such as alcohol and tobacco.

Except for the Nicobarese, all of the other tribes have suffered in varying degrees. The Great Andamanese were the first community to come into contact with the British, followed by the Onge of Little Andaman. Both of these communities have suffered immensely from the ill effects of this contact and the interaction that followed.

The other two negrito communities, the Jarawa and the Sentinelese have scrupulously avoided contact with the outside world and have even used violent means to do so. This isolation, however, changed for the Jarawa by 1998. They for the first time in their history, begun to voluntarily come out of their forest home to interact and mingle with settlers living in the areas bordering their territory (Sekhsaria, 1998). They have broken a centuries-old isolation and their future can only be a matter of speculation.

As a result of this isolation and aggression against colonization, these communities have retained their freedom and original identity (Whitaker, 1985; Paul, 1992; Sekhsaria, 1998). In this respect, the Jarawa and the Sentinelese have succeeded where the Onge and the Great Andamanese failed. Epidemics of diseases such as pneumonia in 1868, measles in 1877, influenza in 1896 and syphilis killed the Great Andamanese by the hundreds. They had no resistance to these diseases, which they contracted from the outsiders. The Great Andamanese are on the brink of extinction as an independent race of people: their population–estimated to be around 5,000 individuals in the early part of the 19th century–today comprises merely 28 individuals (Whitaker, 1985; Saldanha, 1989).

The Onge–inhabitants of Little Andaman–have fared only marginally better. From 600 individuals in 1901 the population has fallen to 101 individuals today (see Table 1). Whereas the population of the Great Andamanese declined because of the various epidemics, the Onge are suffering because of the destruction of their forests and the imposition of a way of life that is alien and insensitive to their own. The destruction on Little Andaman goes on even today, pushing the Onge further and further towards oblivion. The environmental history of the Onge is a classic case of modern development schemes causing the destruction of both vast areas of forest and of communities that depend on them for their survival. The trends described below apply to the other islands in the Andamans and serves as a lesson for their future as well.

DEFORESTATION ON LITTLE ANDAMAN ISLAND: A CASE STUDY

Little Andaman is the southernmost island of the Andaman group of islands with an area of 730 km^2. The island is generally flat. Only the central and southern portions of the island are undulating to hilly, with the island's highest point rising to 156 m above sea level (Box 1). The island is the only home of the Negrito Onge tribe, which has lived there for centuries. Today the community consists of a small group of only 101 individuals (Saldanha, 1989; Anonymous, 1995) (Box 2).

Colonization of Little Andaman Island

Though the history of the settlements and the timber extraction operation in the Andaman Islands in general is more than a century old, Little Andaman remained completely untouched till very recently. The 1960's saw the massive colonization program planned for the Andaman and Nicobar Islands. Until that time, the entire island of Little Andaman had belonged only to the Onge, who were also known to travel in their dugout canoes to nearby islands and occasionally to Port Blair (Whitaker, 1985). A massive development and colonization program was undertaken as a part of which thousands of mainland Indians, refugee families from erstwhile East Pakistan (now Bangladesh) and Tamils from Sri Lanka were settled here.

An 'Inter-Departmental Team on Accelerated Development Programme for Andaman and Nicobar,' set up by the Ministry of Rehabilitation, Government of India, produced a report in 1965. The report, popularly referred to as 'the Green Book' (Anonymous, 1965), prescribed the route for the development of the islands in general and Little Andaman Island in particular. The committee proposed a variety of schemes for the development of the islands under the broad categories of strategy, agriculture, animal husbandry, forest, industry, fishery, water, transport, health, and colonization and manpower.

BOX 1. Chronology of Little Andaman Island.

1867   Attack by the Onge on the British ship Assam valley.
1881   Little Andaman is visited by British officer MV Portman.
1901   Population of the Onge: 672.
1921   Population of the Onge: 346.
1951   Population of the Onge: 150.
1952   Little Andaman is visited by Italian Anthropologist Cipriani to study the Onge.
1957   Declaration of Onge Tribal Reserve including the islands of Little Andaman, Rutland, Cinque, Brothers, and Sister.
1961   Population of the Onge: 129.
1965   Report by the 'The Inter Departmental Team on Accelerated Development Programme for the A&N Islands' Ministry of Rehabilitation, Government of India.
1969   366 East Pakistan families settled in Little Andaman.
1970   Survey for the establishment of red oil palm plantation.
1970   First sawmill is set up on the island; Annual intake 2000 m³ of timber.
1971   Population of the Onge: 112.
1972   First amendment to the Tribal reserve.
1973   165 Nicobari families settled in Harminder Bay area on the island.
1974   Forest Department (FD) assessment of timber productivity of the island's forests.
1975   FD initiates work on the oil palm plantation.
1975   Creation of the first 160 ha of oil palm plantation.
1976   Creation of the Andaman Adim Janjati Vikas Samiti (Andaman Tribal Welfare Society).
1976   Presentation of Andaman and Nicobar Forest Plantation and Development Corporation(ANFPDC) proposal for Logging and Forestry operations in Little Andaman.
1977   The ANFPDC starts functioning.
1977   Second amendment to the Tribal reserve.
1977   118 families from mainland settled on Little Andaman during 1977-1979.
1981   Population of the Onge: 100.
1983   Study of the Onge by anthropologist, Vishvajit Pandya.
1990   Master plan for the development of the tribes of the Islands by SA Awaradi.
1991   Final Amendment to the Tribal reserve.
1991   Population of the Onge: 101.
1996   Patenting controversy related to Onge knowledge.
1996   Supreme Court of India ruling on forests.
1999   Case filed in the Calcutta High Court (Port Blair Circuit Bench) asking for stoppage of all timber extraction operations in Little Andaman.
1999   Intervention filed in the Supreme Court of India in the matter of the deforestation on Little Andaman island.
2000   Supreme Court of India orders the stoppage of felling of 'naturally grown trees from the forests of the entire Andaman and Nicobar Islands.'
2002   Supreme Court accepts report of a special commission set up and issues landmark orders related to various aspects of the forests and indigenous peoples of the islands.

BOX 2. Traditional Knowledge of the Onge.

In the short sighted rush to exploit the forests of the Onge, certain other critical aspects of Onge culture have been completely ignored. This includes the vast storehouse of their traditional knowledge of the plants and animals that are found here. The Onge have knowledge that could prove to be a boon to the whole of mankind; it could be a new food, plant, or a cure for a disease that modern medicine has failed to fight. There is proof of this already, though only very minimal ethno-botanical research has yet been carried out here.

The Onge, for example, use a particular plant *Orphea katshalica* ('tonjoghe' in Onge language) in the process of extracting of honey from the hives of the dangerous rock bees, *Apis dorsata*. The leaves of the plant are chewed and the juice is smeared all over the body before they climb the tree with the hive (Dutta et al., 1983, Basu, 1990). There is some ingredient in the juice that immobilizes the bees and makes them harmless. The Onge keep chewing the leaf as they climb up the tree and on reaching the hive, spit some of the juice on the bees. The method is so effective that the bees are made to move away and only that part of the comb is cut away that has the honey–an entire process that is harmless to humans; to the bees and as safe and efficient as anything can be. Even more recent was the 'discovery' by the Indian Council of Medical Research (ICMR), of a plant used by the Onge, which could possibly have the cure for the dreaded, and often fatal disease of cerebral malaria. This discovery, unfortunately also got involved in an ugly patenting related controversy, obscuring in the process its importance and great potential value (Kothari, 1996, Mc Girk, 1998). Far from respecting and learning from this knowledge, policies are being implemented which will eventually destroy all before it can be documented even.

The Onge as the original inhabitants, have the first right over this land but not many are willing to concede this. The political will, too, is absent. If anything, the weight of political support is on the side of the settlers, as is evident in the statement made in 1990 by one of the most prominent and influential politician from the islands, ". . . . Job seekers (settlers) who have come (to) the island are now serious contenders for the allotment of house sites and agricultural land. Since the political system goes with the number, no political party is in a position to contradict their demands" (Paul, 1992). The Onge with a population of 101 individuals, certainly have no chance of being heard.

It is clear that the Onge are completely dependant of the forests for survival, and only if this fact is given the importance it deserves, can the forests and the Onge be saved. As things stand today, the future for both seems bleak.

The island of Little Andaman was specially earmarked for a Rehabilitation and Resettlement (R&R) program for a number of reasons: it was a relatively large island, it had few inhabitants (only the Onge), and it had valuable natural resources, particularly timber. The suggestions made for Little Andaman included clear felling of half of the island's forests, settlement of 12,000 settler families on the cleared land, creation of plantations of coconut, areca, and palm oil, and use of the felled timber for wood based industries such as sawmills and plywood factories.

The Onge Tribal Reserve

The entire island of Little Andaman was declared a tribal reserve for the Onge in 1957, well before the colonization of the island had reached the planning phase (Anonymous, 1957). It is now clear that the presence of the tribal reserve was completely ignored and that all of the development proposals and activities violated the rights of the Onge. The forests were assigned no value, except for the timber that could be extracted from them. Indeed, they were considered useless wastelands that needed to be cleared, reclaimed, and aggressively developed. Although the Green Book emphasized that its proposals were provisional, it initiated two processes that would bring unprecedented social and environmental destruction: the settlement of thousands of outsiders and the introduction of a large timber extraction operation to support these settlers.

Forest Destruction on Little Andaman Island

Over the last 35 years, roughly 30% of the island of Little Andaman has been taken over by outsiders for settlements, agriculture, timber extraction operations and plantations (Anonymous, 1957; Whitaker, 1985; Saldanha, 1989).

In 1970, the first sawmill on the island was built and large areas were cleared for the establishment of the settlements. As proposed in the development plan, an area was also cleared for the establishment of a red oil palm plantation. At the same time, the Andaman and Nicobar Forest and Plantation and Development Corporation (ANFPDC) was created to manage the red oil palm plantation and timber extraction operations (Anonymous, 1976). In the 1986 the ANDPDC decided not to expand the plantation, and fortunately that moratorium has stood (Bhatee, 1986; Anon., 1997).

In the late 1970s certain legal changes were made to the boundaries of the Onge Reserve so as to allow the development activities that were suggested for the island (Anonymous, 1957). About 20,000 hectares were also handed over to the ANFPDC for timber extraction and creation of the red oil plantation. Seventy percent of the island, however, was retained as a tribal reserve for the Onge. On paper, at least, this was and continues to be exclusive Onge land on which no settlement, logging activity or trespassing is allowed. Unfortunately, this status and the protection accorded have remained only in name and not in practice.

Illegal Logging on Little Andaman Island

Investigative research conducted in 1998 revealed a number of gross violations of the law in the extraction of timber from the forests of Little Andaman island:

First, 19,600 ha of land that was no longer tribal reserve was handed over by the Andaman and Nicobar Forest Department (FD) to the Andaman and Nicobar Forest Plantation and Development Corporation (ANFPDC), which was the sole agency responsible for timber extraction. In 1976, the ANFPDC presented its Project Report for Logging and Marketing of timber from the forests of Little Andaman. The report estimated that a total of 60,000 ha of the island were available for logging and that 60,000 m³ of timber could be extracted annually from 800 ha. This estimation itself was a clear violation of the Onge tribal reserve: the reserve covers 52,000 ha of the island's total 73,000 ha, so how could 60,000 ha be made available for logging without encroaching on the reserve lands?

Second, The Corporation should have limited its operations to the 19,600 ha that had been leased out to it. With 1,600 ha under red oil palm plantation, the actual area available for logging was even less, at 18,000 ha. This meant that the Corporation should have logged only 18,000 m³ of timber from an area of 240 ha annually. The average for the actual logging over two decades of operation was, however, much higher, at 25,000 m³ of timber from an area of 400 ha annually.

Third, it was later found that the transfer of land from the Forest Department to the ANFPDC itself was not legally valid. Though the land had been leased out in 1977, a lease agreement was signed only in 1987, with the clause that it would apply with retrospective effect from 1977 for thirty years. This was pronounced as unacceptable by the Calcutta High Court in an order passed in 1996 which stated, ". . . we are at a loss to understand how under the law of contract or under any law for the time being in force the Government could grant a lease in 1987 with retrospective effect from 1977. The granting of the lease with retrospective effect by the State authorities in favor of a Corporation is not permissible under the law" (Calcutta High Court Order, 1996).

Fourth, a working plan for timber extraction operations, as required under Indian law, was never prepared for the island of Little Andaman. Logging operations continued in spite of a 1996 order of the Supreme Court of India that asked for the stopping of all logging in the absence of a working plan.

Finally, in another shocking violation, it was found out that the Corporation had even logged from within the Onge tribal reserve on the island, making a mockery of the law and also the rights of the Onge. Maps available from the ANFPDC and the Forest Department have logging coupes dated 1990 onwards that are clearly marked within the tribal reserve (Sekhsaria, 1999).

Consequences of Deforestation

Deforestation and the developmental policies have adversely impacted the ecology of the island and the surrounding oceans. Populations of both, the Andaman Wild pig *Sus scrofa andamanensis* and the slow moving sea mammal Dugong *Dugong dugon* have declined because of poaching and habitat destruction. The salt-water crocodile, a reptile found in large numbers in the island's extensive mangrove creeks has seen a similar decline (SANE, undated; Pande et al., 1991; Pandya, 1993; Tambe and Acharya, 1997, Andrews, 2000).

Sand mining for construction activities has destroyed many of the island's beaches, increasing erosion. In some areas, the beaches have simply vanished. Endangered sea turtles such as the Olive Ridley *Lepidochelys olivacea* and the Green sea turtle *Chelonia mydas*, which nest on the beaches here, are now reported in much smaller numbers (Soundarajan et al., 1989). Excessive runoff from deforested areas chokes the coral reefs in the waters surrounding the island (Soundarajan et al., 1989).

The Fate of the Onge

In addition to these ecological impacts, the clearance of land for settlements and the loss of forests to logging have had a direct impact on the Onge. They have been driven away from their preferred lands and have been forced to move deeper into the forest (Paul, 1992). With excessive poaching of their food sources like the wild pig, survival is becoming increasingly difficult for them. Simultaneously, the Onge have had to face the onslaught of an alien modern culture that is highly insensitive to their traditional way of life (Awaradi, 1990).

Logging operations have also opened up the island to further development. Roads that were bulldozed into the forest to carry out the logging operations further opened inaccessible forests areas to settlers. Not only have these roads facilitated greater poaching, but they have also enabled settlers to move further and further into the forest to establish settlements.

The Andaman Adim Janjati Vikas Samiti (AAJVS, The Andaman Tribal Welfare Board) of the A and N administration has a welfare scheme for the Onge. Unfortunately, this program makes them dependant on the aid offered and does little to support and encourage them to lead their traditional and independent lives. Aid provided by AAJVS has introduced foods such as rice and sugar that were never a part of the Onge diet. At one time it also included 250g of tobacco for each adult, as a welfare measure (Awaradi, 1990). Attempts are also being made through the various welfare measures to induce the Onge to give up their nomadic way of life in favor of one that is much more settled. The main aim appears to be to settle the Onge so that they no longer need the entire island to roam and use for survival. Consequently a much larger chunk of the forests, which cannot be exploited at the moment, will become available.

The settlers have also introduced vices such as alcohol, to which many Onge are now addicted. This has made them much more susceptible to exploitation, and the settlers have made the best possible use of the situation. In exchange for the ubiquitous bottle, popularly known here as 180 (after the standard 180 ml size of a bottle of alcohol), the Onge give away to the settlers precious resources such as honey, resins, ambergris and turtle eggs.

Many experts had argued that unless drastic steps are taken, including the termination of logging operations on the island and the halting of this cultural assault, the Onge will not survive as an independent group of people. The Onge can be saved only if the destruction of their forest and the colonisation of their lands are stopped.

The Legal Intervention: 1999-2002

Based on the findings of these large-scale violations and illegalities in Little Andaman, three NGOs–the environmental action group, Kalpavriksh, the Port Blair based Society for Andaman and Nicobar Ecology (SANE), and the Mumbai based Bombay Natural History Society (BNHS)–filed a writ petition before the Calcutta High Court, Port Blair Circuit Bench asking for the stoppage of all logging operations in Little Andaman.

The local administrations and the Forest Department denied the allegations made in the petitions to the court. The Forest Department and local government argued before the judges that the petition had invoked an order of the Supreme Court and therefore the matter could only be argued there. The High Court agreed, and the petitioners approached the Supreme Court of the country in 1999, through an intervention filed in an ongoing case related to forests-Writ Petition (Civil), 202 of 1995, TN Godavarman Thirumulkpad v. The Union of India and Others-that had and continues to have implications for forests in the whole of India (Sekhsaria, 2002).

In the first two years of the case, little happened in either the High Court or in the Supreme Court. Life on the islands too went back to normal after the initial shock and shake-up of the initial hearings in the High Court. The Supreme Court finally heard the case on October 10, 2001 and issued a stay on the felling of naturally grown trees in the entire area of the Andaman and Nicobar Islands. In another order passed on November 23, 2001, the court further banned the conversion of timber as well, effectively stopping the operation of all timber-based industries like plywood and saw mills in the islands. The court also appointed a commission to look into the state of the forests and other related matters in the islands and to submit its report in six weeks time.

The report that was submitted to the court on February 18, 2002, made 47 recommendations, which included among others stoppage of all timber extraction in Little Andaman Island and the shutting down of the ANFPDC, the main violator in Little Andaman (Singh, 2002). The report made a number of other significant recommendations that will have far-reaching implications in the Andaman and Nicobar Islands. The court's earlier recommendations and orders were met with considerable local opposition from those with vested interests in the further development of the islands. At present, the issue of indigenous forest rights and State and para-statal forest development remains hotly contested.

In a hearing held on May 7, 2002, the Supreme Court of India finally accepted the recommendations of the Shekhar Singh Commission thereby beginning a new chapter in the history of the islands. A law is only as good as its implementation, so environmental and indigenous rights NGOs must help ensure that the local administration implements the Supreme Court's orders fully and in their intended spirit.

CONCLUSIONS

Government actors have been shown to be part of the illegal logging problem in the Andaman and Nicobar Islands. The forestry industry itself would not be viable without massive subsidization by the Indian government through low stumpage fees and nominal transportation costs. The efforts by the Indian government to "develop" Little Andaman beginning in the 1960s despite pre-existing land rights of the Onge suggests that some parts of the Indian government were acting in bad faith to exacerbate the problem. Further, it is clear from the role of the local forestry department that they have been largely cooperative in the efforts of private sector actors to expand operations despite the indigenous rights that preclude these activities.

At the same time, the Indian judicial system shows signs of providing a valuable check against some of these irregularities and illegalities performed by the other sectors of government. While the final outcome of the efforts of the three NGOs involved remains inconclusive, the initial decisions of the Supreme Court of the country sound promising. Local organizations have played a central role in this process by conducting "action-oriented" research and identifying the exact nature of illegal activities. These groups have effectively carried out a "citizen suit," forcing the government to abide by its own laws regarding indigenous land rights.

Unfortunately, however, it appears that a court decision alone may not be sufficient. Although forestry activities (both legal and illegal) have been presently stopped in the islands following the orders of the court, the court orders also included a number of other measures related to the rights of the indigenous peoples and the protection of forests on the islands. The local administration has not implemented some of these other important orders, so the struggle to fully realize the rights of the islands' indigenous peoples continues.

REFERENCES

Andaman and Nicobar Forest Department. 1993-94. Forest Statistics 1993-1994: Andaman and Nicobar Forest Department. Port Blair, India.

Andaman and Nicobar Administration. 1995. Island wise statistical outline. Andaman and Nicobar Administration. Port Blair, India.

Andaman and Nicobar Foerst Plantation and Development Corporation, Ltd. 1997. Brief note of the Andaman and Nicobar Forest Plantation and Development Corporation, Ltd., Port Blair, India.

Andrews, Harry V. 2000. Impact Assessment of the Little Known Little Andaman Island, Andamans, India, Newsletter of the Irula Tribal Women's Welfare Society. 12 (200): 52-83.

Andrews, Harry V. and Vasumathi Sankaran, Eds. 2002. Sustainable Management of Protected Areas in the Andaman and Nicobar Islands, Andaman and Nicobar Islands Environmental Team, Port Blair; Indian Institute of Public Administration, New Delhi and Flora and Fauna Internation, Cambridge.

Anonymous. 1957. Andaman and Nicobar Islands (Protection of Aboriginal Tribes Regulation), 1956 and rules framed thereunder; Andaman and Nicobar Administration.

Anonymous. 1965. Report by The Inter Departmental Team on Accelerated Development Programme for A&N Islands; Ministry of Rehabilitation, Government of India.

Anonymous. 1976. Project Report for Logging, Marketing, Forest Plantation and Natural Regeneration in Little Andaman and North Andaman Islands; Andaman and Nicobar Islands Forest Plantation and Development Corporation (ANFPDC, Ltd.), Port Blair.

Anonymous. 1991. Facilities and Incentives for the Development of Small Scale Industries in A&N Islands; Directorate of Industries, Andaman and Nicobar Islands, Port Blair.

Anonymous. 1998. Personal Communication with officials from the private timber industry in the islands.

Awaradi, S.A. 1990. Master Plan (1990-2000) for the Welfare of Primitive Tribes; Tribal Welfare Department, A&N Administration.

Bhatee, S. 1985-86. Cultivation of Red Oil Palm in the Andamans; ANFPDC, Ltd., Port Blair.

Calcutta High Court Order. 1996. Judgement of the Court dated 30/05/1996 in the case FMAT NO. 3353 of 1995, Sushil Dhali vs. The Range Officer, Revenue Range, ANFPDC, Hut Bay, Little Andaman.

Dutta, T.R., R. Ahmed, and S.R. Abbas. 1983. The discovery of a plant in the Andaman Islands that tranquilises *Apis dorsata*; Bee World 64 (4): 158-163.

Indian Census Commission. 1998. Voters list, 1998 General elections. Office of the Register General, India. New Delhi, India.

Kothari, A. 1996. Protect Onge Tribal Knowledge and Rights, a note on the patenting issue regarding the plant Onge use which might have a cure for cerebral malaria.

McGirk, T. 1998. 'Gene Piracy,' Time Magazine, Nov. 9, 1998.

Paul, M. 1992. Report on the Current Status of the Negrito Tribes of the Andamans; Janvikas, Pune and Andaman and Nicobar Islands Environmental Team (ANET), Port Blair.

Pande, P., A. Kothari, S. Singh. 1991. Directory of National Parks and Sanctuaries in the Andaman and Nicobar Islands, Indian Institute of Public Administration, New Delhi.

Pandya, V. 1993. Above the Forest, A Study of Andamanese Ethnoanemology, Cosmology and the Power of Ritual; Oxford University Press.

Saldanha, C.J. 1989. Andaman, Nicobar and Lakshadweep–An Environmental Impact Assessment; Oxford and IBH Publishing Co. Pvt. Ltd.

Sekhsaria, P. 1998. Jarawa Excursions; Frontline magazine, Chennai.

Sekhsaria, P. 1999. A people in peril. Frontline, July 7, 1999.

Sekhsaria, P. 2002. Logging off, for now. Frontline, January 18, 2002.

Singh, S. 2002. Report of the Commission set up under the orders of the Supreme Court on the Status of Forests and other allied matters in Andaman and Nicobar Islands, Volume 1.

Society for Andaman and Nicobar Ecology. Undated. Dugong, Society for Andaman and Nicobar Ecology (SANE). SANE Awarenes Series. Port Blair, India.

Soundarajan, R., R. Whitaker, S. Acharya. 1989. An investigation into the effects of siltation, logging, blasting and other human derived damage to corals in the Andaman and Nicobar Islands; Indian National Trust for Art and Cultural Heritage (INTACH)–A&N Chapter.

Tambe, S. and S. Acharya. 1997. 'Onge Wild Pig and Poachers'; A note submitted to the Lieutenant Governor (LG) of the A&N Islands for the 'Resource Augmentation of the Onges by the Captive Breeding of the Andaman Wild Pig'; (SANE).

Whitaker, R. 1985. Endangered Andamans; World Wildlife Fund–India and MAB India, Department of Environment, Government of India.

Combating Forest Corruption:
The Forest Integrity Network

Aarti Gupta
Ute Siebert

SUMMARY. This article describes the strategies and activities of the Forest Integrity Network. One of the most important underlying causes of forest degradation is corruption and related illegal logging. The Forest Integrity Network is a timely new initiative to combat forest corruption. Its approach is to form global multi-stakeholder coalitions between the civil, private and public sectors and to create synergies between organizations fighting corruption, like Transparency International, and actors

Aarti Gupta is currently a Visiting Fellow with the Technology and Agrarian Development Group (TAO) of Wageningen University in The Netherlands.

Ute Siebert is the current FIN Program Manager and Coordinator at the TI Secretariat in Berlin.

Given that FIN represents a collective effort of numerous institutions and individuals, many of the ideas in this article have been developed in conjunction with FIN supporters, particularly Pierre Landell-Mills and John Spears, Co-Chairs of FIN's Steering Committee. Other FIN supporters and members who have contributed in multiple ways include Ken Rosenbaum, Ajit Banerjee, Nigel Sizer, Nalin Kishor, and Maritta Koch-Weser. For support and input from the TI Secretariat in Berlin, the authors would like to thank Peter Eigen, Hansjörg Elshorst, David Nussbaum, Carin Norberg, Veronique Lerch, Jessie Banfield, Jessica Berns, Juanita Olaya, and Johannes Seybold. For further information about FIN, please contact: Ute Siebert, Program Manager Forest Integrity Network, Transparency International, Otto-Suhr-Allee 97-99, 10585 Berlin, Germany (E-mail: info@forestintegrity.org), FIN-webpage: http://www.transparency.org/fin. Aarti Gupta can be reached at: aartiguptabiermann@yahoo.com.

[Haworth co-indexing entry note]: "Combating Forest Corruption: The Forest Integrity Network." Gupta, Aarti, and Ute Siebert. Co-published simultaneously in *Journal of Sustainable Forestry* (The Haworth Press, Inc.) Vol. 19, No. 1/2/3, 2004, pp. 337-349; and: *Illegal Logging in the Tropics: Strategies for Cutting Crime* (ed: Ramsay M. Ravenel, Ilmi M. E. Granoff, and Carrie A. Magee) The Haworth Press, Inc., 2004, pp. 337-349. Single or multiple copies of this article are available for a fee from The Haworth Document Delivery Service [1-800-HAWORTH, 9:00 a.m. - 5:00 p.m. (EST). E-mail address: docdelivery@haworthpress.com].

concerned with forest conservation and sustainable use. The article describes FIN's current activity in analyzing the relevance of Transparency International's corruption fighting strategies for the forest sector and its future plans to produce an operational Forest Corruption Fighters' Tool Kit. FIN also aims to develop conceptual frameworks for cross-country comparisons and compile replicable best practices in tackling forest corruption. *[Article copies available for a fee from The Haworth Document Delivery Service: 1-800-HAWORTH. E-mail address: <docdelivery@ haworthpress.com> Website: <http://www.HaworthPress.com> © 2004 by The Haworth Press, Inc. All rights reserved.]*

KEYWORDS. Forest degradation, forest corruption, illegal logging, Forest Integrity Network, multi-stakeholder coalitions, Transparency International, Forest Corruption Fighters' Tool Kit, best practices

INTRODUCTION

Persisting degradation of the world's forests is one of the most pressing global environmental and social challenges of the new millennium. In many countries, illegal timber production ranges from 35 to 94% (Greenpeace, 2001; EIA and Telapak, 2001; Contreras-Hermosilla and Vargas, 2002, cited in Contreras-Hermosilla, 2002; Magrath and Grandalski, in press; WWF, 2003; Siebert and Elwert, 2004). Corruption in allocation and use of forest resources constitutes a pernicious and hard-to-address hurdle to making the sustainability transition. This article describes an innovative attempt to bring different actors together to address forest sector corruption–the Forest Integrity Network (FIN), hosted by Transparency International (TI), the leading global anti-corruption organization.

Section 2 of the article assesses the scale of the problem of forest corruption. Section 3 then describes FIN's coalition building approach and the gaps in combating corruption that it seeks to fill. Section 4 focuses on FIN's history, and Section 5 explores the synergies between FIN and the activities of TI. Section 6 provides an overview of current and proposed future FIN activities. The article concludes by emphasizing the urgent need to work together in multistakeholder coalitions to improve forest governance worldwide.

FOREST CORRUPTION: THE SCALE OF THE PROBLEM

Degradation of the world's forest resources is one of the most pressing ecological and human development challenges facing the planet today. Forest degradation impacts the daily life of hundreds of millions of people around the world. Increasing insecurity of access to forest goods and services impacts the poorest of the poor most severely, since they rely on such goods and services for their subsistence and survival. Forest degradation also has catastrophic consequences for the earth's ecosystems, given that forests are major reservoirs of biological diversity, regulators of the hydrological cycle and storers of carbon.

One of the most important underlying causes of forest degradation is corruption and related illegal logging. Forest corruption has many manifestations, ranging from fraudulent logging concessions, to log smuggling and illegal logging, to the laundering of illicit proceeds, fraud, tax evasion and illegal trade. In Papua New Guinea, for example, an estimated 193 million Kina in revenue were lost to landowners and the government in 1993 (Duncan, 1994, cited in Callister, 1999). The Brazilian government estimates that 80% of all Amazonian timber stems from illegal sources (Greenpeace, 2001).

In Northwestern Russia, World Wide Fund for Nature (WWF) estimates that 35% of timber is illegally harvested (WWF, 2003). In the Philippines in the early 1990s, US$ 800 million were lost annually to illegal forest activities (Dauvergne, 1997, cited in Callister, 1999). According to the Environmental Investigation Agency (EIA) and Telapak, 73% of log production in Indonesia and 50% in Cameroon is based on illegal logging (EIA and Telapak, 2001). In Bénin, West Africa, an estimated 90% of timber is cut illegally (Siebert and Elwert, this volume). In Bolivia, 80-90% of all forest clearing is illegal (Contreras-Hermosilla and Vargas, 2002, cited in Contreras-Hermosilla, 2002). Magrath and Grandalski, (in press) estimate that 94% of the timber production in Cambodia is illegal.

Overall, the World Bank estimates that lost revenues in taxes and royalties due to forest corruption are in the order of US$ 10 billion per year minimum (World Bank, 2002). Of course, this does not include the enormous ecological and social costs of corruption-related illegal logging. In the search for global sustainability, therefore, corruption remains one of the hardest challenges to address. Reformers at all levels–national and international, official and non-governmental, public and private–see their efforts to tackle corruption frustrated by lack of political will and information, as well as by lack of common methodologies, appropriate tools of analysis and concerted action. Combating forest-sector corruption urgently requires multi-strategy, multi-stakeholder and multi-level action in order to be successful in the short and long run. FIN is one response to this need.

THE NEED FOR BUILDING COALITIONS IN THE FIGHT AGAINST CORRUPTION

FIN is an innovative and timely new initiative to combat corruption in use of the world's forests. It is unique in that it is the only international network that focuses exclusively on curbing corruption in the forest sector. FIN stresses the need to build coalitions in order to make corruption-fighting initiatives inclusive of all concerned stakeholders. In its inclusive multi-stakeholder approach, FIN aims to incorporate actors in the civil, public and private sector, both national and international. In particular, FIN seeks to bring together two groups of civil society organizations (CSOs) which have so far operated separately from each other: environmental CSOs concerned with forest conservation and sustainable forest use and CSOs engaged in combating corruption, such as Transparency International (TI).

In taking up the challenge of combating forest corruption, FIN aims to become a comprehensive source of information about replicable strategies to combat forest corruption. It seeks to identify gaps in available knowledge about the nature of forest corruption and ways to combat it. Since it is action and results-oriented, it brings together ongoing field experience, as well as thematic and case study research in order to facilitate country and local level action to curb forestry corruption. FIN's success will be judged by whether effective action is actually taken within countries to curb forest corruption.

HISTORY OF THE FOREST INTEGRITY NETWORK

The idea of FIN was first conceived by Peter Eigen of TI and Maritta Koch-Weser of the International Union for Nature Conservation (IUCN) in mid-2000, as an excellent way to combine TI's corruption fighting experience with that of civil society groups concerned with sustainable use of forests. At a meeting in May 2000 at Harvard University, a group of NGOs, donor organizations, academics, and specialists in forestry, conservation, and related disciplines expressed a shared determination to rein in corruption and ensure that forest resources are used more sustainably.

FIN was born from this shared sense of urgency about the need for action to combat corruption in the forest sector, with the objective of seeking peaceful, effective, and long-lasting solutions to the problem of corruption in use of forest resources. The meeting was sponsored by TI, with support from Harvard University's Center for International Development, IUCN, and the World Bank. A follow-up meeting, hosted by the World Bank in Washington DC on 29-30 November 2001, was attended by a broad spectrum of interested stakeholders. The participants agreed on the role, modalities and governance of FIN at this meeting, as well as on the key activities to be undertaken in its initial phase.

FIN is now hosted by Transparency International and has a small coordinating Secretariat at TI headquarters in Berlin, Germany. FIN seeks to accomplish its goals through building up a multi-stakeholder membership, and through drawing on expertise from all parts of the globe. It is thus guided by an expanding group of advisors and resource persons, as well as a Steering Committee, consisting of representatives from civil society groups, the private sector, multilateral organizations, and academia.

SYNERGIES BETWEEN FIN AND TRANSPARENCY INTERNATIONAL

Over the last decade, TI has helped to put corruption firmly on the international agenda. Going into its 10th anniversary, TI is now poised to explore the relevance of its corruption-fighting tools and strategies in specific issue-areas, with forest corruption and environmental sustainability a key concern. At the same time, CSOs long struggling with the problem of sustainable forest use increasingly emphasize the need to address corruption in this area and are therefore interested in TI's corruption fighting strategies.

TI is a global movement devoted to fighting corruption by creating coalitions with other CSOs, business and governments. The organization has developed tools for identifying and combating corruption. These include the well-known Corruption Perception Index (CPI), which ranks countries on the basis of perceived levels of corruption, and the Bribe Payers Index (BPI), which ranks major exporting economies according to the propensity of their companies to bribe foreign officials.

Over the last decade, TI has grown to include more than 80 national chapters, which implement TI's mission on the ground, and develop innovative country-based initiatives to fight corruption. FIN seeks to draw on and involve TI national chapters in the fight against forest sector corruption, partly through fostering synergies with environmental civil society groups active in the same countries. The local knowledge and experience of TI national chapters in fighting corruption is invaluable to combating forest sector corruption.

FIN'S CURRENT AND FUTURE ACTIVITIES

Since participants at the November 2001 meeting agreed on FIN's mandate and modalities, FIN has been engaged in a number of activities to broaden awareness of the problem of forest corruption and expand its membership. The network was invited to make presentations at six international forest governance meetings between 2002 and 2003. These were organized by multilateral organizations, the World Bank, the UN, academic institutions, and coalitions of civil society actors and the private sector, and took place in Rome (Italy), Phnom Penh (Cambodia), Atlanta, New Haven, and Washington DC (USA),

Casablanca (Morocco) and Seoul (South Korea). At all these meetings, the need for FIN was reconfirmed (see also www.transparency.org/fin/activities).

As a central activity, FIN is also in the process of developing a web-based knowledge center which brings together seminal documents on forest corruption. As part of its information synthesizing and dissemination role, the network has produced five newsletters. In October 2002, for TI's Annual Meeting with its national chapters, FIN compiled and circulated a comprehensive 'Information Note' on the activities of civil society groups and other actors in combating forest corruption (FIN, 2002a). Both at the TI Annual Meeting in 2002 and in 2003, a number of TI national chapters expressed strong interest in working to address forest corruption.

A key objective of FIN is to monitor and compile promising and innovative efforts to combat forest corruption around the world. Such monitoring will feed into the process of defining future on-the-ground activities and cross-national initiatives to be undertaken jointly by FIN members. One output of such monitoring and other FIN activities will be a compilation of best practices in fighting forest corruption worldwide.

Below, four FIN activities are described in more detail. These include: (a) an analysis of the relevance of TI's corruption fighting tools to dealing with forest sector corruption, as part of a project funded by the Programme on Forests (PROFOR); (b) a future Forest Corruption Fighters' Toolkit; (c) a future Sourcebook of Best Practices to combat forest corruption; and (d) conceptual frameworks for comparative analysis.

FIN-PROFOR Project: Relevance of TI's Tools to Combating Forest Corruption

PROFOR is a multi-donor initiative to improve forest governance, hosted by the World Bank. PROFOR is currently supporting a project on 'Improving Governance to Ensure Sustainability of Forests and Livelihoods' for which FIN is producing an in-depth analysis of the 'Relevance of Transparency International's Corruption-Fighting Approaches for Combating Forest Corruption'. This study explores whether and how TI's general tools and strategies to deal with corruption can be adapted to deal with forest sector corruption.

TI's general corruption fighting philosophy is based on information generation, awareness raising and coalition-building, involving public and private actors and civil society. The organization's tools and strategies have an emphasis on (1) civil society (such as awareness rasising through surveys and indices, monitoring, and oversight), (2) public sector (transparency in contracting, public hearings, price comparisons), and (3) private sector ('Integrity Pacts,' facilitating development of codes of conduct, business principles).

The TI Corruption Perception Index and the Bribes Payers' Index are powerful tools in generating information and raising awareness of the problem of corruption, at national and international levels. Due to their negative rankings in such indices, many governments have agreed to anti-corruption programs in the recent past. TI has also used citizen feedback surveys to capture experiential data about corruption in different sectors of public life. A detailed report entitled 'Corruption in South Asia' offers insights from surveys and reveals that the police is seen as one of the most corrupt institutions in South Asia, making law enforcement reform a key challenge (http://www.ti-bangladesh. org/). Other examples are surveys on vote-buying and corruption in the private sector by the TI national chapter of Brazil (http://www.transparency.org/ surveys/brazil_survey.html).

Indices and surveys can also be of critical relevance to the forest sector, given the current lack of quantitative, qualitative and comparable measures of forest corruption. With the increasing propensity of donors to link development aid to good public governance, quantitative measures and rankings of corruption levels are important tools with which to pressurize governments and the private sector to aim for better forest governance. In the forest sector, indices can be powerfully deployed to capture not only perceptions of corruption or bribery, but also levels of transparency and vulnerability. Thus, for example, a forest sector 'transparency index' could point to the forms of corruption in the sector and identify countries and companies who are leading the way in good forest governance. A forest sector 'vulnerability index' could demonstrate impacts of forest sector corruption on different segments of society (Gupta, 2003).

In carrying out other civil society functions, such as monitoring and oversight, the TI movement has also relied on citizen watch initiatives, which enhance transparency in public, private and non-profit transactions. TI-Bolivia, for example, has monitored actions of municipal governments in order to enhance accountability of public officials and oversee dispersion of the municipality's financial resources (Transparency International, 2002a). Other initiatives have monitored public procurement activities and the role of the media in election campaigns (Transparency International, 2002a).

In the forest sector, citizens and especially forest user communities can play a similarly key role in monitoring and protecting forests for sustainable use. Forest community initiatives can monitor adherence or abuse of forest laws, the system of awarding logging concessions, and the exploitation process. Through public hearings, they can seek to enhance transparency and accountability in the awarding of logging concessions.

The FIN-PROFOR study also looks at other TI tools with a public sector emphasis, such as public hearings and price comparisons. In public contracting, public hearings involving public sector, private sector and civil society representatives serve to enhance transparency. In 2002, TI national chapters from Latin America, for example, held nine national workshops to produce national risk maps of corruption in public contracting (Transparency International, 2002c). Price comparisons have facilitated more efficient government expenditure by making information on purchase prices available across governmental agencies.

Transparency in public contracting and the construction of risk maps in the forest sector in different national contexts can be important future activities for FIN. The maps could document risks or vulnerabilities in awarding or monitoring forest concessions. They could also focus on illegalities in regional and global forest trade flows.

Another set of TI corruption fighting tools are tailored to corruption in or involving the private sector. 'Integrity Pacts' were developed for public contracting processes involving private parties. The object of such a process may be the construction of buildings, consultancy, operation of services, issuing licenses and concessions (Transparency International, 2002b). The Integrity Pact works through a binding no-bribery and no-collusion agreement between procurement agencies and bidders. It seeks to level the playing field for private sector bidders by assuring that all bidders will not bribe and that public officials will not demand bribes. Public officials, in turn, are assured that bidders will not offer bribes or collude. 'Integrity Pacts' raise the psychological hurdle for the potentially corrupt and facilitate law enforcement (Siebert and Olaya, 2003). Appropriate disciplinary or criminal sanctions enforce the pact and arbitration is used as a conflict resolution mechanism (Transparency International, 2002b). Civil society (NGOs, TI national chapters) plays a key role in monitoring and overseeing the implementation of the Integrity Pact. Both governments and private enterprises can be winners in such a pact: governments can reduce the high cost of corruption in public procurement, privatization or licensing, while private enterprises can save bribing expenses. In Argentina, Colombia, Ecuador, Italy, South Korea and in Pakistan, Integrity Pacts have demonstrated their effectiveness (Transparency International, 2002b).

Since involvement of the private sector is key in effectively fighting forest sector corruption, Integrity Pacts can be a powerful tool for reducing corruption in forest concession awarding systems. They are applicable in all cases in which different stakeholders must comply to a single set of rules and where deviation is expected (Siebert and Olaya, 2003). The challenge remains in generating the political will to undertake Integrity Pacts, as well as in creating incentives for private companies to participate in them.

Another tool to enhance good corporate governance is adoption by socially responsible companies of codes of conduct or business principles to combat bribery. Such codes and principles increase the awareness of corruption and encourage the development of preventive strategies. Some codes consist of detailed rules regarding the conduct of employees and the corporation's standards towards third parties, other codes merely state fundamental ethical principles. TI has been particularly effective in this area, as evident from its role in facilitating the Wolfsburg Anti-Money Laundering Principles and (together with Social Accountability International) the Business Principles to Combat Bribery.

The relevance for the forest sector is obvious. Some forest industry leaders are currently open to voluntary adherence to a set of basic principles, and could take the lead in combating forest corruption among private sector actors. Business principles in the international financial sector are of special relevance to fighting forest sector corruption. The Wolfsberg Anti-Money Laundering Principles, for example, consist of principles of sound business conduct in international private banking, and have been adopted by 11 international banks. Given that money laundering is an integral part of illegal logging and illegal timber trade, it must be addressed by facilitating transnational cooperation between financial institutions.

Finally, TI offers an overarching anti-corruption approach, the 'National Integrity System' (NIS). This system approach is presented in the TI Sourcebook, supplemented by a documentation of best anti-corruption practices worldwide and regularly updated on the TI web-site (Transparency International, 2000). The NIS consists of various 'pillars of integrity' which ensure good governance: Institutions such as the executive, legislature and judiciary, the private sector, media and civil society organizations, as well as their practices and interrelationships. The NIS stresses TI's holistic approach and the necessity to examine each of the institutions and their practices when addressing corruption in a given country. Strengthening the NIS involves transparency in public procurement, a free media, free access to information, an independent judiciary, an elected legislature and a responsible business sector. TI is undertaking detailed country-specific studies to assess the NIS in practice in different countries (approximately 20 country studies are available so far). The NIS approach can be used as an analytical framework in order to systematically explore the current impacts of all government institutions, actors of the private sector, civil society organizations and their practices on the forest sector. The NIS approach can guide the analysis of the kinds of institutions and practices that need to be in place in order to contain forest corruption (Siebert and Olaya, 2003).

The FIN-PROFOR study will result in an in-depth analysis of the relevance of many of the TI tools, described above, to combating forest corruption. It will, therefore, provide the conceptual and empirical basis for identifying and implementing corruption fighting strategies in various countries, ideally in partnership with TI national chapters and FIN members. Experiences gained from subsequent in-country activities will feed into the production of an operational Forest Corruption Fighters' Tool Kit, which seeks to compile effective and replicable approaches to combating forest corruption developed worldwide.

A Forest Corruption Fighters' Tool Kit

As a longer-term goal, FIN proposes to compile, in a user-friendly and disseminable format (web-based, in hardcopy and through a CD-ROM), tools and strategies that civil society and other stakeholder groups are developing to address corruption in the forest sector, as well as innovative new strategies that might emerge from collaborative work. Such a Forest Corruption Fighters' Tool Kit would be modeled along the lines of the newest edition of TI's general Corruption Fighters' Tool Kit, which is a compilation of 'civil society experiences and emerging strategies' in combating corruption globally (Transparency International, 2002a).

A Forest Corruption Fighters' Tool Kit would not be limited to civil society actions, but would seek to emphasize the multi-stakeholder approach to addressing forest corruption. Such a Tool Kit could be organized according to the areas where tools to combat forest corruption are most needed. For example, the Tool Kit could focus on approaches to generating reliable and useful information on the extent of forest concessions, the bidding process, awarding of forest concessions, etc. (through surveys, indices). It could also compile methods of information dissemination and awareness raising, as well as strategies to engage civil society, local communities, the private and public sector in combating corruption (through citizen watch initiatives, integrity pacts, business principles, codes of conduct, and independent monitoring of awarding logging permits, adherence to logging permits, forest certification, etc.). In addition, it could include strategies to promote adoption and implementation of appropriate legal frameworks at national, regional and global levels, and methods to assess law enforcement capacity and gaps in different national contexts. Finally, it could include methods and experiences with development of indicators to monitor and measure progress towards improving forest sector governance.

Producing such a Forest Corruption Fighters' Tool Kit will serve a very important information synthesizing and disseminating function that is not being performed by any other group or network. Such a synthesis of available tools will add value to information being generated by actors in different contexts. Compiling such a Tool Kit requires conceptualization and assessment workshops, and involvement of experts and leaders from civil society and other actors dealing with forest corruption. The aim is to highlight replicable tools, and to develop an evolving product, which can constantly be added to.

A Source Book of Best Practices

The logical consequence of the Forest Corruption Fighters' Tool Kit would, eventually, be the identification and dissemination of a set of best practices in the area of combating forest corruption. Identifying and developing best practices would require further conceptualization of the problem of forest corruption and comparison of empirical experiences. Compiling best practices can be modeled on TI's Source Book on Best Practices, which is accessible by internet and regularly updated (Transparency International, 2000). Again, it will be modified to address the particular challenges of forest sector corruption and corruption in natural resource allocation and use.

Conceptual Frameworks for Cross-Country Comparisons

In parallel with a Forest Corruption Fighters' Tool Kit, and Sourcebook of Best Practices, FIN also proposes to develop conceptual frameworks for cross-country comparisons in analyzing and combating forest corruption. It aims to do so by drawing on its members' activities in different contexts and comparing empirical data generated through in-country activities. The need for comparative conceptual frameworks has been emphasized by FIN supporters from the outset, given the present lack of comparative frameworks which are critical to identifying replicable best practices in combating forest corruption (FIN, 2002b).

CONCLUSION

FIN's goal, ultimately, is the reduction of forest corruption. It seeks to enhance awareness of forest corruption as a key hurdle to forest conservation and sustainable forest use. FIN aims to become a comprehensive clearing-house of information on forest corruption, including policy research and frameworks for comparative analyses. It seeks to develop and disseminate innovative ideas to fight forest corruption, drawing on the experiences of TI as well as organizations concerned with sustainable natural resource use world-wide. The network has the central objective of building a firm basis for coalitions to work together in sharing experiences and identify best practices in combating forest corruption. FIN responds to the timely and urgent need to address the relationship between corruption and natural resource misuse. Hence, it also serves as a model for addressing other areas of corruption and environmental unsustainability, such as in the issue-areas of climate change, hazardous waste, chemical (mis)-management and unsustainable use of the world's water resources.

REFERENCES

Callister, D. 1999. Corrupt and Illegal Activities in the Forestry Sector: Current Understandings, and Implications for the World Bank Forest Policy. Discussion Draft, May 1999. Available at: *http://Inweb18.worldbank.org/eap.nsf/2500ec5f1a2d9bad 852568a3006f557d/01590e1de4ec81d34.*

Contreras-Hermosilla, A. 2002. Policy and Legal Options to Improve Law Compliance in the Forest Sector. In: pp. 43-91. Food and Agricultural Organization of the United Nations (ed.). Proceedings. Reforming Government Policies and the Fight against Forest Crime, Rome, 14-16 January 2002. FAO, Rome, Italy.

Environmental Investigation Agency (EIA) and Telapak. 2001. Timber Trafficking. Illegal Logging in Indonesia, South East Asia and International Consumption of Illegally Sourced Timber. EIA and Telapak, London.

Forest Integrity Network. 2002a. Progress towards Containment of Forest Corruption and Improved Forest Governance. An Information Note prepared for Transparency International's Annual General Meeting (AGM), October 2002, Berlin. Available at: *http://www.transparency.org/fin.*

Forest Integrity Network. 2002b. Forest Integrity Network (FIN). Background and Concept Paper, Berlin. Available at: *http://www.transparency.org/fin.*

Greenpeace. 2001. Partners in Mahogany Crime. Amazon at the Mercy of 'Gentlemen's Agreements.' Greenpeace, Amsterdam.

Gupta, A. 2003. TI Tools and Approaches to Combating Corruption: An Overview Document prepared as input into the FIN-PROFOR study on 'Relevance of TI Tools and Approaches to Combating Forest Sector Corruption.' Berlin. Available at: *http://www.transparency.org/fin.*

Magrath, W. and R. Grandalski. 2002. Policies, strategies, and technologies for forest resource protection. In: T. Enters, P.B. Durst, G.B. Applegate, P.C.S. Kho, and G. Man (eds.). Applying Reduced Impact Logging to Advance Sustainable Forest Management. Food and Agriculture Organization of the United Nations. Bangkok, Thailand. Available at: *http://www.fao.org/DOCREP/005/AC805E/AC805E00.HTM.*

Siebert, U. and G. Elwert. 2004. Combating Corruption and Illegal Logging in Bénin, West Africa: Recomendation for Forest Sector Reform. *Journal of Sustainable Forestry*, 19(1/2/3): 337-349.

Siebert, U. and J. Olaya. 2003. The Forest Integrity Network (FIN) and the Relevance of TI Corruption Fighting Tools. Power Point presentation in the workshop "Forestry and Corruption," 11th Anti-Corruption Conference, 25-28 May 2003. Seoul, South Korea. Available at: *http://www.transparency.org/fin/activities.*

Transparency International. 2002a. The Corruption Fighters' Tool Kit. Civil Society Experiences and Emerging Strategies, Berlin. Available at: *http://www.transparency. org/toolkits/index.html.*

Transparency International. 2002b. The Integrity Pact. The Concept, the Model and the Present Applications: A Status Report, Berlin. Available at: *http://www.transparency. org/building_coaltitions/integrity_pact/index.html.*

Transparency International. 2002c. The TI Narrative Report 2002. Berlin (unpublished).

Transparency International. 2000. TI Sourcebook. Confronting Corruption: The Elements of a National Integrity System, Berlin/London. Available at: *http://www. transparency.org/sourcebook/index.html.*

World Bank. 2002. A Revised Strategy for the World Bank Group. The World Bank, Washington, DC, USA.

World Wide Fund for Nature (WWF). 2003, April 3. Significant part of Russian timber exported to Europe might be illegal, WWF warns. Retrieved June 15, 2003, from *http://www.panda.org/news_facts/newsroom/press_releases/*

Other Websites:

http://www.ti-bangladesh.org/
http://www.transparency.org/surveys/brazil_survey.html

SYNTHESIS

Illegal Logging in the Tropics:
A Synthesis of the Issues

Ramsay M. Ravenel
Ilmi M. E. Granoff

SUMMARY. Illegal logging is a policy problem that has arisen from the failure of protected area and sustainable forestry remedies for deforestation. Nevertheless, illegal logging in some parts of the world is a centuries-old problem rooted in social conflicts over access to forest resources. We synthesize the papers in this volume, which analyze illegal logging issues in a variety of countries–Indonesia, Vietnam, India, Cameroon, Bénin, and Mexico–and from a variety of analytical perspectives–anthropologic, economic, political science, legal, and conservationist. We describe illegal logging activities and analyze the roles of key stake-

Ramsay M. Ravenel is an independent consultant in forestland investment (E-mail: rravenel@aya.yale.edu).
Ilmi M. E. Granoff is a consultant in international conservation and development policy (E-mail: ilmi.granoff@aya.yale.edu).
Address correspondence to: Ilmi M. E. Granoff at the above address.

[Haworth co-indexing entry note]: "Illegal Logging in the Tropics: A Synthesis of the Issues." Ravenel, Ramsay M., and Ilmi M. E. Granoff. Co-published simultaneously in *Journal of Sustainable Forestry* (The Haworth Press, Inc.) Vol. 19, No. 1/2/3, 2004, pp. 351-371; and: *Illegal Logging in the Tropics: Strategies for Cutting Crime* (ed: Ramsay M. Ravenel, Ilmi M. E. Granoff, and Carrie A. Magee) The Haworth Press, Inc., 2004, pp. 351-371. Single or multiple copies of this article are available for a fee from The Haworth Document Delivery Service [1-800-HAWORTH, 9:00 a.m. - 5:00 p.m. (EST). E-mail address: docdelivery@ haworthpress.com].

holders such as governments, the private sector, local communities, and non-governmental organizations. We summarize the major policy recommendations presented by the authors in this volume, which include: broad anti-corruption initiatives beyond the forest sector, information-based approaches, conventional regulation and enforcement, and market mechanisms. Illegal logging is complex. It involves broad networks of power and influence that operate across many geographic, socio-political and economic contexts. It is not simply a question of poor villagers with chainsaws. Effective interventions must identify the key players and their supporting networks of racketeers and timber launderers and give prosecutors the wherewithal to overcome their ability to manipulate political connections to evade conviction. *[Article copies available for a fee from The Haworth Document Delivery Service: 1-800-HAWORTH. E-mail address: <docdelivery@haworthpress.com> Website: <http://www.HaworthPress. com> © 2004 by The Haworth Press, Inc. All rights reserved.]*

KEYWORDS. Illegal logging, deforestation, tropical forest conservation, tropical forest management, forest policy

INTRODUCTION

Academic and Policy Context

Illegal logging in tropical forests has become a hot topic in conservation and development policy in recent years. Major international bodies such as the G8, the World Bank, and the United Nations Forum on Forests have identified illegal logging as a serious threat to forests around the world (G8, 1998; Callister, 1999; UNFF, 2002).

The discourse on the illegal logging problem has succeeded and to some degree grown out of its forest policy predecessor–deforestation. Deforestation captured international attention in the 1980s and 1990s and was often framed in Malthusian rhetoric of population growth and exhaustion of finite resources. Protected areas (State or indigenous) and sustainable forest management emerged as solutions to the deforestation problem. Indeed, protected area coverage increased from 8.8 million km^2 to 18.8 million km^2 from 1982 to 2003 (Chape et al., 2003) and laws mandating sustainable forest management made it onto the books of many countries. However, both of these approaches rely on sound, scientifically based policy interventions and a reliable legal framework.

Sustainable forestry laws in many countries mandate silvicultural practices that may appear to be light on the land–and thus sustainable–when in fact they "high-grade" the forest and irrevocably take commercial and shade intolerant species out of the ecosystem (Ashton and Peters, 1999; Frederiksen and Putz, 2003). This issue of forest degradation is a subtler question of "what is in the forest" rather than simply "where is the forest?" Experience in many countries suggests that misguided "sustainable" forestry has degraded the commercial viability of forestlands. As commercial viability declines, so does the profitability of legal logging. Illegal logging–reducing diameter limits, harvesting more species, prematurely reentering regenerating stands, crossing cut block boundaries, and/or logging in protected areas–becomes increasingly attractive. As illegal logging degrades forest resources, pressure mounts to convert forestlands to non-forest uses such as monocrop plantations. As this process develops, the distribution of forestlands across the landscape shrinks. This process renders remaining forestlands–which are typically protected areas–increasingly isolated and threatens the integrity of their ecosystems (Curren et al., 1999).

The temptation of higher profits through illegal logging is not unique to degraded resources: companies operating in unlogged forests often log illegally too. Unfortunately forest management laws, sustainable or otherwise, are often flouted without recrimination. In the international policy community, governments long considered forest degradation, deforestation, and biodiversity conservation to be sovereign domestic issues. In the late 1990s, however, these problems came to be recognized as symptoms of systematic illegal logging. By this time the international community had grown acutely concerned with the issue of governance (CITE?). In this context, this new framing of forest conservation problems in terms of illegal logging has gained traction.

A number of other issues have increased awareness of the illegal logging problem. Independent certification of sustainable forest management has played an important role in identifying illegal logging as a major policy problem. Certification arose in the mid-1990s as a non-governmental market-based approach to addressing deforestation and forest degradation. Environmental groups hoped to give consumers of forest products the power to discriminate between products that could document the sustainability of their production methods and those that could not. One of the basic principles of forest certification is that forest managers obey national laws. In countries where corrupt and/or under-funded forestry departments are lax in their oversight of forest operations, certification audits represent the only monitoring of basic compliance with forest laws.

Forests play a critical role in global carbon cycles, so the emergence of global climate change as a major international environmental issue in the 1990s has brought increasing attention to illegal logging issues. For example, the Indonesian forest fires of 1997–caused by the synergistic effects of a prolonged El Nino dry season and illegal forest clearing–made mainstream news headlines and raised the profile of the illegal logging problem. In 2002, George W. Bush introduced his "President's Initiative Against Illegal Logging" alongside a major initiative on global climate change, calling illegal logging "a practice that destroys biodiversity and releases millions óf tons of greenhouse gases into the atmosphere" (Bush, 2002).

Illegal logging in protected areas has highlighted both the exhaustion of timber stocks in production forests and the impacts of illegal logging on biodiversity conservation. In Indonesia, for example, decades of timber exploitation have left non-protected areas bereft of timber. With nowhere else to go, opportunistic wood dealers have turned to illegally logging in protected areas, where wood costs only as much as the requisite bribes and operational costs are typically dramatically lower. The Environmental Investigation Agency and Telapak-Indonesia have exposed extensive and politically connected illegal logging in the Tanjung Puting National Park in Indonesian Borneo. Tanjung Puting is one of the few remaining areas of viable habitat for orang-utans (*Pongo pygmaeus*) (Rijksen and Meijaard, 1999; EIA-Telapak, 1999).

Finally, technological improvements in remote sensing have made it much easier and cheaper to detect illegal forest activities. The World Resources Institute's Global Forest Watch program, for example, builds in-country capacity to monitor forest concession holders via satellite imagery and field surveys in the major forested countries of Central Africa, South America, and Southeast Asia, and in Russia, Canada, and the U.S. (GFW, 2003).

Despite the recent flurry of interest in the illegal logging problem, it is important to note that it is the attention and current construction of the problem that is new. Conflict over access to forest resources typically results from differing opinions over what is legal and what is illegal in the forest. People have fought over these issues for centuries. In many countries, conflicts over forest resources trace back to the colonial era (Peluso, 1992; Maathai, 2002; Obidzinski, 2003).

Description of the Conference

In 2002, the Yale Chapter of the International Society of Tropical Foresters (ISTF) hosted a conference in 2002 on "Illegal Logging in Tropical Forests: Ecology, Economics & Politics of Resource Misuse" in order to synthesize existing knowledge on the issue and to identify concrete policy approaches. The conference was designed as an interdisciplinary event, and included speakers from a variety of academic disciplines–such as political science, economics, law, and ecology–and non-governmental organizations. The conference organizers also sought global representation and included work from Asia (India, Indonesia, and Vietnam), Africa (Cameroon), South America (Brazil) and Central America (Mexico). As a result, the event successfully brought together scholars capable of providing a nuanced analysis of the nature and impact of illegal logging, and activists and policy-makers able to provide strategies for addressing illegal logging in the field and in broader legal and political fora.

DEFINITIONS OF ILLEGAL LOGGING

In the Woods

Variously called "forest crimes" or "illegal forest activities," in the simplest of terms illegal logging is the felling of trees in violation of laws and regulations. The question arises, however, who determines the laws and regulations and how well do they reflect accepted social norms and values? These questions point to the broader socio-political context in which illegal logging must be understood-a subject discussed in detail in a later section of this paper.

Illegal logging "in the woods" involves (1) forest practices in otherwise legal forestry operations that violate regulations, and (2) the felling of trees in unapproved areas. Forestry laws typically require timber companies to follow a variety of rules concerning permitting, logging methods, and transportation of forest products. Logging violations could include exceeding approved harvest intensity and diameter limits, thereby diminishing the quality of the residual stand; operating in restricted zones within approved operating areas (e.g., steep slopes, wetlands, riparian zones); felling protected tree species or destroying habitat for endangered plants and animals; and ignoring the boundaries of approved harvest areas. Violation of these types of regulations may be willful and profit-driven, or may be the result of poor training and oversight.

Logging in unapproved areas is in some ways a more severe problem that occurs in both production forests and protected areas. Illegal logging in unapproved areas of production forests may be knowingly carried out by concession holders or by third parties such as crews from nearby villages that may or may not have the consent of the concessionaire. These local crews may have access to heavy machinery or they may only have access to chainsaws. In some cases, concession holders exploit the illegal logging problem by hiring third parties to illegally log within their area in order to increase production without directly violating the law. In situations where tenure is ambiguous, both concession holders and local logging crews justify their illegal logging by arguing, "if we don't cut it, they will." Such logging may take place in primary or selectively logged forest and in active or inactive concessions. In either case, illegal logging in these areas reduces the future economic viability of those production forests and thereby increases the likelihood that that forest will be cleared and converted to another land use that will not provide comparable ecosystem values and services.

In light of the widespread illegal logging in production forests and the extent of forest clearing and conversion, illegal logging in protected areas is something of a *coup de grace* for biodiversity conservation. Protected areas in many once-remote regions of the world are becoming increasingly island-like in their lack of a surrounding matrix of forest. Indeed the matrix has flipped: whereas clearings previously represented holes in a vast matrix of forest cover, forests are increasingly islands in a matrix of non-forest land uses. Island biogeography theory suggests that islands tend to feature higher species extinction rates than continents and that smaller islands support fewer species. With protected areas becoming more like islands than continents, illegal logging threatens the viability of their plant and animal communities and the integrity of their ecosystems. While the most remote areas of the Congo and Amazon Basins as well as parts of Papua New Guinea may not yet suffer from this problem, it describes the pattern of land development in much of the world.

Out of the Woods

Once timber has been illegally harvested, the illegality continues as it is transported and brought to market. The illegality is "over-determined"–that is, multiple links in the chain of custody are unlawful or of questionable legality. Illegal wood is typically laundered by forging documents, mixing it with legal wood, and in some cases illegally exporting it in violation of export bans. Traders of illegal wood are also likely to avoid taxes and pay bribes to regulators and/or judges, either on a case-by-case basis or through more systematic clientelism or patronage. Traders in timber species regulated by the Convention on Trade in Endangered Species (CITES) commonly manipulate the documentation process, undermining its intended purpose (U.S. CITES Management Authority, 2003) and in some cases "legalizing" illegally harvest wood (Blundell and Rodan, 2003). CITES Appendix II, which allows for regulated trade, only lists two major timber species, mahogany (*Swietenia macrophylla*) and ramin (*Gonystylus bencanus*) (CITES, 2003).

Illegality can also extend to the financial backers of illegal operations. These investors may or may not be aware of the illegality of the companies they invest in. Some critics argue that investors and lenders should require companies to demonstrate the legality of their wood and fiber supplies and should be held accountable for any oversight. A recent report on human rights violations at pulp mills on the Indonesian island of Sumatra chides government officials, company managers, and international lenders for their ignorance of severe abuses (Human Rights Watch, 2003). In Latin America, the absence of market accountability has created disincentives to adopt extensive silvicultural understanding of mahogany in on-the-ground management (Blundell and Gullison, 2003).

Illegal activities often accompany legal logging operations that provide a means of laundering illegal timber and the movement of illegal timber is often abetted by forgery or other questionable means of bringing timber to market. Most forestry crimes are by nature on the periphery of the traditional state apparatus. This peripheral status permits state actors themselves to become involved with little concern of oversight. There is thus often some governmental complicity at the local and sometimes even the national level. Like other highly profitable large-scale illegal activities, illegal logging in some cases involves organized crime networks.

In summary, illegal logging involves a complex array of activities, interactions and transactions among a broad range of actors. Who these actors are and the history of the power politics that govern their interactions are discussed below.

STAKEHOLDERS

Government

While many stakeholders have vested interests in illegal logging, questions of legality inherently reflect a state-centered perspective. In the prevailing discourse, a government controls forest resources in public trust, and is charged with regulating them accordingly. Although the legitimacy of the state can be questioned, a pragmatic approach assumes that the state has a role to play in managing forest resources for the public good. Policy interventions, therefore, focus on reforming the state, and improving upon its ability to create policies that maximize the social welfare derived from forests. In this context, other stakeholders are significant in their ability to shape law, to conform to it, and to influence the state's ability to carry out its responsibility.

State forestry ministries are obviously one of the most important government stakeholders in addressing illegal logging issues. They may be necessary for reforms but they are hardly sufficient. However, comprehensive reforms rely on the participation and support of other agencies such as the judiciary, finance (taxation), police, customs, land planning, etc. These agencies support not only enforcement of forest laws in the field, but also help regulate the distribution of forest products along the production chain. In some cases, for example, illegal logging reforms might even be more easily and effectively implemented by focusing on the transportation of finished products (Rosenbaum, this issue). More importantly, in the absence of good governance in these other areas isolated reforms in the forest sector alone are unlikely to succeed.

Measures taken to address illegal logging must not only reach across governmental agencies but they must also reach through the many layers of a federalist government from top to bottom. A weak central government might make any number of reforms but if the will to implement them at the regional and local level is lacking, little change will come. During the conference proceedings, some panelists focused their critiques on national government actors while others focused on the role of local governments. Effective reforms must specifically address the role of each level of the governmental hierarchy.

Private Sector

In most of the countries examined during the conference, private sector forestry companies are granted leases to manage state lands. Vietnam, which features a state-owned forestry company, is the exception (McElwee, this issue). In the Andaman and Nicobar Islands of India (Seksaria, this issue), private forestry companies might be considered quasi-statal since they rely so heavily on subsidies for the transportation of their products back to mainland India. Given the weak governance seen across the board, working directly with the private sector should be considered an alternative approach to affecting change.

Private sector stakeholders in tropical forest management include more than just the companies operating in the forest. The marketing and distribution of forest products along the "chain of custody" involves a much broader range of interests, including domestic wood traders, sawmills of many varieties, pulp mills, international importers and exporters, furniture manufacturers, large-scale retailers, home-builders, etc. Environmental groups such as the Rainforest Action Network, Greenpeace, and others have targeted major retailers of tropical forest products in the U.S. such as The Home Depot, putting pressure on them to only buy wood that has been certified as sustainably produced. The Home Depot has committed to giving preference to certified wood, when available, and to ensuring that its wood does not come from endangered forests. Although they have stopped short of forcing suppliers to undergo full certification (as global furniture retailer IKEA has), they have required suppliers to have their chain of custody audited to ensure that no illegal wood from endangered forests will find its way to Home Depot stores. One compelling argument for this strategy is that it focuses on actors in developed countries, where the vast majority of the world's industrial roundwood is consumed. However, China is also a major global consumer of roundwood and Chinese companies are unlikely to respond to attacks on their brand images–to the extent that they rely on them–anytime in the near future.

Local Communities

Because tropical forests are on the margins of the traditional state apparatus, so too are the local communities that inhabit them. In discussing the roles
and perspectives of local people in illegal logging, panelists were quick to de-
flect blame from local communities. Local people often participate in illegal
logging when they face few alternatives for generating cash income, when
given the incentive by outside actors, and when they believe land is not being
effectively managed by the state on their behalf. Often poor people *from other
areas* are brought in to do the logging, altering community structure and incentives in relation to local forests (Dudley, this issue).

Marginalized rural people–local or otherwise–rarely enjoy much if any of
financial benefits of illegal logging activities. They are typically hired as laborers for illegal logging operations and rarely capture much if any of the profits. More likely, they will borrow money to buy their food for their time in the
woods and to leave with their families while they are gone-only to earn barely
enough to pay off their debts before they return home (Hiller et al., this issue).
While in some cases local people may be running the chainsaws, it is more
likely their boss or their boss' boss who is both orchestrating the operation and
capturing its profits. These investors and wood traders control the access to
markets for illegal wood and are typically players in local, regional, and sometimes international networks of economic and political power and influence
(McCarthy, 2000).

Local communities often have a strong vested in interest in the sustainable
management of forest resources. Indeed, when illegal logging initiates a trend
toward forest clearing and conversion, local communities bear the brunt of the
resulting economic and ecological costs–e.g., loss of water quality and quantity, habitat for game species, and other non-timber forest products. Nevertheless, many community-based initiatives have failed because they did not
address the networks of power that ultimately control these activities. Local
communities may not be the most important perpetrator of illegal logging activities, but they can be valuable partners in localized monitoring and enforcement programs.

Non-Governmental and International Organizations

Non-governmental organizations (NGOs) and international organizations
are an increasingly important stakeholder in the illegal logging arena. NGOs
are often the only institutions with both the interest and the capacity to pursue
citizen suits in those cases where the judicial system provides a legitimate recourse. Often NGOs have better information than those government institutions responsible for monitoring and enforcement. In a number of cases,
valuable partnerships have been formed based on the capacity of NGOs to pro-

vide reliable monitoring, supplementing the enforcement powers of the government. Examples of this include Greenpeace-International in Brazil, Telepak in Indonesia, and Global Witness in Cambodia and Cameroon. NGOs are also an important means by which marginalized actors affected by illegal logging can be represented in policy-making processes, especially when the state is perceived by some other stakeholders as antagonistic.

There is a countervailing danger, however, to NGOs providing services that are traditionally the responsibility of the state: in some cases they face less public scrutiny than the government, and necessary funds for the appropriate agency may get diverted to other governmental activities with the assumption that the NGO is providing a consistent and reliable source of the service. Panelists working in Vietnam and Kenya each noted during the conference proceedings that when NGOs become too dependent on the government for resources, access or credibility, they may be afraid to criticize government actions or failures to act. In the developing world context, NGOs can also create a "brain-drain" on the public service. If NGO or international organization jobs are seen as more lucrative or higher profile, the most qualified public servants are likely to pursue them. Also international organizations, because of their reliance on the funds of a foreign public, may seek to conduct activities that meet their interests instead of local or national ones.

ILLEGAL LOGGING IN SOCIO-POLITICAL CONTEXT

In one of the panel discussions during the conference, Ken Rosenbaum noted that "illegal activities are not simply legal problems; they are social problems with legal, political, economic, moral, social, and historical facets." It is in this spirit that we explore the social and political history of forest use in the developing countries in which much of the illegal logging of tropical forests occurs.

All of the countries examined closely in the conference were colonized at some point in their history. Colonization is both a shared experience and a widely varied one. Dutch (Indonesia), English (India), French (Vietnam, Cameroon), Spanish (Mexico) and Portuguese (Brazil) colonial governments varied greatly in their treatment of indigenous people and resources. After independence, all of these countries saw the establishment of centralized state power and control over valuable resources. In most cases forest resources were nationalized and local access to them was formally restricted. In the context of undemocratic neocolonial governments–some admittedly more so than others–the declaration of certain forest uses as legal and others illegal did not reflect accepted social norms. Forest laws were instead designed to secure political patronage more than reflect accepted social norms, especially at the local level. Many traditional and locally accepted forest uses became illegal in order to make way for legal but often destructive industrial logging. Pamela McElwee noted during the conference proceedings that such nationalization of resources was significantly *not* unique to socialist governments, but was common across regimes of all types. Legality in this sense was a state concept that articulated its control over forest resources, regardless of its intent or ability to maximize social welfare from them.

Much of the debate over illegal logging takes this socio-political history for granted. Illegal logging is often framed as a problem related to poverty, lost tax revenues, state capacity, and economic development. The history described above puts the current illegal logging discourse into context and explains why this problem is not so easily "solved."

Many forest-rich governments have abetted abusive forest activities, ostensibly in the interest of national development. Forest resource exploitation has often provided governments with the means to achieve other political objectives–much to the detriment of the resource itself (Ascher, 1999). Concessions are routinely given out to bad actors who regularly exceed legal production levels, cut outside the boundaries of their licensed areas, and otherwise plunder these nationalized assets. Insufficient state capacity enables this widespread mis-management of forest resources to occur. All too often, no pretence of national interest is even made, and because of forestry's peripheral status, government office-holders are directly enriched by their complicity.

Forest policy reforms enacted in recent decades may have been acceptable to the state as long as no one had to live up to them. Such was the case for many years, but now that productive natural forests are increasingly scarce and the world is now increasingly able to watch, regulations that were never intended to do their job must suddenly do their job after all. Where "legality" may have once been a weapon of the state to secure control over forest resources, it is now being used against the state in international fora to demonstrate the state's failure to do its job.

POLICY RECOMMENDATIONS

Although it is the state's role to create and enforce forest laws, the state is often at some level also complicit in or even an active perpetrator of forest crimes. Illegal logging as addressed in these proceedings is thus largely an issue of corruption. While laws often exist that protect public or state owned timber resources from being misused for private gain, governmental capacity may be insufficient to regulate others, or even itself, from benefiting from forest crime. An initial question to ask, therefore, is whether illegal logging is best addressed through policy measures targeted at government forest sectors, or whether illegal logging is a symptom of a broader corruption problem that must be addressed across government sectors. Policy prescriptions aimed at preventing illegal logging should begin by addressing corruption. Devoting resources to government programs without sufficiently addressing the involvement and complicity of governmental actors can not only risk wasting valuable resources, but also risks supporting bad actors.

Policy prescriptions that target the forest sector focus on aiding governments in their ability to create and enforce forest laws. Priority targets include protected areas, concessionaires, foreign investors and operators, and companies that sell–unwittingly or otherwise–illegal wood. Many regulatory reforms to date have favored command-and-control approaches to illegal logging and such approaches will undoubtedly continue to play a significant role in addressing the problem. This section reviews some of the key issues for forest policy reforms that target illegal logging.

The pervasive illegal logging occurring in protected areas in many tropical countries suggests that current regulations in those countries do not provide a sufficient disincentive. Measures that increase the perceived cost of forest crime-either by better enforcing existing regulations, increasing penalties, or directly affecting criminals' perceptions of the likelihood of capture-relative to the expected benefit of successfully carrying out the activity have the best chance of reducing illegal logging.

Further, given that the illegality of forestry activities typically involves multiple intervention points, other aspects of illegal logging remain under-regulated as well. Illegal logging requires not only concealment in the forest, but also the ability to conceal the source of the timber as it is transported and marketed. Enforcement is usually most effective through careful analysis of records of forestry operations, transportation and marketing by a forestry expert. For example, laundering of illegally harvested timber through legal timber concessions appears to be a common practice. Timber laundering, in addition to being subject to prosecution as a separate offense, is an activity that is identifiable sometimes using public records alone. Careful monitoring of legal forestry operations through remote sensing, on-the-ground inspection, and systematic forest inventories can render them less useful as cover for illegal operations.

Market-based mechanisms are seen by many policy-makers as a progressive, efficient, and effective alternative to traditional command-and-control regulation. It is often overlooked, however, that a baseline level of rule of law, often in the form of effective command and control regulations, is required as a pre-condition for markets to operate efficiently. Indeed, markets themselves are necessary for effective market-based policy mechanisms. Timber markets in the tropics are often insufficiently developed to allow for any serious consideration of more progressive market-based approaches. As economist Nalin

Kishor observed during the conference proceedings, start-up costs can be high, so market-based policy options will not be effective in all situations.

Despite these words of caution, it is important to consider market-based mechanisms as an effective policy tool in those contexts where markets are reasonably well developed. In some cases independent certification can function as a surrogate for state enforced rule of law. Pressuring and rewarding retailers to use certified timber can create economic incentives for forestry operations to improve their conduct, regardless of the state's will or capacity to enforce regulations. Taxing activities that are illegal or unsustainable, such as harvesting of rare or uncommon species, is also an important economic policy instrument–although such fees run the risk of encouraging smuggling. Changing local incentives to commit forest crimes may also be effective when illegal logging activities are diffuse and rely on many local actors. Dudley (2002) provides a systems model of local illegal logging practices in Indonesia that may be used to effectively analyze local incentives and examines how polices might alter them. Akella (2002) evaluates illegal logging on private lands in Brazil and demonstrates that simple models of enforcement economics may be used to analyze how existing regulations actually effect the decision-making of perpetrators of forest crimes.

Any policy measure requires information, and information is often a critical missing ingredient with respect to efforts to curb illegal logging. Successful enforcement of existing forestry laws requires successful monitoring, and most market mechanisms contain an information component as well as some form of certification. Monitoring efforts should track what goes on in the forest, the chain-of-custody of forest products, and government diligence–either by independent branches of government or independent actors. Many NGOs have already taken this approach, conducting "action-oriented" research intended to assist governments in preventing illegal logging, to police government activities, and sometimes to do both at the same time. Greenpeace International, for example, has effectively become the main illegal logging monitoring body in the Amazon for Brazil's environment agency, IBAMA (Adario, 2002). In India, Kalpavriksh conducted investigations leading to a judicial injunction of logging in the Adaman and Nicobar Islands (Sekhsaria, this issue). Partnerships between NGOs and government agencies are most important when NGOs have a stronger access to areas than the state. They appear to be most successful when roles can be clearly delineated and specified, when NGOs have sufficient trust domestically to be effective partners, and when NGOs have access to media outlets that will ensure the government's cooperation when forest crimes are politically sensitive.

In general, effective regulations and policies must have feedback mechanisms. As Dudley's systems model of village-level illegal logging demonstrates, changes to one aspect of large interconnected systems can have many impacts–some predictable, others less so. Regulatory reformers should be careful to allow for future revisions so that regulations can adapt to any unintended consequences. Reformers also need not fret over plagiarism. It is a way of life for legal drafters, Ken Rosenbaum joked: "We look for how other countries solve this problem and we steal that language."

Finally, new approaches to problem analysis may help improve existing policy considerations. Cross-country comparison, categorization and analysis of failures and success may yield significant similarities between countries. Further intra-country comparisons may be useful in identifying successes that are replicable under similar regulatory frameworks.

Authors and presenters who spend a lot of time in the field repeatedly underscored the importance of approaching illegal logging as a system and not just as isolated crews of poor villagers with chainsaws. Effective interventions must identify the key players and their supporting networks of racketeers and timber launderers and give prosecutors the wherewithal to overcome their ability to manipulate political connections to evade conviction.

CONCLUSIONS

As the papers in this volume attest, illegal logging is complex. It involves broad networks of power and influence that operate across many geographic, socio-political and economic contexts. The rural poor are often implicated as a major perpetrator of illegal forest activities, but a closer look reveals that they are typically pawns in a much larger game. Tracking the beneficiaries of the corrupt status quo is more likely to lead to the real culprits than pointing a finger at the man with the chainsaw.

For better or worse, illegal logging is "over-determined": it relies on many other illegal activities once the wood leaves the forest. The downside of this condition is that because these activities typically involve financial transactions, a great many people benefit from illegal logging and may sabotage measures intended to curb it. Conversely, careful analysis of this system may yield strategic points of intervention that may be simpler and cheaper to implement and may produce results.

As a number of authors indicated, illegal logging is not a new problem; rather, it is merely the latest incarnation of age-old conflicts over forest resources. As the global economy spreads its influence to the marginalized forests of the world, so too does global information technology. The power dynamics playing out in remote forests are now visible to a global community of stakeholders. Because illegal logging is such an embarrassing and direct failure of governance, as long as governments fail to control it they will have to yield a seat at the decision-making table to environmental interests, many of which are foreign. Yielding to smear campaigns targeted at the end retailers of products from endangered forests, private sector interests are also pressuring governments to clean up their act by withdrawing from countries where they cannot find suppliers who can guarantee the legality of their production methods. It is for these reasons that the illegal logging debate represents the latest chapter in the historic struggle for control of forest resources.

REFERENCES

Ascher, W. 1999. Why governments waste natural resources: Policy failures in developing countries. Johns Hopkins University Press. Baltimore and London.

Asthon, M.S. and C.M. Peters. 1999. Even-aged silviculture in tropical rainforests of Asia. Journal of Forestry 97: 14-19.

Blundell, A.G., R.E. Gullison. 2003. Poor regulatory capacity limits the ability of science to influence the management of mahogany. Forest Policy and Economics 5: 395-405.

Blundell, A.G., B.D. Rodan. 2003. Mahogany and CITES: Moving beyond the veneer of legality. Oryx 70(1).

Bush, G.W. 2002. President announces clear skies and global climate change initiatives. The White House, Office of the Press Secretary. Washington, DC. Available at: http://www.whitehouse.gov/news/releases/2002/02/20020214-5.html.

Callister, D. J. 1999. Corrupt and Illegal Activities in the Forest Sector: Current Understandings and Implications for the World Bank Forest Policy. Draft for discussion. Forest Policy Implementation Review and Strategy Development: Analytical Studies. The World Bank Group, Washington, DC, USA.

Chape, S., S. Blyth, L. Fish, P. Fox and M. Spalding (compilers). 2003. 2003 United Nations List of Protected Areas. IUCN, Gland, Switzerland and Cambridge, UK and UNEP-WCMC, Campridge, UK. Ix + 44pp.

Convention on International Trade in Endangered Species (CITES). 2003. Appendices I, II, and III, valid from 16 October 2003. CITES Secretariat, United Nations Environment Programme. Geneva, Switzerland. Available at: http://www.cites.org/eng/append/appendices.shtml.

Dove, M.R. and D.K. Kammen. 1997. The epistemology of sustainable resource use: Managing forest products, swiddens, and high-yielding variety crops. Human Organization 56(1): 91-101.

Dudley, R.G. 2002, March. A system dynamics examination of the willingness of villagers to engage in illegal logging. Paper presented at the conference titled: Illegal Logging in Tropical Forests: Ecology, Economics and Policy of Resource Misuse. Yale University School of Forestry and Environmental Studies, New Haven, CT USA.

EIA/Telapak Indonesia. 1999. The final cut: Illegal logging in Indonesia's orangutan parks. London: Environmental Investigation Agency and Telapak Indonesia.

Fredericksen, T.S. and F.E. Putz. 2003. Silvicultural intensification for tropical forest conservation. Biodiversity Conservation 12: 1445-1453.

Hiller, M.A., B.C. Jarvis, H. Lisa, L.J. Paulson, E.H.B. Pollard, S.A. Stanley. Recent trends in illegal logging and a brief discussion of their causes: A case study from Gunung Palung National Park, Indonesia. Paper presented at the conference titled: Illegal Logging in Tropical Forests: Ecology, Economics and Policy of Resource Misuse. Yale University School of Forestry and Environmental Studies, New Haven, CT, USA.

Human Rights Watch. 2003. Without remedy: Human rights abuse and Indonesia's pulp and paper industry. Human Rights Watch-Indonesia. Washington, DC. Available at: http://hrw.org/reports/2003/indon0103/.

G8. 1998. Action Programme on Forests. Released at the Foreign Ministers Meeting, London, England, May 9, 1998. Available at: http://www.library.utoronto.ca/g7/foreign/forests.html.

Global Forest Watch. 2003. Project description and resources. World Resources International, Washington, DC, USA. Available at: http://forests.wri.org/project_text.cfm?ProjectID=58.

Maathai, W. 2002. Closing remarks at the conference titled: Illegal Logging in Tropical Forests: Ecology, Economics and Policy of Resource Misuse. Yale University School of Forestry and Environmental Studies, New Haven, CT, USA.

McCarthy, J.F. 2000. 'Wild logging': The rise and fall of logging networks and biodiversity conservation projects on Sumatra's rainforest frontier. CIFOR Occasional Paper No. 31. Center for International Forestry Research. Bogor, Indonesia.

Peluso, N.L. 1992. Rich forests, poor people: Resource control and resistance in Java. Los Angeles: University of California Press.

Rijksen, H.D. and Meijaard, E. 1999. Our vanishing relative. the status of wild orangutans at the close of the twentieth century. Kluwer Academic Publishers, Dordrecht, Netherlands.

Rosenbaum, K. 2002, March. Illegal actions in the forest sector: A legal perspective. Paper presented at the conference titled: Illegal Logging in Tropical Forests: Ecology, Economics and Policy of Resource Misuse. Yale University School of Forestry and Environmental Studies, New Haven, CT, USA.

Tacconi, L., K. Obidzinski, J. Smith, Subarudi, I. Suramenggala. 2002, March. Can 'legalization' of illegal forest activities reduce illegal logging? Lessons from East Kalimantan. Paper presented at the conference titled: Illegal Logging in Tropical Forests: Ecology, Economics and Policy of Resource Misuse. Yale University School of Forestry and Environmental Studies, New Haven, CT, USA.

United States CITES Management Authority. 2003. Report on *Swietenia macrophylla* as a major importer (United States of America). Report for the second meeting of the Mahogany Working Group of the Convention on International Trade in Endangered Species of Wild Fauna and Flora, Belem, Brazil, 6-8 October 2003 (MWG2 doc 10.2). Available at: http://www.cites.org/common/prog/mwg/MWG2/E-MWG2-10-02-US.pdf.

United Nations Forum on Forests. 2002. Report on the Second Session (22 June 2001 and 4 to 15 March 2002). United Nations Economic and Social Council. Official Records, 2002. Supplement No. 22. United Nations, New York. Available at: http://www.un.org/esa/forests/documents-unff.html#2.

ILLEGAL LOGGING IN THE TROPICS:
ECOLOGY, ECONOMICS AND POLICY OR RESOURCE MISUSE

Speaker List

Paulo Adario–Greenpeace Brazil
Brazil and IBAMA: A Case Study*

Anita Akella–Conservation International
Enforcement Economics and the Fight Against Forest Crime: Lessons Learned
from the Atlantic Forest of Brazil*

Antonio Azuela–Universidad Nacional Autónoma de México
Illegal Logging and Local Democracy: Between Communitarianism and Legal
Fetishism

Lisa Curran–Yale School of Forestry and Environmental Studies
Illegal Logging: An Introduction to the Issue*

Richard G. Dudley–Independent Consultant.
A System Dynamics Examination of the Willingness of Villagers to Engage in
Illegal Logging

Timothée Fomete–Dschang University, Cameroon
Social and Environmental Costs of Illegal Logging in a Forest Management
Unit in Eastern Cameroon

Aarti Gupta–Forest Integrity Network of Transparency International
Combating Forest Corruption: The Forest Integrity Network

Nalin Kishor–The World Bank Group
Does Improved Governance Contribute to Sustainable Forest Management?

Wangari Mathaai–Yale School of Forestry and Environmental Studies
Closing Remarks*

Pamela McElwee–Yale School of Forestry and Environmental Studies
You Say Illegal, I Say Legal: The Relationship Between 'Illegal' Logging and
Land Tenure, Poverty, and Forest Use Rights in Vietnam

Edward H. B. Pollard–Independent Consultant
Recent Trends in Illegal Logging and a Brief Discussion of Their Causes: A
Case Study from Gunung Palung National Park, Indonesia

Kenneth L. Rosenbaum–Sylvan Environmental Consultants
Illegal Actions and the Forest Sector: A Legal Perspective

Pankaj Sekhsaria–Kalpavriksh (environmental NGO in India)
Illegal Logging and Deforestation in Andaman and Nicobar Islands, India:
The Story of Little Andaman Island

Luca Tacconi–Center for International Forestry Research
Can 'Legalization' of Illegal Forest Activities Reduce Illegal Logging? Lessons
from East Kalimantan

*Papers presented but not submitted for publication

Video Presentation

Syahirsyah Syamsuni Arman–World Wildlife Fund Indonesia
Threatened Future: 2001 Documentary of Illegal Logging in West Kalimantan's
Tropical Forests

Panel Discussion Moderators

Benjamin Cashore–Yale School of Forestry and Environmental Studies

Lisa Curran–Yale School of Forestry and Environmental Studies

Wangari Mathaai–Yale School of Forestry and Environmental Studies

Andrew Shrank–Yale University Department of Sociology

Papers Published But Not Presented at the Conference

Ramsay M. Ravenel–Yale School of Forestry and Environmental Studies
Community-Based Logging and De Facto Decentralization: Illegal Logging
in the Gunung Palung Area of West Kalimantan, Indonesia

Ramsay M. Ravenel and Ilmi E.M. Granoff
A Synthesis of the March 2002 Conference on Illegal Logging in the Tropics:
Ecology, Economy, and Politics of Resource Misuse

Ute Siebert and Georg Elwert–Free University of Berlin
Potentials and Recommendations for Combating Corruption and Illegal Logging
in the Forest Sector of Bénin, West Africa

Wynet Smith–World Resources Institute
Undercutting Sustainability: The Global Problem of Illegal Logging and Trade

Wynet Smith–World Resources Institute
The Role of Monitoring in Cutting Crime

THE YALE INTERNATIONAL SOCIETY OF TROPICAL FORESTERS CONFERENCE SERIES

People in Parks: Beyond the Debate. April 2-3, 2004.

Organized by Ines Angulo, Robin Barr, Ellen Brown, Janette Bulkan, Jeremy Goetz, Ilmi Granoff, Iona Hawken, Alder Keleman, Christian Palmer, and Maria Vargas.

Publication to be announced.

Ecosystem Services in the Tropics: Challenges to Marketing Forest Function. April 5-6, 2003.

Organized by Ilmi Granoff, Iona Hawken, Margaret Francis, Rebecca Ashley, Elizabeth Shapiro, Ben Hodgden, Richard Chavez, Fulton Rockwell, Abdalla Shah, and Daniela Cusack.

Not Published.

Illegal Logging in the Tropics: Ecology, Economics and Policy of Resource Misuse. March 29-30, 2002.

Organized by Ramsay Ravenel, Ilmi Granoff, Citlali Cortes-Montano, Abdalla Shah, Barbara Bamberger, Daniela Cusack, and Carrie Magee.

Published as a special issue of the Journal of Sustainable Forestry, Volume 19, Numbers 1/2/3, 2004. Co-published as Ravenel, Ramsay M., Ilmi E.M. Granoff and Carrie A. Magee (eds.). Illegal Logging in the Tropics: Strategies for Cutting Crime. Haworth Press. 2004. http://www.haworthpress.com/web/ JSF.

Transboundary Protected Areas: The Viability of Regional Conservation Strategies. March 30-31, 2001.

Organized by Marc Stern, Uromi Goodale, Ashley Lanfer, Cheryl Margoluis, Matthew Fladeland, and Carrie Magee.

Published as a special issue of the Journal of Sustainable Forestry, Volume 17, Number 1/2, 2003. Co-published as Goodale, Uromi M., Marc J. Stern, Cheryl Margoluis, Ashley G. Lanfer, and Matthew Fladeland (eds.). Transboundary Protected Areas: The Viability of Regional Conservation Strategies. Haworth Press. 2003. http://www.haworthpress.com/store/product.asp?sku=J091

War in Tropical Forests: New Perspectives on Conservation in Areas of Armed Conflict. March 31-April 1, 2000.

Organized by Steven Price, Uromi Goodale, Silvia Benitez, Julie Velasquez Runk, Christie Young, Caroline Kuebler, Matt Fladeland, Omari Ilambu, Keely Maxwell, Harry Bader, Ashley Lanfer, Aya Hirata, Erika Mark, Upik Djalins, Edgardo Gonzalez, Brenda Torres, and others.

Published as a special Issue of the Journal of Sustainable Forestry, Volume 16, Number 3/4, 2003. Co-published as Price, Steven V. (ed.) War in Tropical Forests: Conservation in Areas of Armed Conflict. Haworth Press. 2003. http://www.haworthpress.com/store/product.asp?sku=J091

The Private Sector and Tropical Forest Stewardship. Spring 1999.

Organizers: Robert Kenny, Steven Price, Jon Wagar, Richard Karty, Silvia Benitez, Emile Jurgens, and others.

Not published.

The Ecotourism Equation: Measuring the Impacts. April 12-14, 1996.

Organized by Elizabeth Malek-Zadeh, Joseph Miller, Robin Sears, Alexandra Grinshpun, Kelly Keefe, Christina Cromley, and others.

Published as Malek-Zadeh, Elizabeth and Joseph Miller (eds.). The Ecotourism Equation: Measuring the Impacts. Yale School of Forestry & Environmental Studies Bulletin Number 99. Yale School of Forestry & Environmental Studies, New Haven, CT. 1996. Available at http://www.yale.edu/forestry/publications/

Local Heritage in the Changing Tropics. February 1995.

Organized by Karen Beard, Rachel Byard, Ronald Cherry, Greg Dicum, Gary Dunning, Andreas Eicher, Cesar Flores, Derek Halberg, Chris Hopkins, Andrea Kivi, Lydia Olander, Richard Payne, Cathryn Poff, Walter Salazar, Robin Sears, Austin Troy, Sudha Vasan.

Published as Dicum, Gregory (ed.). Local Heritage in the Changing Tropics: Innovative Strategies for Natural Resource Management and Control. Yale School of Forestry & Environmental Studies Bulletin Number 98. Yale School of Forestry & Environmental Studies, New Haven, CT. 1995. Available at http://www.yale.edu/forestry/publications/

*Timber Certification: Implications for Tropical Forest Management.
February 5-6, 1994.*

Organized by Jennifer O'Hara, Mirei Endara, Ted Wong, Chris Hopkins, Cynthia Caron, and Paul Maykish.

Published as Timber Certification: Implications for Tropical Forest Management: Proceedings from a conference hosted by the Student Chapter of the International Society of Tropical Foresters, Yale School of Forestry and Environmental Studies, New Haven, Connecticut, 5-6 February 1994. O'Hara, Jennifer, Mirei Endara, Theodore Wong, Chris Hopkins, and Paul Maykish (eds.). Yale Reprographic and Imaging Services, New Haven, CT.

Himalayan Conservation and Development. October 17-18, 1992.

Organized by Sally J. Loomis and Maria J. Enzer.

Published as Conservation for Development: A Himalayan Perspective. Loomis, Sally J. and Maria J. Enzer (eds.). Yale School of Forestry & Environmental Studies. New Haven, CT.

Index

"A System Dynamics Examination of the Willingness of Villagers to Engage in Illegal Logging," 3

AAC. *See* Annual Allocable Cut (AAC)

AAJUS. *See* Andaman Adim Janjati Vikas Samiti (AAJVS)

Access, as factor in increase in illegal logging in Gunung Palung National Park, Indonesia, 203-205

Accountability, voice and, defined, 62

Adario, P., 367

Adat foret, 219

ADB. *See* Asian Development Bank (ADB)

Africa, illegal logging and trade in, 15-18,16t

Akella, A.S., 273,363,367

Alerce, illegal logging of, 20

Amazon, illegal logging and trade in, 19

America(s), illegal logging and trade in, 18-21,19t

Andaman Adim Janjati Vikas Samiti (AAJVS), 331

Andaman and Nicobar Forest and Plantation and Development Corporation (ANFPDC), 329,330,332

Andaman and Nicobar Forest Department, 329

Andaman and Nicobar Islands annual extraction of timber in, 323,324t

biogeography of, 321

colonial history of forestry operations in, 322

development of modern timber-based industry in, 324-325

environmental history of, 321-326

illegal logging and deforestation in, 319-335. *See also* Little Andaman Island, illegal logging and deforestation in

indigenous tribal communities in, 321-322,323t

post-independence history of, 323-324,323t,324t

tribal communities in, current threats to, 325-326

Andaman Tribal Welfare Board, 331

ANFPDC. *See* Andaman and Nicobar Forest and Plantation and Development Corporation (ANFPDC)

Angelsen, A., 66

Annammite Mountains, 104

Annual Allocable Cut (AAC), 303

Annual area tax system, in Cameroon, 168,169b

Annual cutting right certificate, 166

Arcview 3.1 Geographic Information System, 189

Arman, S.S., 368

Army, Zapatista, 93

Articulation

in community-based logging in Gunung Palung Area of West Kalimantan, Indonesia, 230-231

in illegal logging in Gunung Palung
Area of West Kalimantan,
Indonesia, 230-231
Asian Development Bank (ADB), 99
Asian Free Trade Zone, 106
Asia-Pacific region, illegal logging and
trade in, 10-15,11t
Auzel, P., 4,153,160
Azuela, A., 3,81,367

Bali Ministerial Forest Law
Enforcement conference, 12
Bangkok Post, 104
Banjir kap, 141,142,144
Barber, C.V., 224
Basic Agrarian Law of 1960, 218
Basic Forestry Law of 1967,
218.219,221
Bassila Region, 243
Belle, A., 3,55
Bénin, West Africa
illegal logging and trade in, 16t,18
illegal logging in, combating of,
239-261. *See also* Illegal
logging, in Bénin, West
Africa, combating of
Bierschecnk, T., 255
Binh Thuan Department of Agriculture
and Forestry, 122-123
Biodiversity Conservation Network, 215
Biodiversity Conservation Strategy, 102
Blair, A., 322
BNHS. *See* Bombay Natural History
Society (BNHS)
Bolivia, illegal logging and trade in,
19t,20
Bombay Natural History Society
(BNHS), 332
BPI. *See* Bribe Payers Index (BPI)
Brazil, illegal logging and trade in,
18-20,19t
Brazilian Amazon, 3
Bribe Payers Index (BPI), 341,343

British Columbia, illegal logging and
trade in, 19t,21
British Columbia Forest Watch
program, 311
Bureau for Regional Oriental
Campaigns, 272
Bush, G.W., Pres., 354
Business Environment Risk
Intelligence, 63
Business Principles to Combat Bribery,
345

Calcutta High Court, 330
Port Blair Circuit Bench, 332
Cam Xuyen SFE, 125
Cambodia
illegal imports from, into Vietnam,
105
illegal logging and trade in, 10,11t,
13,14
Cameroon
annual area tax system in, 168,169b
illegal logging and trade in, 16-17,
16t
Canada, illegal logging and trade in,
18,19t,21
Canadian International Development
Agency (CIDA), 99
Cannon, C.H., 209,224
"Can't 'Legalization' of Illegal Forest
Activities Reduce Illegal
Logging? Lessons from East
Kalimantan," 4
Capital markets, penetration of
in community-based logging in
Gunung Palung Area of West
Kalimantan, Indonesia,
230-231
in illegal logging in Gunung Palung
Area of West Kalimantan,
Indonesia, 230-231
Cashore, B., 368
CBFM. *See* Community-based forest
management (CBFM)

CCU. *See* Central Control Unit (CCU)
Center for International Development, of Harvard University, 241,340
Center for International Forestry Research (CIFOR), 12
Central African Republic, illegal logging and trade in, 16t,17
Central America, illegal logging and trade in, 19t,20-21
Central Control Unit (CCU), 158
CGI meetings. *See* Consultative Group on Indonesia (CGI) meetings
Charcoal, role on national and international market, 246
Chatham sawmill, 324
Chelonia mydas, green sea turtle, on Little Andaman Island, 331
Chile, illegal logging and trade in, 19t,20
China, illegal logging and trade in, 10-11,11t,12,13
CIDA. *See* Canadian International Development Agency (CIDA)
CIFOR (Center for International Forestry Research), 12
CITES. *See* Convention on Trade in Endangered Species (CITES)
Citizen enforcement, allowing, in combating illegal activity, 283-284
Civil society organizations (CSOs), 340
"Combating Corruption and Illegal Logging in Bénin, West Africa: Recommendations for Forest Sector Reform," 4
Command State system, 255-256
Commodity Trade Statistics (Comtrade), 298
Common pool resources, 91-94
Communitarianism, legal fetishism and, 81-96
Community(ies)

local, in synthesis of issues in illegal logging in the Tropics, 359
tribal
   in Andaman and Nicobar Islands, current threats to, 325-326
   indigenous, of Andaman and Nicobar Islands, 321-322, 323t
Community logging operations, financing of, 227-228
Community participation
   in illegal logging in Gunung Palung National Park, Indonesia, 197-199,198f,198t
   in study of recent trends in illegal logging in Gunung Palung National Park, Indonesia, 189-190
Community-based forest management (CBFM), 182
"Community-Based Forest Management of Buffer Zone Forests at Gunung Palung. National Park, West Kalimantan," 215
Community-based logging
   de facto decentralization and, 213-237. *See also* Illegal logging, in Gunung Palung Area of West Kalimantan, Indonesia
   in Gunung Palung Area of West Kalimantan, Indonesia
   articulation in, 230-231
   implications of, 228-231
   penetration of capital markets in, 230-231
   political economy of, 223-228
   reentry in natural production forests, 228-229
   transforming traditional forest management systems, 229-230

"Community-Based Logging and De
    Facto Decentralization:
    Illegal Logging in the
    Gunung Palung Area of West
    Kalimantan, Indonesia," 4
Comtrade, 298
Concession logging systems, local
    appropriation of, 225-227
Congo Basin countries, illegal logging
    and trade in, 15,16t
Conservator of Forests, 279
Construction wood, role on national
    and international market, 246
Consultative Group on Indonesia
    (CGI) meetings, 138
Contreras-Hermosilla, A., 254,271
Control of corruption, defined, 62
Convention on Trade in Endangered
    Species (CITES), 356
Coordination, lack of, failures of legal
    system related to, 274-275
Corrupt practices, illegal activities and,
    253-254
Corruption
    in Bénin, West Africa, combating
        of, 239-261. *See also* Illegal
        logging, in Bénin, West
        Africa, combating of
    concepts and perceptions of, 250-252
    control of, defined, 62
    defined, 61
    facilitation of, change of structural
        conditions and, 258-259
    in forest sector, 250-254
    forestry, on international agenda,
        241
'Corruption in South Asia,' 343
Corruption Perception Index (CPI),
    341,343
Costa Rica, illegal logging and trade
    in, 19t,20
CPI. *See* Corruption Perception Index
    (CPI)
Crime(s)
    forest, nature and impact of, 56-58

reduction in, monitoring in, 293-317.
    See also Monitoring, in
    cutting crime
CSOs. *See* Civil society organizations
    (CSOs)
Curran, L., 367,368

Data, trade, for monitoring in cutting
    crimes, 298-302,299t. *See
    also* Trade data, for
    monitoring in cutting crimes
Dawkins, H.C., 226
Dayak, 216,218
DC. *See* Deputy Commissioner (DC)
De facto decentralization,
    community-based logging
    and, 213-237. *See also* Illegal
    logging, in Gunung Palung
    Area of West Kalimantan,
    Indonesia
de Souza Santos, B., 90
Deacon, R., 66
Decentralization
    de facto, community-based logging
        and, 213-237. *See also* Illegal
        logging, in Gunung Palung
        Area of West Kalimantan,
        Indonesia
    in Indonesia, illegal logging and,
        231-233
Decision 286/TTg, 117
Deforestation
    on Andaman and Nicobar Islands,
        India, 319-335. *See also*
        Little Andaman Island, illegal
        logging and deforestation on
    models of, 66,68
    in Vietnam, 98-99
Democracy, local, illegal logging and,
    81-96. *See also* Illegal
    logging, local democracy and
Democratic Republic of Congo (DRC),
    15
Democratic Republic of Vietnam
    (DRV), 101-102

Department of Forestry, 141,147
Department of Industry, 206
Department of Industry and Trade
    offices, 197
Deputy Commissioner (DC), 279
Detection, failures of legal system
    related to, 273
Development, security and, dual
    functions of, 221
DFO. *See* Divisional Forest Officer
    (DFO)
DFRN. *See Direction des Forêts et*
    Ressources Naturelles
    (DFRN)
Didia, D.O., 66
Direction des Forêts et Ressources
    Naturelles (DFRN), 247
Dispute, resolution, poor, 268-269
District Forestry Office, 143
Divisional Forest Officer (DFO), 279
Djeukam, R., 153
"Does Improved Governance
    Contribute to Sustainable
    Forest Management?", 3
Dove, M.R., 227,230
DRC. *See* Democratic Republic of
    Congo (DRC)
DRV. *See* Democratic Republic of
    Vietnam (DRV)
Dual Functions of security and
    development, 221
Dudley, R.G., 3,31,34f,363,367

East Kalimantan
    illegal logging in, 12,137-151
    illegal trade in, 12
Eastern Cameroon, illegal logging in
    forest management unit in,
    social and environmental
    costs of, 153-180. *See also*
    Illegal logging, in forest
    management unit in Eastern
    Cameroon, social and
    environmental costs of

Easup Forestry-Agriculture-Industry
    Union, 113
Economic(s), as factor in increase in
    illegal logging in Gunung
    Palung National Park,
    Indonesia, 206-207
Economic processes, 83-85
Economist Intelligence Unit, 63
Economy, political, of
    community-based logging in
    Gunung Palung Area of West
    Kalimantan, Indonesia,
    223-228
Ecosystem Services in the Tropics:
    Challenges to Marketing
    Forest Function, 369
EIA. *See* Environmental Investigation
    Agency (EIA)
Eigen, P., 340
El Nino, 353
Elwert, G., 4, 239,368
Emergency Planning and Community
    Right to Know Act of 1986,
    294
Endangered Species Act, 266,284
Enforcement
    citizen, allowing, in combating
        illegal activity, 283-284
    external, 257-258
    lack of, of general laws applying to
        forests and forest products,
        failures of legal system
        related to, 275
    poor, 271
Environmental Intelligence Agency,
    310
Environmental Investigation Agency
    (EIA), 272,311,312,339,354
Estimated regression equations, 78-79
Estonia, illegal logging and trade in,
    21-22,22t
European Bank for Reconstruction and
    Development, 63
Europe-Russia, illegal logging and
    trade in, 21-22,22t

Eurostat, 302
External enforcement, 257-258

Fairhead, J., 243
FAO. *See* Food and Agricultural
    Organization (FAO)
FAOSTAT, 9
Far East, illegal logging and trade in,
    21,22t
FBO value. *See* Free On Board (FOB)
    value
FERN, 23
Feteke, F., 153
Fetishism, legal, communitarianism
    and, 81-96
Field investigations, for monitoring in
    cutting crimes, 309-312
FIN. *See* Forest Integrity Network
    (FIN)
FIN_PROPOR project, 342-346
FIPI. *See* Forest Inventory and
    Planning Institute (FIPI)
5 Million Hectare Program, 115
5MHRP. *See* National Five Million
    Hectare Reforestation
    Program (5MHRP)
FMU(s). *See* Forest management units
    (FMUs)
FMU 10 030, 172,175-176,177,178
Fomete, T., 153,367
Food and Agricultural Organization
    (FAO), 246,298,302
Forest(s)
    adat, 219
    Indonesia, decline of, 32
    logged-over, spatial heterogeneity
        in, 224-225
    natural production, reentry in,
        228-229
Forest and Environment sector
    program, 175
Forest corruption, combating of
    FIN in, 337-349. *See also* Forest
        Integrity Network (FIN)

need for building coalitions in,
    339-340
recommendations for, 254-258
Forest Corruption Fighters' Toolkit,
    342,346
Forest crimes, nature and impact of,
    56-58
Forest Crimes Monitoring Project, 24
Forest Department, 330
Forest Department accounting officer,
    276
Forest destruction, on Little Andaman
    Island, 329
Forest Integrity Network (FIN), 241,
    337-349. *See also* Forest
    corruption, combating of
activities of, 341-347
described, 337-338
history of, 340-341
introduction to, 338
scale of problem, 338-339
TI and, synergies between, 341
Forest Inventory and Planning Institute
    (FIPI), 99
Forest land
    defined, 107-108
    in Vietnam, laws toward,
        107-112,111f
Forest land allocation, in Vietnam,
    goal of, 110
"Forest Law Enforcement and
    Governance," 241
Forest Management Boards, 110
Forest management units (FMUs), 155
    in Eastern Cameroon, illegal
        logging in, social and
        environmental costs of,
        153-180. *See also* Illegal
        logging, in forest
        management unit in Eastern
        Cameroon, social and
        environmental costs of
Forest politics, national, 246-249
Forest product(s), future demands for,
    in Vietnam, 107

Forest Product Harvest Concession, 220
Forest protection, in Vietnam, problems in, 116-118
Forest Protection Department, 107
  in Vietnam, 115-123,119f
Forest Protection Management Boards (FPMBs), 115
Forest Protection Officer, 118
Forest resource(s), 242-244
  in Eastern Cameroon, allocation of, 156-157
Forest resource development, in Indonesia, 183-185
Forest Resources Protection and Development Act of 1991, 107-108
Forest Science Institute, 98
Forest sector
  corruption in, 250-254
  illegal actions in
    combating of, strategies for, 276-284
      allow citizen enforcement, 283-284
      increase transparency, 281-283
      make laws simple and direct, 277-280
      respect local rights and interests, 276-277
      setting priorities for action, 284-286
    failures of legal system related to, 265-276
      implementation-related, 268-276
      lack of coordination, 274-275
      lack of enforcement of general laws applying to forests and forest products, 275
      lack of government oversight, 275-276

law-related, 265-268
poor dispute resolution, 268-269
poor enforcement–related, 271
prevention-related, 272
unfair application of law, 269-270
weak administration, 274
legal perspective on, 263-291
  detection in, 273
  introduction to, 264-265
  suppression in, 273
strategies for, in combating illegal activity, set meaningful penalties, 280-281
illegal activities in, 250-254
Forest sector reform, in Bénin, West Africa, recommendations for, 239-261. *See also* Illegal logging, in Bénin, West Africa, combating of
Forest use rights, illegal logging in Vietnam and, 97-135. *See also* Illegal logging, in Vietnam, land tenure, poverty, and forest use rights and
Forest users, 244-245
Forestry, in Vietnam, laws toward, 107-112,111f
Forestry Code, 22
Forestry corruption, on international agenda, 241
Forestry Science and Technical Association, in Vietnam, 98-99
Forestry sector commodity chain, corrupt and illegal activities in, 295,297f
Forests Monitor, 269
Foundation for the Phillipine Environment, 311

FPMBs. *See* Forest Protection
    Management Boards
    (FPMBs)
Free On Board (FOB) value, 168,170
Freedom House, 63
Fuelwood/charcoal, role on national
    and international market, 246

G8, 352
Gabon, illegal logging and trade in,
    16-17,16t
Gaharu, 223
General Wood & Veneers, 245,246
Georgia, illegal logging and trade in,
    22,22t
German Social Forestry Development
    Project, 218
German Technical Cooperation (GTZ),
    242
Ghana, illegal logging and trade in,
    16t,17
Ghanian Forestry Department, 17
Global Forest Watch, 24
Global Forest Watch Cameroon,
    306,308
Global Forest Watch Canada network,
    311
Global positioning system (GPS), 189
Global Witness,
    14,17,24,105,159,269,270,
    274,276,310,311,360
GOI. *See* Government of Indonesia
    (GOI)
Governance
    defined, 61
    improved, in sustainable forest
        management, 55-79. *See also*
        Sustainable forest
        management, improved
        governance in
    measuring quality in, 61-63
    quality of
        development dividend of
            improving, 65-66,67f

empirical measurement of,
    63-65,64f
role of, first cut at, 68
unbundling of, 61-63
Government, in synthesis of issues in
    illegal logging in the Tropics,
    357-358
Government effectiveness, defined, 62
Government of India, 323,325
    Ministry of Rehabilitation, 326
Government of Indonesia (GOI), 57,184
Government oversight, lack of, failures
    of legal system related to,
    275-276
Government Regulation 34,142
Government Regulation No. 62/1998
    Concerning Delegation of
    Part of Government's Affairs
    in the Sector of Forestry to
    the Region, 219
GPS. *See* Global positioning system
    (GPS)
Granoff, I.M.E., 1,5,351,368
Great Andamanese, 322,325,326
"Green Forest Town," 113
Green sea turtle *Chelonia mydas,* on
    Little Andaman Island, 331
Greenpeace, 258,275,310
Greenpeace Brazil, 19-20
Greenpeace International, 360,363
Gresham's Law, 57,73
GTZ. *See* German Technical
    Cooperation (GTZ)
Gulung Palung Community-Based
    Forest Management project,
    232
Gunung Palung area of West
    Kalimantan, Indonesia,
    illegal logging in, 213-237.
    See also Illegal logging, in
    Gunung Palung area of West
    Kalimantan, Indonesia
Gunung Palung Community-Based
    Forest Management project,
    215,216,218,219-220

Gunung Palung National Park,
     Indonesia, 4
  illegal logging in, recent trends in,
     181-212. *See also* Illegal
     logging, in Gunung Palung
     National Park, Indonesia,
     recent trends in
  map of, 186f
Gupta, A., 5,337,367
Guyana, illegal logging and trade in,
     19t,20

Habibie, B.J., 219
Hak Pemanfaatan Hutan (HPH), 221
Hak Pemungutan Hasil Hutan
     (HPHH), 220
Harvard University, 186,241,340
  Center for International
     Development of, 241,340
  Laboratory of Tropical Forest
     Ecology of, 215
Hiller, M.A., 3,4,181
*Himalayan Conservation and
     Development*, 371
Home Depot, 258
Honduras, illegal logging and trade in,
     19t,20-21
HPH. *See* Hak Pemanfaatan Hutan
     (HPH)
HPHH. *See* Hak Pemungutan Hasil
     Hutan (HPHH)

IBAMA, 20,310,363
IDA framework. *See* Institutional
     Development Analysis (IDA)
     framework
IKONOS, 307,309
Illegal actions, forest sector and, legal
     perspective on, 263-291. *See
     also* Forest sector, illegal
     actions and, legal perspective on
"Illegal Actions and the Forest Sector:
     A Legal Perspective," 5

Illegal forest activities
  corrupt practices and, 253-254
  in forest sector, 250-254
  legalization of, illegal logging
     reduction due to, 137-151
     introduction to, 138-140
     study of, methods in, 140-141
  management of, 143-146,144t-146t
Illegal logging
  in Bénin, West Africa, combating
     of, 239-261
     introduction to, 240
     study of, methodology in, 242
  changes in, 32-33
  decentralization in Indonesia and,
     231-233
  definitions of, 138,355-357
  described, 203
  economics of, timber markets and,
     146-149
  in forest management unit in
     Eastern Cameroon
  incidence of, 176-177
  sanctions imposed on, 177-178
  social and environmental costs
     of, 153-180
     background of, 155-159
     decreased timber resource
        availability, 157
     forest resources, allocation
        of, 156-157
     forestry policy and
        legislation, 158
     introduction to, 154-155
     legal prescription, 178
     level of financial resources
        allocated to forestry
        sector, 157
     monitoring and control of
        activity in forest
        sector, 158-159
     permanent and non-permanent
        forest, 155-156
     study of
        assessing impact in, 163

assessing of spatial extent
in, 160-161,164f
assessing value of damage
in, 162,165f,166f
detecting and spatial
extent in,
163-165,166t
detecting by mapping
logging roads in,
160,162f,163f
discussion of, 173-178
effects of illegal harvest
on local economy
in, 168,169b,170
estimating significance of
illegal harvest in,
165-167,167t,168t
estimating value of
illegally extracted
logs in, 168,169t
infractions in, 172
legal provisions on
sanctions of illegal
logging in, 172
legal sanctions in, 173
methods in, 159-163,
160f-166f
results of, 163-173,
166t-169t,169b,171f
SFM in, 170-171,171f
site of, 159-160,160f,161f
in Gunung Palung Area of West
Kalimantan, Indonesia,
213-237
articulation in, 230-231
described, 215-218,217f
implications of, 228-231
introduction to, 214-215,216f
penetration of capital markets in,
230-231
political economy of, 223-228
site of, 215-218,217f
in Gunung Palung National Park,
Indonesia

ecological impacts of,
199-202,199f,201f-203f,201t
historical background of,
183-184
increase in, factors contributing
to, 202-207
access, 203-205
economics, 206-207
overview of, 187-188
recent trends in, 181-212
case study, 200-202,
201f-203f,201t
study of
aim of, 182-183
area of, 184-187,185f,
186f
community participation
in, 189-190,197-199,
198f,198t
design of, 188-189
historical and current land
use, 186-187
introduction to, 182-188,
185f,186f
logging rates in, 189-190
markets in, 196-197
methods in, 188-189
profit levels and wealth in,
190-193
results of, 189-202,194f,
194t,196t,198f,198t,
199f,201f-203f,201t
socio-economic
organization of
logging, 193-196,
194f,194t,196t
on Little Andaman Island,
319-335,329-330. *See also*
Little Andaman Island, illegal
logging and deforestation in
local democracy and, 81-96
introduction to, 82
out of woods, definitions of,
356-357

reduction of, legalization of illegal
forest activities in, 137-151
in socio-political context, 360-361
spatial extent of, assessment of, in
Eastern Cameroon,
160-161,164f
trade and, global problem of, 7-30
in Africa, 15-18,16t
in Americas, 18-21,19t
in Asia-Pacific region, 10-15,11t
in Europe-Russia, 21-22,22t
introduction to, 8-9
study of
discussion of, 22-24
methods in, 9
results of, 10-22,11t,16t,19t,
21t
in the Tropics
conference and proceedings–
March 29-30, 2002, 3-5
case studies, 4
interventions, 4-5
problems description, 3
synthesis, 5
theoretical approaches, 3-4
described, 2-3
introduction to, 1-6
panel discussion moderators,
368
papers published but not
presented at conference, 368
speaker list, 367-368
synthesis of issues, 351-366
academic and policy context,
352-354
government-related, 357-358
international organizations–
related, 359-360
introduction to, 352-355
local communities–related,
359
NGO-related, 359-360
policy recommendations in,
361-364
private sector–related, 358

stakeholders in, 357-360
video presentation, 368
types of, tools for detection and
monitoring of, 295,296t
in Vietnam
corruption and collision in,
118-123,119f
land tenure, poverty, and forest
use rights and, 97-135
perceptions of, 98-100
social justice claims at local
level, 127-129,128f
willingness of villagers to engage
in, system dynamics
examination of, 31-53
decreasing community support
makes more forest available
for illegal operations,
37-39,40f
disappearing forest decreases
community support for good
forest management,
36-37,39f
disappearing forests cause
disappearing jobs,
39-41,40f,41f
discussion of, 49-51,52f
effect of outside workers,
45,45f,46f
effect of strength of community
rights, 45,47-48,47f,48f
enforcement in, 48-49,49f
full model, 41,42f
illegal workers create more
illegal workers, 36,38f
introduction to, 32-34
limitations of, 50-51
methods in, 34-35,34f
model outcomes, 41-49,43f-49f
model structure in, 35-41,37f-42f
need for income forces villagers
to work illegally, 36,37f
pattern of willingness to work
illegally, 42-44,43f,44f
in woods, definitions of, 355-356

"Illegal Logging and Deforestation in Andaman and Nicobar Islands, India," 5

"Illegal Logging and Local Democracy: Between Communitarianism and Legal Fetishism," 3-4

"Illegal Logging in Forest Management Unit, Eastern Cameroon," 4

"Illegal Logging in the Tropics: A Synthesis of the Issues," 5

Illegal Logging in the Tropics: Ecology, Economics and Policy of Resource Misuse, 369

"Illegal Logging in Tropical Forests: Ecology, Economics & Politics of Resource Misuse," 354

Illegal Logging Monitoring System, 22

Illegal logging sanctions, detecting and dealing with, 176

Illegality
  concepts and perceptions of, 250-252
  defined, 85-91

'Improving Governance to Ensure Sustainability of Forests and Livelihoods,' 342

India
  Andaman and Nicobar Islands of, illegal logging and deforestation in, 319-335. *See also* Little Andaman Island, illegal logging and deforestation in
  Supreme Court of, 320,332,333

Indigenous tribal communities, of Andaman and Nicobar Islands, 321-322,323t

Indonesia
  decentralization in, illegal logging and, 231-233
  forest history in

forest concessions in West Kalimantan, 221-222
legal framework of, 218-221
overview of, 218-223
forest(s) in, decline of, 32
forest resource development in, 183-185
Gunung Palung National Park, 4
illegal logging in, 10,11,11t,12. *See also* Illegal logging, in Gunung Palung National Park, Indonesia, recent trends in
emergence of, 184
willingness of villagers to engage in, system dynamics examination of, 31-53. *See also* Illegal logging, willingness of villagers to engage in, system dynamics examination of
illegal trade in, 10,11,11t,12
Integrated Forest Security Teams of, 272
timber supply and demand for (1997-1998), 302,303t
West Kalimantan, Gunung Palung area of, illegal logging in, 213-237. *See also* Illegal logging, in Gunung Palung Area of West Kalimantan, Indonesia

Indonesian Ministry of Forest, 186
Indonesian Parliament, 141
Indonesian Selective Cutting System, 228
Indonesian Selective Felling System (TPTI), 213
  silvicultural critique of, 222-223
Institutional Development Analysis (IDA) framework, 91,92
Institutionalism, new, 91-94
Integrated Forest Security Teams, of Indonesia, 272
'Integrity Pacts,' 344

'Inter-Departmental Team on Accelerated Development Programme for Andaman and Nicobar,' 326

Interest(s), local, respect for, in combating illegal activity, 276-277

Internal sanctioning, 256-257

International market, role of timber and wood on, 245-246

International Society of Tropical Foresters (ISTF), Yale Chapter of, 1,2,354

International Tropical Timber Agreement, 298,301

International Tropical Timber Organization (ITTO), 9,11t, 12,24,298,300,301,302,303

International Union for Nature Conservation (IUCN), 340

Inter-Secretariat Working Group on Forest Statistics, 302

IPPK (Izin Pemungutan dan Pemanfaatan Kayu), 142,143 logging activities of, 141

IPPK (Izin Pemungutan dan Pemanfaatan Kayu) concessions, management of, 143-146,144t-146t

ITTO. *See* International Tropical Timber Organization (ITTO)

IUCN. *See* International Union for Nature Conservation (IUCN)

Izin Pemungutan dan Pemanfaatan Kayu (IPPK), 142,143 logging activities of, 141

Jalan gerobak, 195,196

Jalan kuda-kuda, 193,195,196,226

Jalan sepeda, 195

Japan, illegal logging and trade in, 11t,13,15

Jarawa, 322, 325

Jarvis, B.C., 181

Johnson, S., 304

Kahn, P., 87

Kaimowitz, D., 66

Kalpavriksh, 5

Kaufmann, D.A., 63,71

Ke Go Nature Reserve, 125,126,126f, 127,128f,129

Kenya, illegal logging and trade in, 16t,18

Kepala rombongan, 191,227

Kiem Lam, 115,116,117,118,119,120, 121,127,128,130

Kiet, V.V., Prime Minister, 104,105

Kishor, N., 3,55,367

Koch-Weser, M., 340

Konfrontasi, 221

Kummer, D., 123

Kuznets hypothesis, 70

Lacandona Reserve, in Chiapas, 93

Lack of coordination, failures of legal system related to, 274-275

Lack of enforcement, of general laws applying to forests and forest products, failures of legal system related to, 275

Lack of government oversight, failures of legal system related to, 275-276

Lam Truong, 112-113People's Committees113

Land Law, 109

Land ownership, shifting of, in Vietnam, 109-112,111f

Land tenure, illegal logging in Vietnam and, 97-135. *See also* Illegal logging, in Vietnam, land tenure, poverty, and forest use rights and

Landsat 7, 308

Landsat 7 ETM images, 160,163f,176

Landsat 7 imagery, 164

Lao Dong, 99,122

Laos, illegal logging and trade in,
    11t,13
Law(s)
    clear standards for, in combating
        illegal activity, 277-280
    failures of, 265-268
    Gresham's, 73
    rule of, defined, 62
    in society, role of, 85-91
    toward forestry and forest land, in
        Vietnam, 107-112,111f
    unfair application of, 269-270
Leach, M., 243
Legal fetishism, communitarianism
    and, 81-96
Legal reformers, rules of thumb for,
    289-291
Legal system, failures of, types of,
    265-276. *See also* Forest
        sector, illegal actions and,
        failures of legal system
        related to
Lepidochelys olivacea, Olive Ridley,
    on Little Andaman Island,
    331
Liberia, illegal logging and trade in,
    16t,18
Lindsay, J., 290
Lisa, H., 181
Little Andaman Island
    chronology of, 327b
    colonization in, 326-328,327b
    deforestation on, consequences of,
        330-331
    described, 326
    endangered sea turtles on, 331
    forest destruction on, 329
    green sea turtle *Chelonia mydas* on,
        331
    illegal logging and deforestation on,
        319-335
        case study, 326-333,327b,328b
        consequences of, 330-331
        introduction to, 320-321

legal interventions for
    (1999-2002), 332-333
    Olive Ridley *Lepidochelys olivacea*
        on, 331
    Onge Tribal Reserve on, 328-329
    Rehabilitation and Resettlement
        (R&R) program on, 328
Local communities, in synthesis of
    issues in illegal logging in the
    Tropics, 359
Local democracy, illegal logging and,
    81-96. *See also* Illegal
        logging, local democracy and
*Local Heritage in the Changing
    Tropics*, 370
Local political power, mystery of,
    94-95
Local rights and interests, respect for,
    in combating illegal activity,
    276-277
Log(s), Red Meranti, 146
Log tracking system, for monitoring in
    cutting crimes, 312-314
Logged-over forests, spatial
    heterogeneity in, 224-225
Logging
    community-based, 213-237
        de facto decentralization and.
            See also Illegal logging, in
            Gunung Palung Area of West
            Kalimantan, Indonesia
    illegal. *See* Illegal logging
    small-scale, recent forest policy
        and, 141-143
    socio-economic organization of, in
        Gunung Palung National
        Park, Indonesia,
        193-196,194f,194t,196t
Logging systems, concession, local
    appropriation of, 225-227
Long, T.D., 121
"Love Song for the Forest Protection
    Officer," 118

Mainardi, S., 66
Malaysia, illegal logging and trade in, 11t,13
MARD. *See* Ministry of Agriculture and Rural Development (MARD)
Market(s)
    capital, penetration of
        in community-based logging in Gunung Palung Area of West Kalimantan, Indonesia, 230-231
        in illegal logging in Gunung Palung Area of West Kalimantan, Indonesia, 230-231
    for illegal logging in Gunung Palung National Park, Indonesia, 196-197
    international, role of timber and wood on, 245-246
    national, role of timber and wood on, 245-246
Mathaai, W., 367,368
McCarthy, J.F., 231
McElwee, P., 4, 97,367
McKean, M.A., 91,92
Meadows, D., 284,287
Melayu, 216-217
Mexican American War, 265
Mexico
    illegal logging and trade in, 19t,21
    Office of the General Attorney for Environmental Protection in, 93
MINEF. *See* Ministry of Environment and Forests (MINEF)
MINEF reports. *See* Ministry of Environment and Forests (MINEF) reports
Minh, H.C., 116
Ministry of Agriculture and Rural Development (MARD), 107,121,122
Ministry of Defense, 121

Ministry of Economy, 274
Ministry of Environment and Forests (MINEF), 158,159,172,173, 175,177,178
Ministry of Environment and Forests (MINEF) reports, 17
Ministry of Finance, 172
Ministry of Forestry, 104,108,115,141, 142,215,220
Ministry of Rehabilitation, Government of India, 326
Monitoring, in cutting crime, 293-317
    field investigation in, 309-312
    framework for, 294-295,295f,296t,297f
    introduction to, 294
    log tracking system in, 312-314
    methods of, 295,296t,297-314,299t, 303t
    paper audits/assessments in, 304-307
    production/consumption data analysis in, 302-304,303t
    remote sensing in, 307-309
    trade data in, 298-302,299t. *See also* Trade data, for monitoring in cutting crimes
Monitoring framework, 294-295,295f,296t,297f
Mount Elgon Forest Reserve, 18
Multicollinearity, 70
Myanmar, 300
    illegal logging and trade in, 11t,12

National Conservation Strategy, 102
National Five Million Hectare Reforestation Program (5MHRP), 106
National forest politics and regulation, 246-249
National Forestry Database System (SIGIF), 158
'National Integrity System' (NIS), 345

National market, role of timber and
wood on, 245-246
National parks, 108,116,227
Natural production forests, reentry in,
228-229
Nature Reserves, 108,116
New institutionalism, 91-94
New Order regime, 218,220
New Order state, delegitimization of,
231
NGO staff. *See* Non-governmental
organization (NGO) staff
NGO Transparency International, 241
NGOs. *See* Non-governmental
organizations (NGOs)
Nguiffo, S., 153
Nicaragua, illegal logging and trade in,
19t,21
Nicobarese, 325
Nicobari, 321
Nigeria, illegal logging and trade in,
16t,18
1927 Forest Act, 269
1927 Forest Law, 270,273
1994 Forestry Law, 177
Article 159 of, 172
NIS. *See* 'National Integrity System'
(NIS)
Non-governmental organizations
(NGOs), 284
staff of, 9
in synthesis of issues in illegal
logging in the Tropics,
359-360
Norm awareness, publicity and
increase in, 256

Obidzinski, K., 137,140
Office of the General Attorney for
Environmental Protection, in
Mexico, 93
Olive Ridley *Lepidochelys olivacea*, on
Little Andaman Island, 331
Olivier de Sardan, J.-P., 255

Onge, 322,325,326
fate of, 331-332
traditional knowledge of, 326,328b
Onge Tribal Reserve, on Little
Andaman Island, 328-329
Ostrom, E., 91

Paper audits/assessments, for
monitoring in cutting crimes,
304-307
Papua New Guinea (PNG),
300,339,356
illegal logging and trade in,
10,11t,15
Paraguay, illegal logging and trade in,
19t,20
Partnership for Governance Reform,
Jakarta-based, 149-150
Patlis, J., 142
Paulson, L.J., 181
Peluso, N., 100,129
Penalty(ies)
meaningful, in combating illegal
activity, 280-281
weak, 268
People in Parks: Beyond the Debate,
369
People's Committees, 108,116,117,
122,123,127
Peoples Republic of Laos, illegal
logging and trade in, 11t,13
Peru, illegal logging and trade in,
19t,20
PGFTR. *See Programme de Gestion
des Forêts et Terroirs
Riverains* (PGFTR)
PGRN. *See Projet de Gestion des
Ressources Naturelles*
(PGRN)
PGTRN. *See Programme de Gestion
des Terroirs et des
Ressources Naturelles*
(PGTRN)
Philip, M.S., 226

Phillipine Department of Natural
    Resources, 14-15
Phillipines, illegal logging and trade in,
    11t,14-15
Pinedo-Vasquez, 226
"Places to Intervene in a System," 284
Plantation production, in Vietnam,
    106-107
PNG. *See* Papua New Guinea (PNG)
Political economy, of community-based
    logging in Gunung Palung
    Area of West Kalimantan,
    Indonesia, 223-228
Political power, local, mystery of,
    94-95
Political stability, lack of violence and,
    63
Pollard, E.H.B., 181,367
Poor enforcement, 271
Poverty, illegal logging in Vietnam
    and, 97-135. *See also* Illegal
    logging, in Vietnam, land
    tenure, poverty, and forest
    use rights and
Power, political, local, mystery of,
    94-95
"President's Initiative Against Illegal
    Logging," 354
Private sector, in synthesis of issues in
    illegal logging in the Tropics,
    358
Privately-Owned Forest and Social
    Forest, 219
Production/consumption data analysis,
    for monitoring in cutting
    crimes, 302-304,303t
Profit levels, of illegal logging in
    Gunung Palung National
    Park, Indonesia, 190-193
PROFOR. *See* Programme on Forests
    (PROFOR)
*Programme de Gestion des Forêts et
    Terroirs Riverains* (PGFTR),
    248

*Programme de Gestion des Terroirs et
    des Ressources Naturelles*
    (PGTRN), 248-249
Programme on Forests (PROFOR),
    342-346
Project Report for Logging and
    Marketing, 330
*Projet de Gestion des Ressources
    Naturelles* (PGRN), 246,249
*Projet Restauration des Ressources
    Forestières dans la Région
    Bassila* (PRRF), 242,243
Provincial Control Units, 158
PRRF. *See Projet Restauration des
    Ressources Forestières dans
    la Région Bassila* (PRRF)

Rainforest Action Network, 258
Ravenel, R.M., 1,4,5,213,351,368
"Recent Trends in Illegal Logging and
    a Brief Discussion of Their
    Causes," 4
Red Meranti logs, 146
Reforestation, in Vietnam, 106-107
Regulation, national, 246-249
Regulatory burden, lack of, defined, 62
Rehabilitation and Resettlement
    (R&R) program, on Little
    Andaman Island, 328
'Relevance of Transparency
    International's
    Corruption-Fighting
    Approaches for Combating
    Forest Corruption,' 342
Remote sensing, for monitoring in
    cutting crimes data, 307-309
"Renovation of State Forest
    Enterprises' Organization and
    Management Mechanism,"
    114
Republic of Congo, illegal logging and
    trade in, 16t,17
Resource(s), common pool, 91-94

Revenue(s), distribution of,
    management of, 143-146,
    144t-146t
Rhee, S., 214,220,231
Right(s), local, respect for, in
    combating illegal activity,
    276-277
"Rome Plows," 101
Rosenbaum, K.L., 5,263,360,364,367
Rotary Club, 87
Roundwood equivalent (RWE), 303
Royal Canadian Mounted Police, 21
Rule of law, defined, 62
Russia, illegal logging and trade in,
    21,22t
RWE. *See* Roundwood equivalent
    (RWE)

SAE. *See* Secretaria de Assuntos
    Estrategicos (SAE)
Sanctioning, internal, 256-257
SANE. *See* Society for Andaman and
    Nicobar Ecology (SANE)
Sarawak Timber Museum, 226
Sea turtles, endangered, on Little
    Andaman Island, 331
Secretaria de Assuntos Estrategicos
    (SAE), 19
Security, development and, dual
    functions of, 221
Sekhsaria, P., 5,283,319
Sekhsaria–Kalpavriksh, P., 368
Sentinelese, 325
SFEs. *See* State Forest Enterprises
    (SFEs)
SFM. *See* Sustainable Forest
    Management (SFM)
Shafik, N., 66
Shekhar Singh Commission, 333
Shompen, 321
Shrank, A., 368
Siebert, U., 4,5,239,243,337,368
Sierra Legal Defense Fund (SLDF),
    306

SIGIF. *See* National Forestry Database
    System (SIGIF)
SLDF. *See* Sierra Legal Defense Fund
    (SLDF)
Smith, J., 137,140
Smith, W., 3,5,7,293,368
Social justice claims at local level, in
    illegal logging, 127-129,128f
Socialist Republic of Vietnam (SRV),
    98
Society, law in, role of, 85-91
Society for Andaman and Nicobar
    Ecology (SANE), 332
Socio-political context, of illegal
    logging in the Tropics,
    360-361
Solomon Islands, illegal logging and
    trade in, 10,11t,15
Sourcebook of Best Practices, 342
    of TI, 347
South America, illegal logging and
    trade in, 18-20,19t
SRV. *See* Socialist Republic of
    Vietnam (SRV)
Stability, political, lack of violence
    and, 63
Standard and Poors, 63
Stanley, S.A., 181
State Forest Enterprises (SFEs),
    101,110,112,121,126
    changing times for, 114-115
State Forest Policy, 154
State Forest Service, 242
State Foundation of the Environment
    in Mat Grosso, 308
State logging companies, in Vietnam,
    role of, 112-115
State Planning Commission, 104
Sterman, J.D., 34
Subarudi, 137,140
Suharto, Pres., 33,213,218,231
Sungai Meliya, 225
Suppression, failures of legal system
    related to, 273
Supreme Court, 177

of India, 320,324,332,333
Suramenggala, I., 137,140
Suriname, illegal logging and trade in,
19t,20
Sustainability, undercutting of, 7-30.
See also Illegal logging and
trade, global problem of
Sustainable forest management (SFM),
58,155,157
described, 170-171,171f
improved governance in, 55-79
data analysis in, 66,68-71,69t
demystification of,
61-66,64f,67f
full econometric analysis,
68-71,69t
sample size in, 68
variables in, 68
Sustainable forest management (SFM)
plans, 59-60
Swedish Western Match Company
(WIMCO), 322,324

Tacconi, L., 4,137,368
Tanh Linh case, 122,123
Tanjung Puting National Park, 354
Tanzania, illegal logging and trade in,
16t,18
Tauke, 227
Telepak, 311,312,339,354,360
Temple, R.C., 320
Tenure, land, illegal logging in
Vietnam and, 97-135. *See
also* Illegal logging, in
Vietnam, land tenure,
poverty, and forest use rights
and
Thailand, illegal logging and trade in,
11t,13
*The Ecotourism Equation: Measuring
the Impacts*, 370
"The Forest Integrity Network," 5
"The Great Barbeque," 270
'The Green Book,' 326,329

*The Private Sector and Tropical Forest
Stewardship*, 370
"The Role of Monitoring in Cutting
Crime," 5
The World Conservation Union
(IUCN), 241
Thu, N.B., 117
TI. *See* Transparency International (TI)
Timber, role on national and
international market, 245-246
*Timber Certification: Implications for
Tropical Forest
Management*, 371
Timber markets, economics of, illegal
logging and, 146-149
Toxic Release Inventory, 294
TPTI. *See* Indonesian Selective Felling
System (TPTI)
Trade data, for monitoring in cutting
crimes, 298-302,299t
challenges in, 300-302
examples of, 298
Traditional forest management
systems, transforming of, in
community-based logging in
Gunung Palung Area of West
Kalimantan, Indonesia,
229-230
*Transboundary Protected Areas: The
Viability of Regional
Conservation Strategies*, 369
Transparency, increasing, in combating
illegal activity, 281-283
Transparency International (TI),
5,338,340,342
FIN and, synergies between, 341
Source Book on Best Practices of,
347
Transparency International's Corruption
Perception Index, 66
Treaty of Guadalupe Hidalgo, 265
Tribal communities
in Andaman and Nicobar Islands,
current threats to, 325-326

indigenous, of Andaman and
Nicobar Islands, 321-322,323t
Tropics, illegal logging in. *See* Illegal
logging, in the Tropics
Tuake, 193,195
2000 FAO (Food and Agricultural
Organization) FRA (Forest
Resource Assessment)
statistics, 9
2001 Ministerial Declaration, at West
Asia Forest Law
Enforcement and Governance
conference, 139

Uganda, illegal logging and trade in,
15-16,16t
"Undercutting Sustainability: The
Global Problem of Illegal
Logging and Trade," 3
United Nations, 270,341
United Nations Economic Commission
for Europe, 302
United Nations Expert panel on Illegal
Exploitation of Natural
Resources and Other Forms
of Wealth in DRC, 15
United Nations Forum on Forests,
8,352
United States, illegal logging and trade
in, 18,19t
United States Forest Service, 313
U.S. Department of Agriculture, 313

Value(s), FBO, 160,168
Van Klaveren, 247
Variable definition, data sources and,
75-76
Vietnam
current forest cover statistics in,
103-105,103f,105f
deforestation rates in, 98-99
forest(s) in, 98

Forest Protection Department in,
115-123,119f
forest protection in, problems in,
116-118
forest resources in, 100-107,103f,
105f
Forestry Science and Technical
Association in, 98-99
future demands for forest products
in, 107
illegal imports from Cambodia in,
105
illegal logging in, 11t,13-14
land tenure, poverty, and forest
use rights and, 97-135. *See
also* Illegal logging, in
Vietnam, land tenure,
poverty, and forest use rights
and
land ownership in, shifting of,
109-112,111f
laws toward forestry and forest land
in, 107-112,111f
plantation production in, 106-107
reforestation in, 106-107
state logging companies in, role of,
112-115
trade in, 11t, 13-14
Vietnam Airlines, 124
Vietnam News Agency, 122
Vietnam News Service, 117
Vietnam War, 101
Villager(s), illegal logging by,
willingness in, system
dynamics examination of,
31-53. *See also* Illegal
logging, willingness of
villagers to engage in, system
dynamics examination of
Vinafor, 105
Violation(s), undetectable, 267-268
Violence, lack of, political stability
and, 63
Voice, accountability and, defined, 62

War in Tropical Forests: New Perspectives on Conservation in Areas of Armed Conflict, 370

Wardojo, W., 272

Weak administration, failures of legal system related to, 274

Weak penalties, 268

West Asia Forest Law Enforcement and Governance conference, 2001 Ministerial Declaration at, 139

West Kalimantan, Indonesia
forest concessions in, 221-222
Gunung Palung area of, illegal logging in, 213-237. *See also* Illegal logging, in Gunung Palung area of West Kalimantan, Indonesia

Wildlands League, 311

WIMCO. *See* Swedish Western Match Company (WIMCO)

Wolfensohn, J.D., 241

Wolfsburg Anti-Money Laundering Principles, 345

Wood
construction, role on national and international market, 246
role on national and international market, 245-246

World Bank, 8,14,21,173,177,241,246, 248,257,308,339,340,341,342, 352

World Development Indicators, 68

World Resources Institute, 300
Global Forest Watch program of, 354

World Trade Organization, 106

World War II, 101

World Wide Fund (WWF) for Nature, 339

Writ Petition, 332

Yale Chapter, of International Society of Tropical Foresters, 1,2,354

Yale International Society of Tropical Foresters Conference Series, 369-371

Yale School of Forestry and Environmental Studies, 2

"You Say Illegal, I Say Legal," 3-4

Zapatista army, 93